Wilhelm Reich, Biologist

WILHELM REICH
BIOLOGIST

JAMES E. STRICK

Harvard University Press

Cambridge, Massachusetts
London, England
2015

Copyright © 2015 by the President and Fellows of Harvard College
All rights reserved
Printed in the United States of America

First Printing

Library of Congress Cataloging-in-Publication Data available from
the Library of Congress
ISBN: 978-0-674-73609-2 (alk. paper)

CONTENTS

Introduction	1
1. Reich's Background, Origins of His Research Program, and Relevant Context	16
2. Reich's Move toward Laboratory Science	64
3. Reich and du Teil: Control Experiments Begin	99
4. An Independent Scientist: The Basic Theoretical and Methodological Features of Sex-Economic Research	146
5. Reich's Theory of Cancer	186
6. Opposition to the Bion Experiments	218
7. SAPA Bions and Reich's Departure for the United States	270
Epilogue	311
Glossary	329
Reich's Bion Preparation Types	333
Dramatis Personae	335
Chronology	339
Periodicals Important in the Norwegian Press Campaign	343
Archives Consulted	345
Notes	349
Acknowledgments	453
Index	455

Wilhelm Reich, Biologist

INTRODUCTION

In 1956 and again in 1960, officers of the U.S. government supervised the public burning of the books and scientific instruments of Austrian-born scientist Wilhelm Reich. This was one of the most heinous acts of censorship in U.S. history, as New York publisher Roger Straus was heard to remark many times over decades afterward, explaining why his firm, Farrar, Straus and Giroux, steadfastly brought all of Reich's published works back into print beginning in 1960. Surely no one felt the injustice of the act more keenly than Reich himself, having narrowly escaped Nazi Germany in April 1933 not long before the Nazis burned his books. Reich eventually came to America in 1939, hoping to find a home where finally his scientific research would be protected from such government-sponsored attacks.

Reich is famous as a brilliant pupil of Freud who developed character analysis and other breakthroughs in analytic technique, as the creator of an innovative synthesis between psychoanalysis and Marxist dialectical materialist philosophy, and as a keen political critic and author of a landmark 1933 analysis of the Nazis' skilled manipulation of the helplessness of the masses, *The Mass Psychology of Fascism*. Having pioneered techniques such as character analysis and working directly with patients' emotions bound up in rigid muscular tensions—what he called "muscular armoring"—Reich is highly regarded among therapeutic and self-help circles as the "father of body therapies," "decades ahead of his time" in that regard. Similarly, Reich campaigned to liberalize divorce laws and laws against abortion and homosexuality, and to spread birth control education and access among the working classes. He was central in the creation of a Sex Political (SexPol) movement in late Weimar Germany, tens of thousands strong. This and his incisive 1936 critique of political reaction on all these fronts in the Soviet Union under Stalin, *Die Sexualität im Kulturkampf* (translated into English in 1945 as *The Sexual Revolution*), have led to his being widely regarded as one of the most important thinkers who foresaw and acted as midwife to the sexual revolution of the 1960s. In the 1968 student uprisings, Reich's writings were considered seminal as a roadmap for dealing with reactionary tendencies in both political and sexual revolutions; Reich was a big name in many American and European intellectual circles. However, one group for whom that was never true was scientists.

This book is not primarily about Wilhelm Reich the psychoanalyst, the social and political thinker, and the pioneer of body therapies. It is about Reich's work as a laboratory scientist—at least his first foray into that realm in 1934–1939. To be sure, Reich's work is remarkably of a piece, and one cannot describe even the logic that led him into the laboratory without setting that logic in the context of the larger body of work. There is scholarship on those other parts of Reich's work, though even in the history of psychoanalysis there has usually been silence on Reich or the same old "black boxed" dismissal, and few take his research program seriously. His laboratory science has largely been ignored by scholars. A huge, popular literature demonizes scientist-Reich as a charlatan, a pseudoscientist, at best a victim of mental illness deluded into thinking he was discovering important things.[1] His charismatic personality was so strong, in this view, that several dozen talented students—physicians, biologists, physicists, and other professionals—were hypnotized into sharing his delusions, not least belief in the existence of a specific biological energy (Reich named it "orgone energy").[2] Reich was convinced by later experiments that orgone energy also existed in the atmosphere and in the cosmos beyond Earth. The U.S. Food and Drug Administration (FDA) centered its legal case against Reich on his experimental device, the orgone energy accumulator, and the machinations of the legal system finally led to Reich's 1957 imprisonment for contempt of court charges when a student of his (without Reich's knowledge or consent) violated a federal injunction against shipping accumulators across state lines. Reich died in prison in November 1957. The FDA widely propagated the idea that Reich was a scientific fraud and the accumulator a bogus scam with no scientific validity. The American Medical Association (AMA), the American Psychiatric Association, the Menninger Clinic, and numerous other organizations aided the FDA in its attacks on Reich and in widely propagating the story of Reich as at best a self-deceived, mentally unstable pseudoscientist and at worst a charlatan deliberately out to take advantage of a naïve public.

A central part of the narrative constructed by Reich's enemies is that he fraudulently claimed the orgone energy accumulator could cure cancer, impotence, and a host of other ailments. This patently false statement was first published by a muckraking journalist with Stalinist ties, Mildred Edy Brady, in an April 1947 article in *Harpers Magazine,* followed soon after by a second article in *The New Republic*.[3] The second article, "The Strange Case of Wilhelm Reich," was reprinted in the *Bulletin of the Menninger Clinic* in March 1948. Despite the falsity of these claims—Reich pointed out that every experimental cancer patient he worked with died, for instance—the claims have nonetheless been blindly repeated in almost every single thing written about Reich since that time.[4] These claims were the initial basis for why the FDA began its investigation of Reich. The FDA, the AMA, Karl Menninger, and others picked up this claim—which can only be described as slander—as though it were fact and ran with it, attempting to defame Reich, steer patients away from his therapy, and eventually initiate legal action against his use of the orgone energy accumulator. FDA investigators make clear in their notes they felt they were on the trail of a "sex racket." In this narrative, the accumulator is derisively referred to as "the orgone box."[5]

The dominant narrative, widely circulated in intellectual circles for almost sixty years since Reich's death, insists there is no point in looking more closely at Reich's science because there *was* no legitimate science from Reich—even if one credits him with talent as a therapist and political or social analyst. Thus, the most important contribution the current book makes is to show, by a very detailed examination of published works as well as of Reich's unpublished laboratory notebooks, that the dominant narrative is not correct and that a much more complex and interesting story is involved. These notebooks, and all of Reich's archives, were closed to scholars until November 2007 and thus only recently became available for careful study and analysis. As the title implies, the picture that emerges makes clear that, at least during the period closely examined here, 1934–1939, Reich was doing careful, state-of-the-art research in laboratory biology.

When it comes to Wilhelm Reich, there seem to be only two kinds of audiences discussing his work: cheerleaders and, far more often, repeaters of old false slanders. Both live in echo chambers. The kind of fact-checking usually required—even with the likes of the *New York Times Book Review* or National Public Radio's *All Things Considered*—seems to be waived when it comes to repeating rumors of Reich as pseudoscientist or as "crazy, mad scientist." The Skeptics groups, from Martin Gardner to the present day, have a positively religious zeal (I use the term advisedly) when it comes to attacking Reich. For this reason alone, this story seems ripe for treatment by a professional historian of biology (not only of psychology), but there simply is no such treatment as yet. In addition, even among those in America interested in Reich and even among historians of psychoanalysis, there has been very little understood about his five years in Scandinavia, perhaps partly because of the inaccessibility of primary sources and relative lack of familiarity with those languages. Yet this was just the crucial period when Reich became a laboratory scientist and did the work culminating in the discovery of what he interpreted as a previously unknown biological energy, which he named "orgone energy."[6] Clearly, then, any attempt to evaluate the validity of his later work in America depends on a detailed understanding of the theoretical and experimental work done in Oslo from 1934 to 1939.

The "pseudoscience" category has become a broad, vague, epistemological wastebasket into which all kinds of things get thrown. In Reich's case, sensationalist attacks began in the press even while his experiments were still underway in 1937–1938, when Reich had published very little about the experiments and well before he thought he had discovered orgone energy. The case is far more interesting than "pseudoscience."

In addition to examining Reich's science seriously, this book does several things. Much as Sherwin and Bird's recent biography of J. Robert Oppenheimer, *American Prometheus*, acts as a window into many of the political developments and much of the physics of the 1920s to 1950s, I use Reich's story as a lens that brings into focus much in the politics, psychiatry, and life sciences of the same period.[7] In addition, I explain the logic of Reich's research program and how the bion experiments fit within the overall trajectory of his work. That includes explaining the steps that led Reich from psychoanalysis to physiology and laboratory biology, and eventually to experiments on the origin of life. I discuss the relationship between the sequence of events in unpublished laboratory notebooks and the published reporting of those experiments, a subject of great interest to historians of science in recent years, since pioneering analyses by Frederic Holmes, Gerald Geison, and others. More on this follows. This book also compares Reich's self-conscious and unique use of dialectical materialism in his thinking about biology, especially origin-of-life research, with other biologists such as J. B. S. Haldane, J. D. Bernal, and Alexander Oparin using dialectical materialism to inform their research during the 1930s.

Moreover, I set Reich's bion experiments in the larger context of biology and medicine in the 1920s and 1930s, a time of major conceptual changes in the life sciences engendered by the introduction of powerful new tools such as the ultracentrifuge and the electron microscope and by many other forces. Reich's research program was rooted in many life sciences approaches that were respected minority research traditions in 1930. His work was fully within the tradition of "biomedical holism" in Germany and France well documented recently by Anne Harrington, George Weisz, and Christopher Lawrence. But many of these approaches had become marginalized by 1940 and almost all were considered dramatically out of date by 1950, with the dominant trend toward what would shortly thereafter become "molecular biology." In studying cancer tissue, for example, Reich prioritized observations made on cells in the living state over those on fixed, stained specimens. This was a minority opinion in the 1930s, but with the supremacy of electron microscopy by the early 1950s, it had come to be considered an extreme, unreasonable view.[8] Thus a research program that might seem like a respected minority view in 1936 could seem separated from the mainstream by an unbridgeable chasm a mere fifteen years later, so rapid was change in life sciences during this time.

As a detailed analysis of the bion experiments shows, Reich's experimental methods were in many ways unorthodox. Moreover, Reich created structures he called "bions," with many lifelike properties, from nonliving starting materials. Thus his main claim in the bion experiments—that a naturalistic origin of life is still possible under present-day conditions—was in direct conflict with Pasteur's claim of the impossibility of spontaneous generation. Pasteur's claim had become biological dogma well before the 1930s. However, the common criticisms that Reich's sterile technique was inadequate, or that all he saw was Brownian movement, are shown in this book to be largely groundless.

I also show that other forces—such as strong reaction against Reich's sexual theories and competition for Rockefeller Foundation funding—likely played a role equal or greater in importance than the actual experimental findings in provoking such powerful opposition to the bion experiments from scientific quarters. A single example for this introduction is the following: An official of the U.S. Embassy in Oslo interviewed a number of Norwegian scientists and public figures a decade after the controversy over Reich's experiments; many of them were participants in the 1930s debate. The goal was to assess whether Reich was a legitimate scientist and whether criticisms of his work were based on actual experimental data or, rather, upon prejudice against his sexual theories. In this investigation Reich's two leading critics, cancer researcher Leiv Kreyberg and psychiatrist Johann Scharffenberg, were thought by one of their own colleagues and a prestigious voice in Norwegian science, Harald Sverdrup, to be violently emotionally prejudiced against Reich. Kreyberg in particular was said by Sverdrup to be a volatile personality whose opinion was largely untrustworthy on scientific matters in any situation where he had strong emotions on the topic. The embassy official did not think Reich blameless in the controversy either; nonetheless, it is clear that far more than a simple dispute about microscopy took place.

Thus, explaining away Reich's science by the well-established narrative of Reich as mere pseudoscientist simply fails to hold up under historical scrutiny. And since the bion experiments are the scientific origin of Reich's theory of orgone energy and of Reich's cancer theory, this study suggests that a close reexamination of those supposedly pseudoscientific theories might also prove fruitful.

Interpretations other than Reich's may explain some of what Reich saw, given more recent research in related areas (such as nanoparticles, prions, and cancer viruses).[9] But the evidence suggests Reich clearly saw *something* meaningful; no longer can these experiments be dismissed as careless or incompetent, or as pseudoscience, which is a label sometimes used to dismiss unconventional work out of hand. There is much to be learned in this story about microscopy, about the study of biological material in the living state, about scientific method in a broader sense—even about why a Freudian psychoanalyst would find it logical, indeed compelling, to end up in a physiology laboratory and eventually studying primitive microorganisms.

This is a story full of fascinating contingencies: a left-wing Jewish intellectual famous for radical sexual theories operating in 1930s Lutheran, small-town Oslo, Norway, where all local scientists knew each other and were often related by marriage and where anti-Semitism was common; an outsider competing with the local talent for coveted Rockefeller Foundation grant money—almost the only game in town for large-scale funding for life sciences research during the Depression years; an outsider to professional physiology and microbiology disciplinary circles who did not publish in their journals but, rather, in his own highly eclectic journal alongside political and sociological analysis in what Reich viewed as a new scientific field, named by him "sex-economy."[10] Fascinating comparisons can be made with recent literature on others who self-consciously tried to remain "independent scientists"—James Lovelock and Peter Mitchell being two of the most prominent.[11]

Recent studies that compare a scientist's published work with the messy, detailed daily record from laboratory notebooks have highlighted the way scientists rhetorically create and deploy "rationalized reconstructions" of the path of research. It is evident that polished published works are often used to simplify and render more persuasive the way in which scientists come to some important finding and realize its significance. Like his contemporary Ludwik Fleck, Reich already pointedly criticized this rhetorical sleight of hand in the 1930s. He argued that it denied the public the full understanding of just how difficult and confusing the daily practice of science was, leading them to inflated expectations that all discoveries could be quickly and easily distilled into "sound-bites" they could digest without much effort. Reich seems to have stuck pretty faithfully to the actual sequence of laboratory events in his published work on the bion experiments. And yet, for Reich, intuitive leaps out ahead of data and controls are a significant, indeed a signature, feature of his research style. He defends his reliance on intuition or direct contact with his subject thus: "Admittedly, measurements and replicate experiments still have the last word in science. But when I see an amoeba stretching, and the protoplasm flowing in it, I react to this observation with my entire organism. The identity of my vegetative [i.e., autonomic] physical sensation with the objectively visible plasma flow of the amoeba is directly evident to me. I feel it as something that cannot be denied. It would be wrong to derive scientific theory from this alone, but it is essential for productive research that confidence and strength for strict experimental work be derived from such involuntary, vegetative acts of perception."[12] (The term "vegetative" refers to involuntary activity of the autonomic nervous system; it has nothing to do with plants.) Given a research style in which intuition or "a feeling for the organism" (as Keller has put it in her biography of geneticist Barbara McClintock) plays so significant a role, it is more interesting than ever to compare how these intuitive feelings interact with the data and control experiments from the daily work in the laboratory, particularly for a figure who, like Reich, is credited with such extraordinarily insightful intuition in so many other fields than laboratory science.[13]

In discussing Reich's work and the public reaction to it, it is quite important to remember a distinction Reich himself often emphasized: that between workaday science within well-established boundaries and truly pioneering research that crosses into unknown territory. The two are fundamentally different intellectual enterprises. This difference is also what Thomas Kuhn emphasized in his classic book *The Structure of Scientific Revolutions*, and which Kuhn called "normal science" versus "revolutionary" or "paradigm-changing" science. While he emphasized that normal science, done *within* an existing paradigm, is crucial to scientific progress, Kuhn asserted that revolutionary science, because by definition it violates or fundamentally overturns an existing paradigm, must always, by its very nature, provoke strong opposition—perhaps most of all from the most well-established, credentialed, specialist experts in the scientific community.

Of course, this does not mean that all science attacked by the mainstream is automatically the bearer of the next revolutionary paradigm-shift. Many marginalized figures claim this status; the challenging task—until such a shift occurs—is to be able to make an educated guess about which few of the many challengers really carry some vital new insight and which (most by far) are likely just mistaken.[14] Terms such as "pseudoscience" are not very helpful here, since they are always used by the mainstream *during* a controversy in an attempt to discredit any and all paradigm-challenging claims. Similarly, in biology since the 1920s or 1930s, to label an idea "vitalistic" puts a fatal millstone around its neck, since that label became code among the mainstream for "pseudoscience." The historian often reads "vitalist" since 1945 more like "so unlike what we now believe is the right way to do science (both ideas *and* practices), that it *must* be wrong and seems so quaint as to be lame." Whatever Reich was, this analysis will show that he was no vitalist.

Scientists, then, whose names we associate with revolutions—Copernicus, Newton, Lavoisier, Darwin, Pasteur, Semmelweiss—all, by definition, faced staunch, often irrational resistance to their ideas, not least from the established scientific authorities of their day. I argue that Reich's work on biogenesis in the bion experiments, and certainly the visceral reactions it provoked, need to be understood in this light. Reich was engaged in research that challenged, one after another, numerous deeply entrenched beliefs and practices. Research suggesting a purely naturalistic origin of life has, in particular, usually touched a religious nerve, even sometimes among scientists themselves. Here Louis Pasteur makes an interesting case: for all his brilliant, revolutionary contributions to understanding the role of germs in fermentation and in infectious disease, Pasteur was a staunch Catholic who took his ideas about germs too far and argued famously in public debates that life could never, under any circumstances, arise from nonliving materials, in other words, that spontaneous generation—or a naturalistic origin of life such as most scientists today think plausible—was impossible (this despite Pasteur's private views to the contrary, about which he only hinted—much later—in public). Reich has interactions with the French Academy of Sciences regarding his bion experiments that echo in many fascinating ways those of Pouchet and Bastian with the academy in their disputes with Pasteur over spontaneous generation in the 1860s and 1870s.

As mentioned earlier in this introduction, in the past thirty-five years, historians of science have shown that the study of laboratory notebooks can provide fascinating close-up insight into how a scientist works and how his ideas develop, even on a day-to-day basis. Frederic Holmes and Gerald Geison in particular have shown that differences between the published account (or even the scientist's own recollections later in life) and the detailed sequence of development of experiments can reveal unexpected things. Holmes studied a wide range of figures, from eighteenth-century chemist Antoine Lavoisier to nineteenth-century physiologist Claude Bernard to twentieth-century biochemists Hans Krebs and Matthew Mesehlson, both of whom he was able to interview. Geison exhaustively studied the notebooks of Louis Pasteur, which Pasteur himself on his deathbed instructed his family never to make public. Fortunately for scholars, Pasteur's grandson and last surviving male heir decided in 1964 that such scientific treasures must not be lost; he donated the 102 laboratory notebooks to the Bibliothèque Nationale in Paris. Geison was the first scholar to spend the long hours, weeks, months, and years deciphering Pasteur's tiny, cramped handwriting and thereby to begin reading the new revelations about Pasteur's work contained in those treasures. Deliberate fraud and deception are not the point (usually); rather it is better understanding how scientific ideas form and are shaped in interaction with recording raw observations. In his finished biography *The Private Science of Louis Pasteur*, Geison provides an enlightening discussion of the potential of studying laboratory notebooks.

As a student of Geison's who assisted in the early 1990s with some of the Pasteur notebook studies, I witnessed the value of this approach firsthand. Thus in 2007 I saw the opening of Reich's archive as an opportunity to determine whether his laboratory notebooks had been preserved and, if so, what new insights they might offer into Reich's method of work and thought. In addition, I found that specific technical details related to culture media and other parameters used in the bion experiments might help resolve difficulties researchers have had in recent years replicating crucial parts of those experiments, which were frequently and unproblematically replicated during Reich's time (for example, by Roger du Teil at the Mediterranean University Center in Nice and later by biologists Bernard Grad at McGill University and Sol Kramer at the University of Florida and by physicist Adolph Smith at Sir George Williams [now Concordia] University in Montreal). I discuss the importance of these technical details as each arises in the narrative of the experiments.

I have examined a number of Reich's laboratory notebooks from 1935 to 1939, though it is not yet clear whether all such notebooks were preserved. What is most striking is the complete agreement in the sequence of development and logic of the bion experiments through October 1937, with the account Reich published in his 1938 book *Die Bione* (available now in English as *The Bion Experiments on the Origin of Life*). Reich did not indulge in the kind of rational reconstruction many scientists do in writing up *how* they came to a significant discovery. Reich seems unafraid to directly reveal the groping guesswork that constitutes so much of daily laboratory practice. But developments also actually seem to follow a fairly logical progression, with appropriate control experiments created along the way every time some new finding is sufficiently striking to imply that such controls are needed.

Reich's Movement toward the Laboratory

Some other interesting features of Reich's research style and trajectory are worthy of comment here. In both he is building upon but departing from Freud. Reich's work was from early on characterized by two prominent strands: (1) the concept of simultaneous identity and antithesis and (2) the energy principle. I briefly introduce here these two key items from Reich's intellectual tool kit. The simultaneous identity and antithesis concept helped Reich see that two forces imagined to be opposites (an antithesis) could both simultaneously also be manifestations of a common, underlying unified force (thus, in that sense, they constitute an identity). For example, expansion and contraction—apparent opposites—are unified as phases of the unitary act of pulsation (as in the heartbeat). Another example of how this idea is practically employed is given by Mary Boyd Higgins, the trustee of Reich's estate: "Freud's original assumption was that if sexual excitation is barred from perception and discharge, it is converted into anxiety. But no one knew how this conversion takes place. In wrestling with this question therapeutically, Reich observed the relation of anxiety to the vegetative (autonomic) nervous system. 'There is not a conversion of sexual excitation into anxiety,' he concluded. 'The *same excitation* which appears as pleasure in the genital is manifested as anxiety if it stimulates the cardio-vascular system.' Sexuality and anxiety represent opposite directions of vegetative excitation."[15] Reich did not see the organism as a Rube Goldberg mechanism in which lever A → button B → switch C. He thought the movement of energy, and the direction in which it moved, more likely to be the fundamental driver of emotions and libido.[16] It was movement, and not the underlying details of structure or biochemistry, that most attracted his attention. In the decades leading to recognition of macromolecules and the development of molecular biology, this alone made Reich's approach stand out: he was interested in questions that went against the general tide in life sciences, increasingly so as that tide strengthened by the late 1930s.

Reich believed, as did some other psychoanalysts—like Norwegian Harald Schjelderup, originally trained in physics—that psychoanalysis and the libido theory needed to be put on a firm foundation in laboratory science if they were ever to converge with other work in the experimental tradition in psychology (see Figure Introduction.1). [17] As early as 1923 Reich gave serious thought to Freud's picture of "sending out libido" (or withdrawal into the self, when rejected or disappointed) as analogous to the extension (and retraction) of pseudopodia by an amoeba.[18] Reich found this image so compelling that he told friends it might someday be proven that the extension of an amoeba's pseudopod is physiologically identical to the erection of the penis, "a stretching out toward the world" in a fundamental biological sense.[19]

Figure Introduction.1: Harald Schjelderup (left) and Reich, 1934. Courtesy of Wilhelm Reich Infant Trust.

Hence, Reich eventually desired to study microorganisms in the laboratory, which led serendipitously to the bion experiments after Schjelderup invited Reich in 1934 to work in a laboratory in the Psychological Institute at Oslo University.

Reich's pursuit of the energy principle was already evident in a 1923 publication, "On the Energy of Drives."[20] In Reich's view, the energy principle is the idea that all psychic events must be driven by tangible energy of some measurable form. Since energy must obey the law of conservation of energy, it could be converted among different forms, but no energy would be lost in the conversion. Like Freud, taking the conservation of energy principle seriously, Reich noted that the energy source behind neurotic symptoms increased when patients did not experience any sexual discharge and decreased when they did.[21] "Libido" seemed to be a real, quantifiable something that could potentially be measured in a laboratory. As one of Reich's physician students expressed it,

> Just as Freud had essentially abandoned any expectation of finding a somatic basis for the psycho-neuroses and was inclined more and more to "psychologize biology," the youthful Reich appeared on the psychoanalytic scene, with his enthusiasm for the idea of a "sexual energy." In 1923 he . . . took issue with the neglect of the genital function. . . . This signaled the beginning of his sex-economic approach to neurosis and psychosomatic disease. Thereafter, we are assured of an unrelenting effort to comprehend genitality as a biological function apart from the function of procreation and to establish its relevance in health and disease. . . . While Reich continued to adhere to the libido theory and to emphasize "the quantitative factor in psychic life," Freud was moving away from it and emphasizing his new ego psychology.[22]

Freud found Reich's idea hard to accept and challenged Reich to defend it. "The questions that Freud's revision raised helped Reich refine his own position and defend Freud's original thesis regarding sexuality. Questions of culture, prophylaxis of the neuroses, the negative therapeutic reaction, which Freud was unable to solve in a practical way, found clarification in Reich's theory based on the principle of a biological energy—a principle which he hoped would eventually be demonstrable and measurable."[23] Most importantly, Freud and other analysts objected that they knew plenty of patients

Introduction

who had regular sex yet were still quite neurotic. Clarifying, Reich noted that frequency of sexual intercourse alone was not the key element in prevention of neurotic symptoms; rather, it was whether the sexual contact produced was completely gratifying, whether all the dammed-up excitation was fully discharged. He called the capacity for this "orgastic potency," to distinguish it from, for example, "erective potency" or "ejaculatory potency," the ability (of a man) merely to carry out the sex act. This is discussed further in Chapter 1.

Building on Freud's analogy of undischarged libido acting like water overflowing in a dammed-up river, Reich later put it thus: "The important issue, hygienically is not *that* the patient have coitus, but that the experience *satisfy fully*. I conceived of sexuality as a river, which springs from natural sources and has to follow a certain course. *Sexual release in the sex act must correspond to the excitement which leads up to it.* If one constructs artificial dams which impede the normal course of the river, it will flood its banks and do damage. Sexual energy which is bound up, like quantities of water behind a dam, spills over into . . . sexual aberrations (child seduction, rape, sex murder), insomnia, . . . compulsive behavior, anxiety states, menstrual disorders, impotence, frigidity, premature ejaculation, etc."[24] The "water" driving neurotic symptoms, then, was not original sexual libido *converted* into some other form of energy, in Reich's understanding. It was the same "water" as before, merely turning different mills because of having been diverted out of its natural channel.

In 1929 Reich was still open-mindedly exploring Freud's suggestion that libido might be governed by a chemical substance in the body (perhaps akin to the new hormones being discovered) rather than an energy phenomenon.[25] Reich read widely in the latest research in biological and medical literature on the effects of different chemicals on the autonomic nervous system, for example, choline, whose effects at relieving anxiety had been investigated by two Berlin doctors.[26] He took great interest in the theory of famed Berlin cardiologist Friedrich Kraus that all life phenomena might be rooted in movements of fluid and electrical charges within membrane-enclosed compartments within the body.[27] Reich also closely followed the work of biologist Max Hartmann at the Kaiser Wilhelm Gesellschaft, whose experiments on protist cells (such as marine algae) suggested their gender in reproduction might be determined by relative electrical charge.[28] By 1934, his explorations in this direction led to his formulation of "the basic antithesis of vegetative life."[29] This in turn led Reich to the first physiology experiments he conducted in the laboratory, attempting to measure whether bioelectrical charge was moving or changing at the skin surface—particularly at the erogenous zones—during different emotions. These experiments are described in Chapter 2. Reich's bioelectric experiments and Hartmann's "relative sexuality" theory led Reich to keep open the possibility that libido was a phenomenon of bioelectrical charge.

From Dialectical Materialism to Energetic Functionalism

Reich joined the Austrian Social Democratic Party, but by 1927 left it and joined the Austrian Communist Party. When he moved to Berlin in November 1930, Reich joined the German Communist Party. From the writing of his "Dialectical Materialism and Psychoanalysis," published in 1929, onward for about ten years, Reich considered the Marxist philosophy of dialectical materialism to be a powerful tool of his thinking and research. However, already by late 1934 he had come to realize that most others using the language of dialectical materialism for their ideas—not least his former friend and colleague Otto Fenichel—meant very different things by it than did he.[30] The two central ideas Reich had before encountering Marxist philosophy—the crucial role of the energy principle (especially in the orgasm function) and his concept of simultaneous identity and antithesis—were central to Reich's understanding of a dialectical materialist science. Yet, Reich gradually began to realize, their importance was not appreciated, or perhaps not understood by others who saw dialectical materialism as their lodestar. As Reich moved more and deeper into laboratory science, from 1935 on, this intellectual gap widened. His discoveries in the bioelectric experiments, leading to the discovery of the bions and their effect on cancer cells, and finally in 1939 to Reich's belief that he had discovered orgone energy, affirmed and greatly strengthened Reich's confidence in the orgasm theory and the unity/antithesis concept, as well as in the central importance of a quantitative energy principle in research in psychology and the life sciences. Thus, by the time of the discovery of orgone energy, that breakthrough acted as a final catalyst to convince Reich once and for all that his conceptual framework was now so fundamentally different from what any other researchers meant by "dialectical materialism" that it was misleading and no longer even accurate to call his own thought method by that name. Coining the new term "energetic functionalism" for his approach, Reich captured in the name both of the key conceptual pieces he himself had brought from the outset.[31] And as he later put it, to express his sense of having transcended dialectical materialism, "the energetic functionalism of today [1947] has as much to do with dialectical materialism as a modern electronic radar device with the electric gas tube of 1905."[32]

It seems unlikely that the discovery of orgone energy alone caused the final intellectual break for Reich from the language of dialectical materialism. The break occurred in the context of the Left parties' repeated rejection (even expulsion) of Reich and his ideas from 1933 to 1939. During this time Reich was among the first of those on the left outside the Soviet Union to level a detailed critique at the reactionary policies of Stalin and the Moscow Communist Party.[33] His own social and political thinking by 1939 was moving away from the Communist and Socialist Parties and toward what he called "work democracy." More than that, Reich's break with dialectical materialist philosophy occurred in the context of his much more general break with his European milieu that inevitably resulted from Reich's departure from Europe for America in August 1939 and from the chaos of World War II that quickly followed; the remainder of his career developed entirely in the American context.

The development of Reich's biological work, both theoretically and in the laboratory, from 1934 to 1939 is the subject of this book.[34] Understanding this story will begin to put the reader in a position to evaluate Reich's later scientific work and his eventual confrontation with the U.S. government. First, though, we must take a step or two back, to survey Reich's background and to understand the context of science and medicine from 1900 to the 1930s, particularly in Vienna where Reich trained.

1

REICH'S BACKGROUND, ORIGINS OF HIS RESEARCH PROGRAM, AND RELEVANT CONTEXT

Wilhelm Reich was born in Galicia in 1897; his family soon moved to Bukovina, another eastern province of the Austro-Hungarian Empire, where Reich's father purchased a large farm. Reich grew up on that farm, largely educated by private tutors until age twelve, after which he attended a gymnasium in Czernowitz. Reich later said that early life on a farm, "close to agriculture, cattle-breeding," taking part each year in the growing of crops and the harvest, stimulated his interest in nature and in biology. This interest was cultivated further by a private tutor who guided Reich to keep his own "collection and breeding laboratory of butterflies, insects, plants, etc.," so that, as Reich recalled it, "natural life functions, including the sexual function, were familiar to him as far back as he can remember."[1]

Reich's mother committed suicide in 1910 after revelation of her affair with one of Reich's tutors. His father went into a profound decline and died of tuberculosis less than four years later, shortly before the outbreak of World War I. When war began Russian troops swept through the farm in Bukovina where Reich, age seventeen, was staying trying to keep the family business going. In 1915 he enlisted in the Austrian Army and in 1916 was commissioned as a lieutenant. His regiment was involved in action on the Italian front. Over time, like so many other young men who went excitedly to war, Reich came to see war and military life as mindlessly mechanical. Upon demobilization in 1918 he enrolled at the University of Vienna to study law. But within a few months he felt compelled to switch to medicine, his youthful interest in biology revived. He took advantage of a program offered by the university to veterans, to complete both the bachelor's degree and the M.D. in four years of intensive study instead of the usual six years for the combined degree program. Thus, he graduated as an M.D. in 1922. Having lost all possessions during the war, Reich worked his way through school by tutoring other students in the premedical subjects. He also received somewhat grudging aid during these years from distant relatives in Vienna.[2]

While at the University, Reich and several other students (including Otto Fenichel and Edward and Grete Lehner Bibring), bothered by any lack of systematic study of sexuality in their medical training, organized their own informal seminar on sexology. Reich saw his later work as part of the field of sexual science, not only as psychoanalysis. As part of research for this seminar, Reich in 1920 made contact with Professor Sigmund Freud. Reich found Freud warm and encouraging, and Reich soon became interested in psychoanalysis. Later that year, Reich read his first paper to the Vienna Psychoanalytic Society and became a member of the society in October 1920. By his final year of medical school, 1921–1922, he was a practicing psychoanalyst. In addition, during that final year he "took postgraduate work in Internal Medicine at the University clinics of Ortner and Chvostek at University Hospital, Vienna." After graduating M.D. in July 1922, Reich "continued his postgraduate education in Neuro-Psychiatry for two years (1922–24) at the Neurological and Psychiatric University Clinic under Professor Wagner-Jauregg, and worked one year in the disturbed wards under Paul Schilder." Reich was regarded by Freud and others as a gifted clinician who sought breadth as well as depth: he also "attended at polyclinical work in hypnosis and suggestive therapy" at the Neurological and Psychiatric University Clinic, as well as "special courses and lectures in biology at the University of Vienna."[3]

Reich found Freud's libido theory compelling: Freud had speculated early on, for example, that some chemical substance might underlie the impulses and emotions of sexuality, and he was certain libido would obey conservation laws, that is, that the amount not discharged would quantitatively correspond to the amount and severity of neurotic symptoms that resulted. This made sense to Reich and was corroborated in his experience with his patients. At first he thought the more patients were able to have sex, the more their neurotic symptoms were reduced. As mentioned previously, many older analysts objected that they had many patients who were sexually active but severely neurotic. Reich himself dealt with male patients who bragged of how much sex they had, so at first this seemed to contradict libido theory. However, when Reich inquired in more specific detail, he found those men "were erectively very potent, but ejaculation was accompanied by little or no pleasure, or even the opposite, by disgust and unpleasant sensations."[4] When Reich inquired in similar detail about whether neurotic patients actually felt full gratification in intercourse, he came to the conclusion that the vast majority, both male and female, did not. When some of his patients became capable of complete gratification, then their symptoms did improve. He concluded that the key to relief of neurotic symptoms was "orgastic potency," not merely having sex (in any quantity), but the ability to completely discharge the dammed-up sexual tension in a fully gratifying sex act. He claimed that very few analysts would take the trouble to question their patients in so much detail about the specifics of their sexual experiences. Thus, many remained skeptical of Reich's concept of "orgastic potency." However, all of Reich's future research was based upon it.

As a psychoanalyst, Reich was interested in improving technique as much as in theory. It seemed to him that many analysts were content to just interpret material from the patients' free association and not much concerned with whether patients actually got better. He was one of the founders in 1922 of the Viennese Seminar for Psychoanalytic Therapy. Freud and others recognized Reich as gifted in this area, and Reich rose to be elected leader of the Technical Seminar by 1924. His 1925 book *Der triebhafte Charakter (The Impulsive Character)*, according to psychiatrist and psychosomatic medicine pioneer Theodore P. Wolfe, showed "the necessity of extending symptom-analysis to character analysis, and the possibility of making certain types of schizoid characters accessible to psychoanalytic therapy which had previously been considered as inaccessible."[5]

In 1928, Reich gave a paper that formulated the principles of a whole new approach to psychoanalysis. He argued that, rather than focus on individual symptoms, the analyst should attempt to view the patient's entire character structure as a fundamental neurotic defense. If the patient approached the therapy with a typical attitude of contempt, for example, Reich argued that should be viewed as a central feature of the character structure. He said that the character structure itself should be understood as "character armor," containing the history of the patient's traumas and of how the patient dealt with the world in order to survive.

Reich's therapeutic contribution was widely recognized as significant; however, in the meantime a theoretical rift formed and was deepening between Freud and Reich. In the fall of 1926 Freud, in his book *Hemmung, Symptom und Angst*, put forward major changes in Freud's earlier position about the sources of neurotic anxiety. Freud no longer interpreted anxiety as a result of sexual repression; now he asserted that the anxiety was the cause of the repression. The question of how libido became converted into anxiety—which for Reich had become central—Freud now said was no longer important. For Reich, however, if his character analysis made a useful contribution, it was because it was fundamentally rooted in the therapeutic goal of helping convert the patient's stasis anxiety into genital excitation. "Where it was possible to bring it about, there were good and lasting therapeutic results." But it was precisely in those cases where he did not succeed "in liberating cardiac anxiety and in producing its alternation with genital excitation" that Reich first came to believe that "the character armor was the mechanism which was binding all energy."[6] Freud acknowledged some differences with Reich in July 1927 but assured Reich that psychoanalysis was a science and that, if he (Reich) was right, over time that would become clear.[7]

In addition to trying to improve therapeutic technique, Reich's thought continued further in the direction of trying to understand what the actual biological basis was of libido and of the energy that fueled neurotic symptoms when it was blocked. What was the real, biological nature of psychic and sexual drives, Reich had wondered for some time already. (He had given Freud an early draft to read of his paper "On the Energy of Drives" as early as 1920).[8] Reich sought a model that would account for the buildup of sexual tension and its discharge as a real, tangible "something" in physiology, since clinically it seemed clear to him from his patients that only when the "something" built up was discharged completely did the patients' neurotic symptoms diminish or disappear, at least temporarily. Because the severity of patients' neurotic symptoms appeared to be directly proportional to the amount of undischarged libido, Reich referred to this research as the study of "sex economy." He spoke, for instance in his 1927 book *Die Funktion des Orgasmus*, of the "energy household" of the organism and theorized that the disruption of normal energy buildup and discharge was the source of neurosis. Additionally, as he was drawn further into socialist and communist politics, he came to believe that unhealthy social conditions were producing neurosis on a mass scale and that prevention via social reforms would be far more effective than individual therapy for people already damaged.

Vienna, Biology, and Medicine in the 1920s and 1930s

It is easy to forget how much of modern biology, modern medicine, and modern psychology were all created since 1900—in the case of biology even since 1930. The range of topics under discussion, the ideas viewed as realistic or plausible, the social concerns associated with scientific and medical ideas were beginning to take on the outlines of what we recognize as modern. But many older traditions, discourses, and research agendas still overlapped from the nineteenth century. The world of biology, medicine, and psychology between 1918 and 1922, when Wilhelm Reich trained in medicine at the University of Vienna, was, then, as remarkable for the differences in ideas and practices from today—perhaps more—than for the similarities.[9]

By 1926 or 1927, Reich had become a politically active figure on the left; he saw his scientific work on the etiology of mass-neurosis and, later, of organic disease (cancer, hypertension, and numerous other diseases) as well, to be in the service of clarifying what social and political reforms were needed for the prevention of these terrible public health problems. The working class, in particular, suffered from hysterical, compulsive, and violent behaviors that Reich thought were produced by their living conditions.[10] Many people sleeping in one room, for example, allowed very few to have a gratifying love life undisturbed. As Elizabeth Danto has shown, a significant number of the Freudians—and Reich was prominent among them—felt strong obligations to directly address the mental health problems of the poor, even of those too poor to pay at all. Many of them, like Reich, Otto Fenichel, and Erich Fromm, embraced Marx's analysis of society.[11] Reich had been an early advocate and activist for abortion rights, birth control education programs, and repeal of legal sanctions against homosexuals. Rooted in the "nurture" tradition, Reich was deeply skeptical of genetic determinism, unlike some advocates of eugenics among Red Vienna's socialists, since Reich saw the eugenics movement as primarily a tool of fascism and political reaction, with Nazi "race biology" as the epitome of this trend.[12]

Reich had gone further, however, much to the distaste of Freud and more conservative analysts, writing a pioneering 1929 article that attempted to reconcile dialectical materialism and psychoanalysis. While at first an enthusiast of Bolshevik reforms of marriage and divorce laws, Reich was nonetheless by 1935 one of the first Western supporters of the Soviet experiment to criticize Stalin's reactionary moves.[13] Reich was among those who early on began to use the term "Red fascism" to describe the totalitarian system Stalin had created by the mid- to late 1930s; indeed, Reich's 1933 book *The Mass Psychology of Fascism* was denounced by doctrinaire Marxist politicians (and he himself was declared persona non grata by the Communist Party), while at the same time the book was proscribed for burning by the Nazis and officially disowned by the psychoanalytic establishment in its eagerness to compromise with the new Nazi regime.[14]

In his theoretical work, as well as in the laboratory, Reich strove, as many did in the 1920s and 1930s, to find a path that "steered between the extremes of reductionism and vitalism."[15] To many, living organisms were not machines pure and simple: one could not disassemble the parts and reassemble them with the same result. But neovitalism seemed too extreme in the other direction, declaring that the most essential properties of life were simply inaccessible to modern laboratory investigation. The preceding quote from Loren Graham describes the Marxist philosophy of dialectical materialism as applied to the sciences, showing that it was one of a number of paths by which scientists attempted during these years to find the middle path that combined the strengths of mechanism and vitalism but avoided their excesses. Thus, because of his social reform interests and his scientific investigations, Reich found Marx's social analysis and dialectical materialism attractive in the late 1920s and on into the 1930s. By 1929 he had completed his ambitious theoretical paper, analyzing in great depth the compatibility between dialectical materialism and psychoanalysis. In this paper Reich explored the implications of the dialectical materialist thought method for pursuing Freud's libido theory, a theory Reich found clinically fruitful and theoretically compelling. He began to lay out a research agenda, pointing toward laboratory biology investigations that might test the theory or ground it in tangible, quantifiable biological processes rather than only in psychology or psychoanalytic theory.

Already by 1923 Reich questioned the workability of psychophysical parallelism to account for the interactivity between psychic processes and underlying physiological processes. This was a theory that represented psychic processes as "parallel to" but separate from physiological ones. A prominent psychosomatic researcher at the time, Theodore P. Wolfe, said that many were frustrated by this dualistic concept, that is, that "the 'psyche' did this or that to the 'soma,' [while] . . . there were always others voices which warned . . . the soma also does this or that to the psyche." Wolfe said he and others intuitively groped for a concept of psychosomatic identity, but they could find no such concept in the literature.[16]

Reich was interested in Semon's "mneme" theory, but even more so in Henri Bergson's conception of the interrelation of the psyche and soma.[17] Reich had been a student of Paul Kammerer in his University of Vienna Medical School training, and he was deeply impressed by Kammerer's suggestion, repeated in the 1915 first edition of his textbook *Allgemeine Biologie*, that Kammerer found very probable the existence of a specific life energy.[18]

Let us look in more detail at each of these related ideas.

The Mechanism-Vitalism Debate

For hundreds of years, scientists who studied living things had a basic disagreement. Though it took many different forms, it boiled down to one central question: can living things and how they work be completely explained by, or reduced to, only the laws of physics and chemistry? This idea is sometimes referred to as physical-chemical reductionism in the life sciences. One reading of this claim is that organisms are just complicated physical-chemical machines. This view is often referred to as mechanism, or mechanistic materialism. (Marxists referred to it as "crude mechanistic materialism," to distinguish it from their more nuanced idea of dialectical materialism.)[19]

Or is there something different about living things—not that they disobey the laws of physics and chemistry, but are there elements of life that cannot be explained *only* by the laws of physics and chemistry? Most of all, what is the difference between nonliving matter and living matter? What has changed between the living organism that was here ten minutes ago and the corpse that now lies before us? For many hundreds of years, at least some life scientists took the position that there had to be something more than physics and chemistry involved, because living things had a purposefulness about them as intact organisms that was not understandable in any of the individual parts. This idea goes at least as far back as that astute observer of nature, Aristotle— that the whole design of an intact organism, every aspect of it, seemed to serve the needs of the whole in a coordinated way. One did not have to believe in a supernatural creator (à la Intelligent Design) to think this design implied there is something purposeful about living things and to recognize there was not anything purposeful about the nonliving realm. This position, which resisted the claim that physics and chemistry could explain everything about life, is often called vitalism, though usually by its opponents, at least in modern times. People took many different versions of this position.

The vitalist position had been espoused to varying degrees by William Harvey, discoverer of the closed blood circulation, and early nineteenth-century physiologist Johannes Müller. But by the nineteenth century, nonvitalistic physics and chemistry had made quite a bit of progress in explaining many features of metabolism and biochemistry. Organic compounds had, for example, once been thought to be qualitatively different from inorganic ones, perhaps only capable of synthesis within a living organism. In 1828, however, German chemist Friedrich Wöhler first carried out a complete synthesis of an organic compound, urea, entirely in a test tube. Mechanists, including

Wöhler's colleague Justus von Liebig, considered this a major blow against vitalist beliefs. Liebig went on to demonstrate that inorganic forms of nitrogen, phosphorus, and other chemicals served as perfectly adequate fertilizers for plants, declaring that humus or other "organic" components of soil were not needed in agriculture for plant nutrition and that all of animal and plant metabolism would be explained in purely chemical terms before long.[20]

Similarly, Johann Wolfgang von Goethe believed a vital force necessary to explain plant morphological development. Others, like Swiss botanist Augustin de Candolle, accepted the idea of a vital force, "but it was not the universal creative force in which Goethe believed. The vital force, as de Candolle perceived it, had the same relationship to organisms as gravity has to the planets, and he therefore thought the vital force ought to be analyzed by the methods of science."[21]

Beginning in 1847, a group of German researchers (sometimes called the "physicalists" or the "1847 group"), Hermann Helmholtz, Ernst Brücke, Carl Ludwig, and Emil DuBois-Reymond, also pushed back hard against all versions of the vitalist tradition. Using physical instruments, they studied, for example, the electrical behavior of nerve impulses, assuming that nerve impulses *were* electrical impulses, plain and simple. Their model saw nerves as like the new telegraph wires, carrying messages from the "central government" of the brain to the various "subjects" like the muscles—a very Prussian model, as has oft been noted, not least by Reich.

Thus, by 1850, few defenders of the vitalist position would say that organic chemistry would never be explained by the same basic principles as inorganic chemistry or that animal movement had nothing to do with electricity. Scientists wrestled for some time over whether galvanic electricity, electrochemistry, the electrostatic generator, the electric eel, the current produced by moving magnets, and the current produced by a voltaic pile were actually the *same* phenomenon, though in the end Michael Faraday was convinced he had demonstrated the common identity of them all.[22] Resistance to the mechanistic program could no longer be as sweeping as in the past. By the mid-nineteenth century, vitalist claims said that, while much of living metabolism might be analyzed in purely chemical terms, the key difference between living and nonliving matter still had to be explained in terms of a *nonmaterial* force, often called a "vital force" ("Bildungstrieb," "Lebenskraft") that does not obey the same laws as the forces of the nonliving realm such as magnetism, light, heat, electricity, and so forth.[23] Thus, some of them concluded that questions such as the difference between living and dead matter would simply not be accessible to modern laboratory methods. Not that this was not basically a scientific question—these were scientists—but by arguing for some clear limits on the reach of physics and chemistry, they came to the conclusion that methodologically it was not accessible to solely physical-chemical methods—in principle—because the vital force was of a qualitatively different kind. Vitalistic views varied over the degree to which this basic force was amenable to laboratory investigation.

As a young student in biology and physician in training, Reich said he was extremely interested in this controversy; it had entered a new round in the early twentieth century. Since about 1900, biologist Jacques Loeb had been advocating a very aggressive form of mechanism. Loeb did an experiment demonstrating that an invertebrate egg cell could be stimulated to divide by a mere physical or chemical signal rather than only via fertilization by a sperm cell.[24] Similarly, since 1888, Wilhelm Roux put forward mechanistic ideas about the nature of embryological development and showed they could be investigated by experimental methods, a research program he called "Entwicklungsmechanik" and which was further developed in the early to mid-1890s by Hans Driesch.

Yet by 1899, Driesch had begun to doubt whether the extreme mechanist model of the organism could explain the "harmonious equipotential system" of an early embryo. He felt he could understand each step in development only by introducing a teleological principle into embryogenesis. Because Driesch was a well-respected embryologist who had done important work under the mechanistic model in the 1890s, when he declared in 1899 that he was having real doubts about the mechanistic project, and, switching to philosophy, declared his beliefs after 1900 to be "neovitalism," it was not something that could just be ignored by laboratory scientists. He revived Aristotle's term "entelechy" to refer to a teleological, nonmaterial principle that was an essential distinguishing feature of living matter.[25]

The immensely popular French philosopher Henri Bergson came to somewhat similar conclusions, positing an élan vital that animated living matter and was not necessarily accessible to modern laboratory science.[26] Both this and Driesch's entelechy were quite reminiscent of earlier "vital force" conceptions. To many researchers, however, impressed by the progress of Loeb and of Entwicklungsmechanik, of physiological chemistry, and after 1910 of *Drosophila* genetics, this seemed too reactionary a rejection of the newly evident power of experimental method in biology. Many scientists wrestled with the limits of excessive mechanism or reductionism but did not seek an outright rejection of materialism and recourse to a nonphysical "vital force." Among these in Vienna we may note Freud's colleague, the physiologist and psychologist Josef Breuer, Freud himself, and Darwinian biologist and neo-Lamarckian heredity theorist Paul Kammerer, in whose work Freud took an interest.[27] Breuer's biographer, Albrecht Hirschmüller, makes clear that Breuer's training under physiologist Ewald Hering at Prague was an important source of his antireductionism but also made him skeptical of mystical vital forces: "The basis of [Hering's antivitalist] view was that life was not to be considered as a mechanistic process operating on purely physico-chemical principles, but that attention should be paid to its characteristics of spontaneity, autonomy and purpose, and these general features looked for in all particular manifestations of life."[28] I will quote at length from several sources, including Breuer, to show in detail some of the ideas being suggested to resolve the mechanism-vitalism dilemma.

In an 1895 critique of an essay by DuBois-Reymond, Breuer argued for a much more nuanced kind of vitalist position.[29] Breuer observed that the idea of a "vital force" was antiquated, that the word "force" was now recognized to be ambiguous and was rightly going out of fashion, and that "as the concept and the word 'energy' become more familiar," it was this concept that would more usefully capture the key vital element. He expressed it by way of analogy:

> Let us suppose that the ancient physicists had possessed all our knowledge concerning mechanics and calorifics, whilst their knowledge of electrical phenomena was limited to those few things with which they were in fact familiar. Then they would have been perfectly justified in trying to explain these phenomena in terms of the known facts of mechanics and calorifics; and the presumption would have been perfectly justified even after the attempt had failed. Nevertheless, those people would have been right who considered the attempt to be impossible and declared that those phenomena originated in some other "force" lying beyond the then known bounds of physics. They would not thereby have postulated that the [electrically charged] amber contained a soul; but they would have maintained that the phenomena of attraction and repulsion manifested by the amber were not explicable by means of mechanics or calorifics.[30]

The secret of the energy of life was not inaccessible *in principle* in this view, but only inaccessible now given the tools available at the present day. Breuer expanded, using another analogy:

> My view is that the physicalist school [the 1847 group and their followers] have no more investigated the essence of life than Columbus explored the coast of Japan, but, like him, in striving to do just that they have achieved the most fruitful and important successes and conquests. These frustrated hopes concerning the central problem should therefore not be regarded as misfortunes, but rather as richly rewarded: but it is necessary to recognize them. . . . If we can realize that only the outworks of organic phenomena have fallen to physico-chemical analysis whilst the fortress itself remains untouched, then we must doubt the validity of the assertion that it is only a question of time and of cracking the central problem also by means of physics and chemistry.[31]

Breuer closed with a final comparison: all the mechanics of bird flight, he said, are a purely mechanical problem, including gravity, air resistance, and the structure of the bird's body, wing, and feathers. With a flying machine the entire analysis could be physical and chemical, Breuer argued, "involving only the chemical and calorific factors, [but] this is not the case with birds. Here the motor principle, muscular power, eludes physical interpretation; thus, everything has been solved in terms of physics, except the main problem. It is the same with all biological problems. They involve certain definite elements intelligible in terms of physics, but at just that point where the living creature takes over, this form of explanation ceases to apply. That is why we are vitalists."[32]

(By 1940 and later, Reich would almost certainly refer to Breuer's position here as "functionalism" rather than "vitalism" and argue that Breuer had hereby moved to a more successful middle position, resolving the extremes of mechanism and vitalism. This shares a great deal in common with how Reich felt he had resolved the conflict, and with his position by 1939–1940, which he then began to call "energetic functionalism.")[33]

Against this background, Freud first coined his concept of libido, which he sometimes referred to as possibly a chemical substance but at other times spoke of as a sexual energy. In *Three Contributions to the Theory of Sex*, he said, "'Libido' is the motor force of sexual life. It is a quantitative energy directed to an object."[34] Describing the origins of psychoneurosis via the inability to attain normal sexual gratification, Freud adds that "the libido behaves like a stream, the principal bed of which is dammed; it fills the collateral roads which until now perhaps have been empty."[35] He also described libido as "something which is capable of increase, decrease, displacement and discharge, and which extends itself over the memory traces of an idea like an electric charge over the surface of the body."[36] Freud clearly did not conceive of libido as a vitalist force, but rather of something capable—eventually if not at present—of being measured, quantitatively, in a laboratory.

Physiological chemistry had made strides as well by the last quarter of the nineteenth century, which were beginning to change the language of the debate between mechanists and vitalists. Louis Pasteur argued as early as 1860 that optical activity (of biochemical stereoisomers) is "the only distinct line of demarcation that we can show today between dead and living matter." He believed that "cosmic asymmetric forces" were responsible for this. In an 1899 paper, British chemist F. J. Allen reviewed a number of definitions of life from the previous fifty years. One of Allen's most useful observations is that most of the attempted "definitions of life" up to that point have some merit, but they all tried to find a single key formulation to capture an extremely complex, many-sided phenomenon. "Life," he said, "is too complex to be described in a concise aphorism."[37] Pasteur was pointing out that producing purely the d-glucose enantiomer, solely the l-amino acids, and so forth, is one of the most crucial features of living systems we know on Earth, but as recent work on non-50:50 enantiomers in extraterrestrial amino acids from meteorites has shown, even that formerly sharp line of demarcation has begun to blur into a spectrum.[38]

"Every one of these definitions has some merit," said Allen, "and yet none of them can be said to have hit the mark." Reading them one after another only produces confusion in the mind of the reader, he noted, while a series of even fair approximations ought to help the reader at least begin to envision the thing being described. Further, "Even the best of them is only like a dictionary definition, which amplifies the expression of a word but calls up no idea of the thing to which the word is applied. For instance a man reading in the dictionary that a plough is 'an instrument for turning up the soil' would form no conception of a plough unless he had been acquainted with it beforehand. And so with the most explicit of the foregoing definitions, no one previously ignorant of life would be able to form any notion of it from learning that it is 'the definite combination of heterogeneous changes, both simultaneous and successive, in correspondence with external co-existences and sequences.'"³⁹ The final definition he cited is from Herbert Spencer's 1860 *Principles of Biology*.

After cautioning that previous single definitions of life were inadequate, Allen then plunged into a review of what is known about the biochemistry of metabolism and a discussion of the central role of "energy-traffic" in cells. Allen had first advanced his argument in a talk at the British Association meeting of September 1896; he fleshed it out fully in the 1899 paper. This is in essence the old analogy of Heraclitus, between life and a flame. Life in this view depends fundamentally upon an environment full of reactive chemicals that can combine to release energy, but do so only when catalyzed so that the energy flux may be channeled by the living system. The phenomenon of "trading in energy," Allen asserted, is "the most prominent and perhaps most fundamental phenomenon of life." Living systems carry out oxidations of some molecules, catabolism, in order to capture the energy released and incorporate it into the building up of other molecules or biopolymers, such as proteins, the process of anabolism. "The building up of these large molecules is always accomplished by slow steps; but when formed the said molecules are very . . . *labile*."⁴⁰ He compared such large molecules to a house of cards, because it could be slowly broken down piece by piece, or it could collapse all at once. The larger and more complex the structure, he said, the greater the potential energy it contained and thus, the more susceptible it was to sudden collapse. But this instability was also constructive, since those molecules with high potential energy stored in them were capable of driving life processes. "It should be distinctly understood," added Allen, "that it is not the mere size of the molecules that makes them labile, but rather the manner in which the elements are linked together, and the amount of potential energy which is occluded in the molecule."⁴¹

In his enthusiasm for this approach, ironically enough, notwithstanding his opening caveat, Allen soon arrived at his own pet, too-concise formulation, which he called a law: that "every vital action involves the passage of oxygen either to or from nitrogen."⁴² Nitrogen has generally been "admitted to be indispensable to life," said Allen. But he concluded that nitrogen's role in life had been underrated; that its properties offer clues to the basic nature of life itself. Allen thought all "the phenomena of life (i.e., such life as we know) are the phenomena of nitrogen compounds" and that every phenomenon of life coincides with a change in a nitrogen compound. As he said, "In short, it is to nitrogen that living matter owes its power of exchanging energy, its variability, instability, lability."[43] Nitrogen was "the secret of life" for Allen, much as DNA was later for James Watson.

Allen cited another German, Oscar Loew, who had recently suggested in his book *The Energy of the Living Protoplasm* that the basis of life is continuous atomic motion within large unstable organic molecules. Allen commended the theory for explaining some phenomena, but thought "Dr. Loew does not lay enough stress on the functions of Nitrogen, to which I personally trace so many of the phenomena of life. . . . It is chiefly owing to the instability of Nitrogen compounds that continuous atomic motion is possible."[44] Allen gave priority to N, among the CHON.

Allen said he was impressed with the 1875 theory of the German physiologist Eduard Pflüger, which also pursued the analogy between life and fire (Pflüger titled his paper "Regarding Physiological Burning in Living Organisms"). Pflüger argued that cyanogen (C-N) in the Earth's early atmosphere is the defining unstable compound of living reactions.[45] Pflüger believed that proteins come in two fundamentally different types: dead, or storage protein such as albumen (egg white), and *live* protein, such as made up the catalytically active living protoplasm. Only the latter was composed with its nitrogen in the form of the unstable, labile cyanogen radicle, Pflüger felt. He thought cyanogen formed only when nitrogen compounds come in contact with red-hot coal; he therefore supposed that "life comes from fire and, in the last analysis, was formed at a time when the Earth was a red hot mass."[46] Allen disagreed on this point, saying that in his nitrogen theory one need not assume with Pflüger that only Earth in the state of a "glowing fire-ball" could allow the first origin of life. Allen found it more believable that the circumstances that support life would also favor its origin. In other parts of the universe, "even if there be no other world where nitrogen is the critical element, yet other elements may be in the critical state on the moon, or Mars, or the sun, or even in unknown and unimagined regions of the universe."[47] In other words, other elements than nitrogen might allow the same kind of strained, high-energy bonds with an organic molecule in those settings, Allen suggested.

Allen did accept Pflüger's conjecture that the carbon and nitrogen of living matter were combined as cyanogen. He thought that at least living substances had a near approach to that structure. Summing up his views "as to the structure of the active molecule of living substance," Allen stated that a very large molecule containing a lot of potential energy, especially in the bonds between nitrogen and the rest of the molecule, was the thing that made life possible. Thus, he said, "Death consists in the relaxation of the strained relationship of the nitrogen to the rest of the molecule. When thus 'the silver cord is loosed,' the released groups fall into a state of repose." When all the theories he discussed were summarized, Allen said the key points were the following:

1. Every vital phenomenon is due to a change in a nitrogenous compound . . .
2. There is no vital action without the transfer of oxygen, and the transfer is performed by nitrogen (often assisted by iron).
3. In the anabolic action of light on plants, the nitrogen compounds are affected primarily, and the CO_2 and water secondarily.
4. In the living and active molecule the nitrogen is centrally situated and often in the pentad [valence 5] state. In the dead molecule it is usually peripheral and in the triad state.
5. The oxygen store of the *living* molecule is more or less united with the nitrogen, but passes to some other element at death.
6. The nitrogen of the living molecule is combined in a complex and perhaps changeable manner—the compound resembling in some respects the cyanogen compounds, in other respects the explosives such as nitro-glycerine; other analogies are also traceable.[48]

Allen argued that life's cause might belong to the metaphysical domain, "but . . . until we find very strong evidence to the contrary, we ought to assume that the cause of life is inherent in the universe,—that life is the direct outcome of the properties of matter, energy, etc."[49]

Here we are seeing a transition to a new, biochemical way of talking about the most basic characteristics of life, both structural and metabolic. Robert Kohler has argued that the emergence of biochemistry as a modern science is connected to abandonment of the protoplasm theory between about 1897 and 1905.[50] But this transition was gradual: within this new approach, Allen's attempt to hold on to some single sine qua non of life represents the deep grip of a late nineteenth-century scientific gestalt that we need to look at more closely, in order to understand why biochemical ideas that in some ways sound so modern are still shaped by it. From the 1860s through the first decade of the twentieth century, theories abounded as to the simplest "living unit." From Thomas Graham's colloid, containing "energia," to T. H. Huxley's "protoplasm," to Charles Darwin's "gemmules" of pangenesis, to Ewald Hering and

Ernst Haeckel's "plastidules," to Carl Nägeli's "micelles," to Max Verworn's "biogen," in many ways Allen's focus on nitrogen as *the* crucial "living element" is in this tradition.[51] These theories reflect the fact that from 1860 to 1900, perhaps for some as late as 1918, "explaining life" was interpreted by most investigators to require addressing all or most of the parts of a large, once-unified problem complex. This included generation, development, inheritance, sensation or irritability, and evolution.[52] Reich included a long section positioning his "third way" in the mechanism-vitalism debate in his book on the bion experiments, in which it is clear that for him, as for Bastian, Hering, Breuer, and others, understanding the biological roots of sensation, memory, and drives is central to this problem complex.[53]

The Concept of a Specific Biological Energy

The gradual teasing apart of the strands of this once-unified problem complex, for instance, the separation from about 1910 to 1920 of transmission genetics from embryology/development and its more inclusive notion of "inheritance," greatly altered discussions of the origin of life. Thenceforth also, the biochemistry of cell metabolism was described, as in the work of Soviet biochemist Alexander Oparin by 1924, without reference to any single defining "living unit." Even the notion of a "life substance," still to some extent carried along in the terms like "protoplasm" and "biogen," began gradually fading from view. Allen's paper represents a turning point in certain kinds of discourses about the origin and nature of life, the beginning of a relatively new biochemical language about "energy trafficking," but still entwined for the last time in a widely respected way with an older language about "the active molecule of the living substance." In another example less than ten years later, by 1906 J. Butler Burke was the subject of sarcastic barbs in *Nature*, for a similarly too-vague use of "biogen," in his book describing "radiobes," as Luis Campos has described.[54]

But as we saw above, figures as diverse as Breuer, Pflüger, and Allen all thought seriously about how the new concept of energy might help explain the specific character of the living without recourse to a "vital force." For some, a new kind of "living molecule" that "trafficked in energy" was unsatisfactory, and they sought instead to solve the problem by conceptualizing a *kind* of energy specific to living things (beginning with simple colloids) but obeying the same conservation law as known forms of energy. When between 1906 and 1912 Liverpool biochemist Ben Moore spoke of a distinct "biotic energy," a life energy measurable in the lab (unlike the old "vital force"), physiologist Edward Sharpey-Schafer, in his 1912 British Association for the Advancement of Science (BAAS) Presidential Address, set the new tone for the strong mechanistic program when he criticized this as empty language that should be discarded.[55] Moore responded that Schafer was rather missing his point, saying that "it seemed useless for the biologists and physiologists to shut their eyes to the fact that they had a separate science [from chemists and physicists], that they had a separate type of energy which was in a distinctive state of phenomenon, just as radio-activity was different from the ordinary electric waves."[56]

Moore's 1912 book *The Origin and Nature of Living Matter* developed and defended the idea of a specific biological energy in some detail. Moore pointed out that heat energy and electrical energy, though "mutually transmutable one into the other," are not seen as *identical*. Thus, he asked, why should the energy of the living be seen as identical to (or a mixture of) heat energy and electrical energy, even if these are observed in living organisms? Moore continues, "The position which denies the existence of a form of energy characteristic of life [Shafer's position, for example] is one of peculiar absurdity even for the pure mechanician, which can only be explained as a natural reaction from the entirely different medieval conception of a vital force which worked impossible miracles. As well because of the errors connected with the idea of 'phlogiston' might the present ideas regarding 'energy' as a whole be scouted." Since, Moore concludes, there is a need for some name for an energy specific to living matter itself, he proposes calling it "biotic energy."[57]

Moore was attempting to avoid the extreme of vitalism. But rather than fall back on a purely mechanistic view of physical-chemical reductionism, he and others at this time (c. 1910–1920) saw a specific life energy as a more likely explanation of the phenomena that such "vitalists" as Breuer had been pointing to:

> It is *biotic* energy which guides the development of the ovum, which regulates the exchanges of the cell, and causes such phenomena as nerve impulse, muscular contraction, and gland secretion, and it is a form of energy which arises in colloidal structures, just as magnetism appears in iron, or radio-activity in uranium and radium, and in its manifestations it undergoes exchanges with other forms of energy, in the same manner as these do amongst one another. There are precisely the same criteria for its existence as for the existence of any one of the inorganic

energy types, viz., a set of discrete phenomena; and its nature is as mysterious to us as the cause of any one of these inorganic forms about which we also know so little. When we know *why* hydrogen and oxygen unite to form water, then we shall be near to understanding the balance of organic colloids. In fact, the knowledge may come to us in the reverse order.[58]

In other words, Moore hoped the study of colloids and their specific chemical properties might serve as the bridge that would lead to more fundamental explanatory power in both inorganic chemistry and the living realm. In this, he was carrying on a conceptual tradition begun by the famous English chemist Thomas Graham in 1861 and picked up by numerous scientists interested in the origin of life, not least Henry Charlton Bastian.[59] The progress of colloid chemistry and its relevance for Reich's bion experiments will be discussed further in Chapter 4.

A very similar argument was made for a specific biological energy by experimental biologist Paul Kammerer in the 1915 first edition of his textbook *Allgemeine Biologie*. Kammerer saw himself as a Darwinian, interested, like Darwin, in inheritance of acquired characteristics. Kammerer was no vitalist, identifying strongly with the modern experimental tradition in biology. (Our modern absolute dichotomy between Darwinism and Lamarckism was created by biology textbooks well after 1915. Darwin did not see this as an either-or proposition and felt that both natural selection and Lamarckian "use and disuse" were active forces driving evolution.)[60] I will reserve a discussion of Kammerer's neo-Lamarckian experiments until later and for now focus only on his interest in a specific life energy.

Kammerer's 1915 textbook stated his position as follows:

> Just to prevent the author from being lumped together misleadingly with one of these lines of thought [mechanism or vitalism], let him be allowed to make his general standpoint in the matter clear. Accordingly, neither the mechanistic nor the vitalistic hypothesis can be supported at present with enough certainty so that one might entrust oneself to it blindly. No one can claim that there is a special life-force (vital energy, entelechy) that can sovereignly ignore the law of cause and effect that rules inorganic Nature and thus fall outside the framework of all the other physico-chemical energies; it is just as impossible for someone to *prove* the opposite.... Were I finally also to make known what

I personally think to be most probable, and thereby make . . . an unproven, at present unprovable scientific confession, then I would have to say: the existence of a *specific life energy* [*besondere Lebenskraft*] seems entirely probable to me! That is, an energy that represents neither heat, nor electricity, magnetism, motion (including vibration and radiation), nor chemical energy, nor a mosaic of all of them together, but rather an energy that specifically belongs only to those natural processes that we call "life." But it does not therefore restrict itself to those natural bodies that we call "organisms." Rather it is at least present in the formative processes of crystals. For this reason, to prevent misunderstandings, it would perhaps be better named *"form-energy"* than life-energy. But there would be nothing super-physicalistic about it, even though it could not be identified with any physical energies known to date; no mysterious *"entelechy"* (Aristotle, Driesch), but rather a genuine, natural *"energy."* Only, just as electrical energy is bound to electrical phenomena, chemical energy to chemical transformations, it is bound to the phenomena of life, form-giving, and transformative. Above all, it is subject to the law of conservation of energy: in the appropriate manner changeable into other kinds of energy, the way heat transforms itself into motion and motion into heat.[61]

What did all this mean to Wilhelm Reich? Reich tells us the answer in the 1940 manuscript of what later became his scientific autobiography up to that point.[62] As a young medical student in Vienna beginning in 1918–1919, just decommissioned as an officer from the Austro-Hungarian Army at the close of World War I, Reich says he was fascinated by the mechanism-vitalism debate. The question "What is life?," he claims, had drawn him intellectually for some time. He says that he found Driesch's critiques of the overreach of mechanists compelling but that he thought unconvincing Driesch's attempt to explain living matter by the concept of entelechy. "It gave me the feeling that a gigantic problem was being evaded by way of a word." In that regard he was apparently more attracted to the mechanistic approach, saying, "On the other hand, the concept of the organism working like a machine was more appealing to the intellect; one could think in terms of what one had learned from physics. In my medical studies I was a mechanist and in my thinking rather too systematic."[63]

Reich tells us he also found some pertinent ideas in the work of the vitalist biologist Jakob von Üxküll, particularly his 1921 book *Umwelt und Innenwelt der Tiere*.[64] Similarly, Reich says he found much to admire in Bergson's work: his rejection of teleological finalism about living organisms, "his explanation of the perception of time-duration in mental life and of the unity of the self only confirmed my inner perceptions of the non-mechanistic nature of the organism."[65] Reich says, however, this was mostly a *feeling* rather than a knowledge he could have clearly articulated at the time. But again, he says he saw serious shortcomings with Bergson: he thought his concept of an élan vital reminiscent of Driesch's entelechy, but that neither was "satisfactory as long as it was not tangible, as long as it could not be described or practically handled." Reich believed that the supreme goal of natural science involved *tangibility* and the capacity to *practically handle* the phenomenon under study. He did, however, think that the vitalists seemed to come closer to a correct understanding of life "than the mechanists who dissected life before trying to understand it."[66]

Looking back from 1947, Reich emphasized the mechanistic grounding of Freud's ideas that were so important to him; he wrote, "Somewhere around 1919 my initial functional assumptions linked up with the theories formulated by the science of psychoanalysis, which at the time had not yet abandoned its energy-related orientation, as it has done today. Freud was, I believe, the first researcher in the field of psychology to assume the existence of a 'psychic energy.' . . . At the time, psychoanalytic theory was based on the same principles as classical physics." The science of 1919 considered that in nonliving nature matter and mass were primary and were moved and displaced by "forces." And in parallel, "in the psychological sphere 'amounts of energy' became affixed to static ideas, moving and displacing them. The ideas corresponded to the 'matter,' and the drives corresponded to the 'forces' or 'impulses' of classical physics."[67]

Thus, it seems to me, Reich was not a straightforward vitalist, as has been recently claimed.[68] He was much more inclined to the kind of resolution proposed by Moore and Kammerer. Writing about the late 1930s when he was involved in the bion experiments, Reich said he was attracted to Kammerer's concept of "a force [in nature] which is neither mechanical nor mystical, a force which regulates life without itself being electrical or magnetic or representing any other known form of energy."[69] Reich said he grasped the full meaning of Kammerer's idea only when he discovered this force experimentally as orgone energy in 1939–1940. But, he said, the idea "does not appear to have lain dormant in my unconscious" during the intervening years as he became a psychoanalyst and became interested in trying to put Freud's theoretical "libido" on a solid basis in laboratory science.[70] It bears emphasizing again, however, that this was not a clear agenda or roadmap in Reich's mind from 1919, when he first heard Kammerer's lectures, to set out to find that force. Rather, Reich's initial strong response when Kammerer posited a specific life energy was because it agreed with his own instinctive groping toward resolving the tension of his own conflicting feelings about mechanism and vitalism. The specific life energy concept guided Reich's explorations in at most a half-conscious way from 1919 to 1939. This account of Reich's will be borne out by the story he tells regarding how he found his way into a laboratory at Oslo University in 1935 and how the bion experiments unfolded afterward. Many upon first reading of Reich's concept of a specific life energy immediately think it sounds like a "vital force" and see Reich's ideas through that lens, imagining therefore that he is essentially just reviving *Naturphilosophie* or nineteenth-century Romantic science under a new name. As this discussion has made clear, the specific life energy concept was by definition *not* vitalism. Much as Lloyd Ackert shows that Russian microbiologist Sergei Vinogradskii transformed the cycle of life concept "from a popular, quasi religious holistic view of nature into an experimental method applied in plant physiology, then microbiology" and other fields of research, I demonstrate here that Reich's "specific life energy" enabled and grew into a practical, experimental research program.[71] Similarly, his approach demonstrated new connections between the psychology of emotions and drives, physiology, and microbiology.

Since Reich said he was so struck by the quoted passage in Kammerer's *Allgemeine Biologie*, it is worth noting that in the 1920 second edition of his text, Kammerer *removed* reference to a specific life energy. By 1920, a similar fate overtook Moore's biotic energy concept. Lacking the authority of the BAAS Presidency that Edward Sharpey-Schafer had in dismissing Moore's idea, as well as the publicity apparatus guaranteed to Schafer's address, by 1920 or so Ben Moore's argument for a specific biotic energy was rather drowned out in the changing tide. In addition, after his discrediting in 1926 over neo-Lamarckism, which will be discussed in the following section, Paul Kammerer's support for a specific life energy would likely have done more to discredit the idea than to help it. So it is perhaps not surprising that the concept—never vocally supported by large numbers of scientists—waned by the late 1920s.

Heredity, Teleology, and Moralism in Biology and Psychiatry

Reich said he thought Kammerer a "highly talented biologist," and indeed Kammerer enjoyed a prominent reputation by 1915–1920.[72] Even those deeply skeptical of his interpretation of his data admitted he was unexcelled, almost unparalleled, in his skill in captive breeding of animals, especially amphibians and reptiles. Kammerer claimed to have proven experimentally that certain traits acquired during an animal's lifetime could then be transmitted through heredity to the descendants of that animal. This could be a means, in addition to natural selection, by which evolution might occur, said Kammerer, supporting the basic model of evolution (or "species transmutation") put forward by Jean Baptiste de Lamarck in 1809. The young medical student Reich heard Kammerer lecture on those experiments as well and found them "most interesting."[73] Kammerer, Reich noted, was influenced in his support for Lamarckian inheritance by his collaboration with Viennese physician Eugen Steinach. Steinach's research showed that it was the interstitial cells in the gonads—not those producing sperm or eggs—that were responsible for secondary sex characteristics. Steinach surgically implanted interstitial gonadal tissue in aging animals and showed this led to a new expression of secondary sex characteristics. The success of these experiments also led Steinach to offer a similar procedure, the so-called Steinach operation, to aging Viennese patients who hoped to experience rejuvenation. Freud, in 1923, was among those who tried it.[74] This, too, appealed to the young medical student who was so fascinated with the scientific understanding of sexuality.

Reich was impressed, as well, with the power of environment to modify organisms. He felt Kammerer and Steinach counteracted the too-sweeping claims of Mendelian heredity theory, which many, especially in the eugenics movement, said could explain simply everything from alcoholism to prostitution ("sexual licentiousness"), from criminality and pauperism to the tendency to become a sailor ("thalassophilia").[75] Reich also noted that Kammerer believed in the possibility of abiogenesis under current Earth conditions: "He was a convinced advocate of the natural organization of living matter from inorganic matter, and of the existence of a specific biological energy. Of course, I was not really capable of forming a valid scientific opinion about all this, but I liked these scientific views." Reich claimed he also found it instructive that "Steinach as well as Kammerer were being opposed violently. When I visited Steinach one day, I found him tired and worn out. . . . Kammerer later committed suicide."[76]

One of Reich's biggest problems with the mechanistic approach, especially as exemplified by Mendelian heredity theory and the new gene theory, was that Reich said he found most of that work deeply tied up with moralistic teleology, hearkening back to roots in Cartesian dualism. As Reich expressed it, "The cell was supposed to have a membrane *in order to* better protect itself against external stimuli; the male sperm cell was so agile *in order to* better get to the ovum. The male animals were bigger and stronger than the females, or more beautifully colored, *in order to* be more attractive to the females; or they had horns *in order to* beat off their rivals. The workers of the ants were sexless *in order to* be able better to do the work . . . and 'nature' had 'arranged' this or that in such and such a fashion *in order to* achieve this or that end. In brief, biology was also dominated by a mixture of vitalistic finalism and causal mechanism."[77] Reich claims he found this most pronounced of all when it came to sexuality, which was almost universally supposed to exist solely "in order to procreate." Reich conducted a survey of the sexology literature in the summer of 1919, and he read the resulting paper to the sexological seminar he and a group of fellow medical students had convened on their own. With the sole exception of Freud, all the researchers he surveyed (Forel, Bloch, Moll, and Jung), as Reich saw it, "believed that sexuality was something that at the time of puberty, descended on a human being out of a clear sky. . . . Where it had been before, nobody seemed to know." Further, Reich said, in the sexology and psychology of 1919 as he was introduced to these fields, almost every behavior was explained by arbitrarily invoking the existence of a biological instinct. And often these "instincts" were attributed—just as arbitrarily—to a hereditary factor. Obviously this allowed for easy infusion of ideas that were actually moralistic into what was supposedly objective science. As it appeared to the young medical student Reich,

> There was a hunger instinct, a propagation instinct, an instinct to exhibit, an instinct for power, an instinct for prestige, . . . a cultural and herd instinct, . . . instinct to suffer pain, one for masochism, sadism, transvestitism, etc., etc. In brief, it all seemed very simple. And yet, terribly complicated. One did not know one's way out. The worst of all was the "moral instinct." . . . Sexual perversions were considered something purely diabolical and were called moral "degeneration." So were mental diseases. Whoever suffered from a depression or a neurasthenia, had a "hereditary

taint," in other words, was "bad." . . . One has only to read Wulffen's book on criminality or the psychiatric texts of Pilcz or any of his contemporaries to ask whether one is dealing with science or moral theology. Nothing was known then about mental and sexual disorders; their existence aroused moral indignation, and the gaps in science were filled in with sentimental morality. According to the science of that time, everything was hereditary and *biologically* determined, and that was that.[78]

Any of a number of good recent histories of the eugenics movement, such as Diane Paul's *Controlling Human Heredity*, bears this assessment out clearly.

Reich instinctively rejected such moralistic and metaphysical ideas, which in his mind were masquerading as scientific. He said he searched the existing primary literature in biology, and that he carefully read Mendel and Darwin, to see whether the facts were there to substantiate the rampant biological and genetic determinism. In Mendel, however, he found "on the contrary, much more substantiation for the variability of the process of inheritance than for its alleged wooden uniformity." Similarly, Darwin's theory of natural selection "also corresponded to the realistic expectation that, although life is governed by certain fundamental laws, there is nevertheless ample room for the influence of environmental factors. In this theory, nothing was considered eternally immutable. . . . Everything was *capable of development*."[79] Because "hard hereditarian" Mendelian biology as he found it in Vienna was so un-self-consciously dripping with moralism (later reinforced in his mind by the excesses to which the Nazi state spurred "race biology"), Reich found Kammerer's experimental evidence for inheritance of acquired characteristics refreshing and appealing.

In 1925–1926 Paul Kammerer engaged in a public debate over his evidence for inheritance of acquired characteristics, particularly in the tiny number of specimens of *Alytes*, the "midwife toad," which had survived World War I in Vienna. At one point, evidence emerged that a critical specimen had been altered by injection of india ink, to appear more in line with Kammerer's theory. It was unclear whether Kammerer had done this; the possibility of a disgruntled laboratory assistant sabotaging him was raised and the matter never finally settled. Kammerer, who had a turbulent personal life, committed suicide not long afterward in the mountains outside Vienna. For most biologists, this has ever since been taken as a straightforward admission of guilt on Kammerer's part. Notwithstanding apologists then and later—most notably Arthur Koestler in the late 1960s and early 1970s—his suicide was read far and wide as the final death-blow for Kammerer's evidence for neo-Lamarckian inheritance. If some left-leaning biologists later took a revived interest because of the work of T. D. Lysenko in the Soviet Union, the discrediting of that work in the West after 1948 drove a still more final nail in the coffin-lid of belief in inheritance of acquired characters.[80]

Despite the discrediting of neo-Lamarckism, and notwithstanding the popularity of eugenics, even among many geneticists, it is worth recalling that before the "modern synthesis" of Darwinian evolution with population genetics in the 1930s through early 1940s, many biologists who took Darwin seriously, and most embryologists, were deeply skeptical that Mendel's work had much relevance for traits of any importance. There were two main causes of skepticism: first, if Mendel was right about "hard heredity," they believed, then his ratios applied only to trivial traits. The central question of those who previously studied inheritance (embryologists) was: how is the entire species body plan transferred with fidelity from one generation to the next? Compared to this question, such minor characteristics as whether a pea was wrinkled or smooth or whether a fruit-fly had bent bristles seemed to pale into insignificance. Even T. H. Morgan, originally an embryologist of *Drosophila*, shared this harsh skepticism about Mendelian transmission genetics until his rather sudden "conversion" in 1910 after discovering the Mendelian pattern of the *Drosophila* white-eye mutation.

But well after the *Drosophila* school, centered on Morgan's "Fly Room" at Columbia during the 1910s and 1920s, produced some spectacular successes with chromosome mapping, the second basic conflict that many biologists saw, with Darwinian evolution, persisted. If Mendel was right, then most traits should come in only one of two discrete forms—in the case of garden peas, smooth or wrinkled, tall or short, and green or yellow. Three forms might occur in the occasional case of co-dominant inheritance. But for Darwinian natural selection to make sense, it was key that most traits of any importance for survival should occur in a broad spectrum of variation. Indeed, Darwin's cousin Francis Galton had been the first to emphasize that most traits of interest—human height, intelligence, and so on—fell in a "bell curve" distribution. Only the completion of the modern synthesis, bringing, for example, the work of the Chetverikov group in the Soviet Union to the West with Dobzhansky's 1937 book *Genetics and the Origin of Species*, as well as with Ernst Mayr's 1942 *Systematics and the Origin of Species*, among other key works, did the mathematical treatment of gene distributions—"population genetics"—finally resolve this apparent incompatibility.[81] Thus, even into the 1930s and early 1940s, Freud and Reich were hardly alone in their skepticism about "hard heredity" and simplistic "beads-on-a-string" models of genes as discrete, hard objects. One did not need to have neo-Lamarckian sympathies to share this skepticism. And there were clear scientific reasons for such skepticism, not only "political" ones, such as revulsion at the excesses of the eugenics movement or of Nazi-supported "race biology."[82] As Eliza Slavet has made clear, for example, Freud's support of neo-Lamarckism was not "just because he was out of date" in modern biology by the 1930s as his biographer Ernest Jones later tried to spin the story. Slavet shows that after Julian Huxley popularized the new Darwinian synthesis in the 1940s, Jones tried in 1953 to put it down as a peculiarity of Freud that Freud realized Lamarckism was disproven but could not himself let go of it intellectually. In reality, Freud was aware of biology's rejection of Lamarckism but considered that a power play, still in 1939 not based on any convincing evidence disproving neo-Lamarckism.[83]

Holism/Organicism

The tide of reductionism swept into life sciences more strongly than ever at the turn of the last century, after the start of Roux's *Entwicklungsmechanik* program and Jacques Loeb's discovery of artificial parthenogenesis. But the inevitable ebb followed, as the extreme reductionism of Loeb and his followers overreached. It was countered by what Garland Allen has called "holistic materialism" in the work of, for example, Henderson and Cannon in physiology; by the "field theories" of Paul Weiss and others in embryology; by a wide range of programs under the name of "theoretical biology," from Ludwig von Bertalanffy to Julius Schaxel; and by another wide range of programs under the banner of "dialectical materialism," taking their charter from Engels's *Dialectics of Nature*, Lenin's *Materialism and Empirio-Criticism*, and other Marxist philosophy of science works. Of course, "holism"—then as now—lent itself to a number of misuses because of its vagueness. It could be used to support Nazi "volkish" ideology and sloppy clinical practice or astrology.[84] Indeed, as Gilbert and Sarkar have shown, the vague and varied ways in which "organicism" was used caused many scientists to conflate it with vitalism and outright rejection of mechanistic methods.[85] Nonetheless, it also captured widespread dissatisfactions with a purely mechanistic approach. Talented scientists in many fields—not least life sciences—sought for such formulations because they recognized the limitations of pure mechanism to account for the phenomena they observed.

As mentioned above, historian Garland Allen distinguished among the highly mechanistic version of materialism deployed by Jacques Loeb and newer, more nuanced materialistic research by physiologists such as Lawrence Henderson and Walter Cannon, an approach Allen dubs "holistic materialism." Allen states that "this change was more than the replacement of a simpler theory by a more complex one."[86] He argues that the change was nothing less than a philosophical move to a new way of looking at the organism by mechanists, as a complex organic whole system, rather than as a mere "cog and gear" type of machine. In many ways, Breuer, Gurwitsch, or others who had previously felt the need to call their position some qualified version of vitalism would now have found themselves comfortable in this new approach; this ought not to be overlooked because of the retrospective label "holistic materialism."[87] Soon, however, researchers thinking along these lines had a philosophical banner under which to gather.

After Alfred North Whitehead's 1925 Lowell Lectures at Harvard, many physiologists who had been working toward a middle path between vitalism and extreme mechanistic materialism saw in his formulation of "organic mechanism" a clear vision of the new philosophy of the organism toward which they had been groping. L. J. Henderson reviewed Whitehead's lectures in 1926 in the *Quarterly Review of Biology*. Whitehead attempted to bridge the gap between vitalism, mechanism, and all positions in between by emphasizing *events*, the importance of process in living organisms, rather than material units or biochemical structure. Henderson, among others, was greatly stimulated by Whitehead's view because, as Allen puts it, "it fit well into his own research program, physical chemistry, where atoms and molecules as material entities are less important than the *interactions*, the processes in which they are involved."[88]

Both Loeb and "holistic materialists" were materialists, after all, Allen emphasizes. Thus, "they believed that biological phenomena were the result of the interaction of atoms and molecules in accordance with discoverable (although not necessarily discovered) physical laws." If one adds "energy" to "atoms and molecules," this certainly applies to Wilhelm Reich as well. But while the mechanists saw the properties of the whole as "derivable from the properties of the separate, individual parts," in Allen's account, "to holistic materialists, the properties of the whole were derivable partly from the properties of the separate parts, but also from their properties when acting in consort."[89] They recognized what today are called "emergent properties," only present in the whole, intact system—and, as Reich came to think, functioning in a way that makes sense only in the context of the whole system.

During the late 1920s and early 1930s, a number of biologists continued striving toward an "organicist" or nonreductionist view of the organism. Much attention was focused particularly on the problem of organization, for example, during embryonic development. Hans Spemann had demonstrated an "organizer" region of the embryo, capable of orchestrating the symphony of movements and developments of its different parts, perhaps via chemical signals. For others, this was still too mechanistic. Austrian embryologist Paul Weiss attempted to create a "field theory," in which development was directed not by chemical signals but by a "field" (a term usually used to describe electromagnetic forces)—presumably an energy or force field.[90] Many biologists saw mitogenetic radiation as a model for such an energy field capable of directing biological processes. Russian biologist Alexander Gurwitsch had discovered in 1922 that actively dividing cells from an onion root tip could stimulate cell division in other tissues brought close to, but not in contact with, the tip in which mitosis was originally active. The separation suggested the possibility that some radiant energy was involved. Experimenting further, Gurwitsch found the effect could be blocked by interposing a glass slide between the "sender" root tip and the "receiver" tissues. A quartz slide, by contrast, would allow transmission of the influence. Since this made it highly unlikely a chemical signal was involved, Gurwitsch coined the term "mitogenetic rays" for the emanations; their transmission through quartz suggested the rays had wavelengths in the ultraviolet range. Dozens of scientists in many labs across Europe and in the United States confirmed the phenomenon and found it could also be detected from actively dividing yeast cells and from cancer tissue.[91] Many other labs claimed to be unable to replicate mitogenetic radiation experiments.

Given Weiss's stature as a scientist (including in the United States after his emigration here) and the widespread excitement over mitogenetic radiation, even as late as 1946–1947 John Tyler Bonner felt compelled to rule out "rays" or "energy fields" as a means of cell-to-cell communication.[92] In his famous experiment demonstrating that release of a chemical compound was the intercellular signal between separate cells of the slime mold *Dictyostelium discoideum*, directing individual amoebae to aggregate together to form a complex, differentiated fruiting body, Bonner felt he must convince a strong school of thought led by Weiss, who still believed some kind of rays or energy fields were the most likely agent of communication between cells.[93]

Julius Schaxel was an experimental embryologist, the last student of Ernst Haeckel, and extraordinary professor at the University of Jena since 1916. In the revolution in Germany that began on November 9, 1918, Schaxel joined the German Socialist Party and at the same time declared a deep crisis in the life sciences that could be cured only by creation of a "theoretical biology," which he sought to begin with a series of *Abhandlungen zur theoretischen Biologie* beginning in 1919.[94] By 1924 he synthesized his Marxist ideas with his attempts to create a theoretical biology, stating that the crisis in biology was the result of a crisis in capitalism. His involvement in socialist university reform efforts caused him to be labeled a "Red professor," an anomaly among the usually very conservative German professoriate. But in some ways, Schaxel's path prefigures that taken by Reich, led by surrounding political events to rethink his academic ideas and to seek a synthesis between Marxism and life sciences.

When, in 1927, Engels's *Dialectics of Nature* first appeared in German, Julius Schaxel had helped with the translation, working at the Marx-Engels Institute on a visit to the Soviet Union from the winter of 1924 to the summer of 1925. Schaxel saw himself as engaged in using biology to produce "fighting knowledge" for "people of the future."[95] As part of a broad socialist reform program, in February 1931, Schaxel's article "Soziale Eugenik und erotisches Kollektiv" declared that in real socialism sex would be like any other form of recreation or sport: personal "love" (a capitalist ownership phenomenon) would wither away.[96] In another article, "Das biologische Individuum," Schaxel made further arguments about the role of dialectical materialism for progress in biology by 1931, even among right-wing biologists, for example, Richard Goldschmidt, "no doubt without [their] becoming conscious of it."[97] After the Nazi seizure of power, Schaxel was forced into exile in the Soviet Union in 1933.

With regard to the problem of embryonic development, some researchers, including Ludwig von Bertalanffy and embryologist Joseph Needham, saw a powerful analogy between the growth of crystals and the growth and development of an embryo. They did not necessarily posit a specific biological energy responsible for both, as Kammerer had done in 1915. Bertalanffy, in a book in Schaxel's *Abhandlungen* series, hypothesized an organizing principle he called the "Gestaltprinzip" or "Ganzheitfaktor." Unlike Driesch's entelechy, "it was totally immanent in matter and could be studied by analytical science."[98] Joseph Needham similarly sought some kind of nonvitalist, "organicist" model of embryology, as did other members who founded the Cambridge Theoretical Biology Club in a meeting at Oxford in 1932. These included embryologist C. H. Waddington and biophysicist/crystallographer J. D. Bernal.

Needham's thought was steered toward organicism by his exchanges with theoretical biologist J. H. Woodger, beginning in 1929. Previously, Needham had also begun to take an interest in left-wing politics, first catalyzed by Britain's General Strike of May 1926 and maturing further in response to the rise of Hitler's fascism. As he immersed more in left-wing politics, Needham began to explore the relevance of the Marxist dialectic for his thinking about a materialistic but nonmechanistic approach to problems in the life sciences.[99] In this, he was paralleled in Britain at the same time by Lancelot Hogben, J. B. S. Haldane, J. D. Bernal, and Hyman Levy, the so-called British scientific socialists.

Needham was spurred to think further about dialectical materialism—the Marxist philosophy of the sciences—in particular, as were all the British scientific socialists, by attending the Second International History of Science Congress in London from June 26 to July 4, 1931. Indeed, Hogben and Needham were on the planning committee for this congress. A team of Soviet scientists attended and gave talks that influenced many in the West to engage seriously (even if often critically) with a Marxist analysis of the history of science. Most famous of these was physicist Boris Hessen's talk "The Social and Economic Roots of Newton's *Principia*." All the Soviet papers, at Hogben's suggestion, were collected in a single volume and published within a week of the congress, on Tuesday, July 7.[100] *Nature* published a detailed review of the entire meeting.[101] Afterward, a "party of scientific and medical men and women" from Britain visited the Soviet Union, organized by the Society for Cultural Relations. The visitors included James G. Crowther and Julian Huxley, who wrote a book about the trip, *A Scientist among the Soviets*.[102] On Thursday, July 2, there was a session at the congress about whether life could be explained purely in reductionist physical-chemical terms. Here B. Zavadovsky and A. F. Joffe supported Hogben and Needham's attack on "vitalism" and idealism in biology. Joffe noted that, contrary to physics always explaining biology, recently "physicists have to use biological methods for the finest measurements. He quoted the experiments of Prof. Gurwitsch, who claims to have discovered 'biological rays,' in support of the closer relationship between physics and biology, which will lead in time, he hoped, to the disappearance of the 'mysterious' vitalistic conceptions."[103] Gary Werskey describes this discussion further: "After Hogben had wondered aloud whether Western fears of social decay had led to a declining enthusiasm for scientific materialism, Zavadovsky felt able to assert that 'these tendencies characterize the general disillusionment of bourgeois society in the problems of material culture.' The Soviet biologist added that, for the dialectical materialist, the dilemma posed by this particular session—could 'life' be explained in purely physical terms—was a false one. To admit that the biological level was qualitatively distinct from the physical merely implied the need for novel experimental tools and a subtler version of the materialist viewpoint."[104] What, then, was dialectical materialism, and how was it used in the debates over mechanism versus vitalism and in research on the origin of life?

Dialectical Materialism

It is sometimes difficult for modern, post–Cold War biologists to recall that there was a time in the 1930s, before T. D. Lysenko had politically bastardized the idea of "dialectical materialist biology," when many brilliant biologists used dialectical materialism in genuinely fruitful ways in their thinking.[105] This included scientists of the highest stature such as J. B. S. Haldane, J. D. Bernal, C. H. Waddington, and Joseph Needham. Even in the Soviet Union, while many scientists felt pressure to give lip service to dialectical materialism, important scientists in many fields—as Loren Graham has shown—did find dialectical materialist thought genuinely helpful in their research. These included Lev Vygotsky in psychology, for example.[106] But of all life science fields, none experienced more significant figures using dialectical materialism in their research than origin-of-life studies. Here, Alexander Oparin in the Soviet Union was a major figure, as were Haldane and Bernal in Britain. Before we look at how they deployed dialectical materialism, it is important to introduce how Reich initially encountered this philosophy and put it to use. Reich, too, claims that he considered dialectical materialist thinking critical to his research, including his laboratory work.

For those who found the mechanistic approach unconvincing, one of the phenomena that most called their attention was the existence of different levels of organization in nature: atom, molecule, colloidal aggregate, living cell, tissue, organ, organism, population, ecosystem, and so on. Each higher level appeared qualitatively different in its behavior from those below it. For Needham and many other biologists, this was a key point at which they felt dialectical materialism could help clarify what, for mechanistic materialism, seemed a puzzle.

Dialectical materialism is a form of philosophical materialism, that is, one that supposes all of nature is composed only of matter and energy. It posits that objective reality does exist, outside the human mind, and that it exhibits lawful behavior that can be understood (though perhaps never totally understood) through scientific investigation, including laboratory experiments. First sketched out by Frederick Engels in his *Anti-Dühring* and in the form of notes, which only later in the 1920s were published as *Dialectics of Nature*, the theory was then elaborated in depth in 1905 in Lenin's book *Materialism and Empirio-Criticism*.[107]

Dialectical materialism embraced many assumptions shared by other "holist" or "organicist" materialist thinkers, as well as by many mechanistic materialists. These included, for example, (a) belief that the universe is material, composed of matter-energy only; (b) that matter-energy contains within itself all factors needed to explain motion, so that no "external mover" need be invoked; (c) changes that occur in matter follow certain regularities or "laws"; and (d) laws of development of matter exist on different levels—thus, chemical systems follow only chemical and physical, not biological, laws, nor do sociological laws apply to single organisms (though biological, chemical, and physical laws do apply). In addition, (e) change is a constant feature of natural systems at all levels; because of this inherent dynamism, truly static systems do not exist. "Matter-energy" here means matter and energy considered as alternative manifestations of a single thing, but interconvertible as in $E = mc^2$.

The use of these ideas among Soviet intellectuals became common only after Stalin consolidated his power in the late 1920s. Dialectical materialist writings emphasized several important "dialectical laws," taken quite dogmatically by some but used rather more flexibly by others. These were, first, the "law of transformation of quantity into quality," which Graham argues was used by sophisticated Marxist thinkers merely as an assertion of nonreductionism from one level of organization to the next. In origin-of-life thinking, for example, for Oparin, this amounted to a claim that a living system was greater than merely the sum of its physical and chemical parts. Oparin noted that once a system reached a new qualitative level—such as the level of a simple living organism—it became subject not only to the laws of the next lower level (physics and chemistry), but also to *new* laws, for example, the law of natural selection. So for many this helped clarify a reason for the qualitative differences of living systems. They were not reducible to *only* the laws of physics and chemistry because they also obeyed additional, *biological* laws. The hypothesis of the existence of a specific biological energy, in Kammerer's sense, could fit very well into this niche within dialectical materialist theory, even though relatively few biologists still mentioned this hypothesis by the 1930s.

Second, the "law of unity and the struggle of opposites," Graham suggests, was an attempt to explain the presence and role of energy in nature. For Oparin in origin-of-life thinking, this law highlighted or suggested that the dynamism of life arises from the fundamental tension between anabolism (chemical reactions building up larger molecules out of smaller ones) and catabolism (reactions that break down larger molecules, usually releasing energy in the process). This dynamism, rather than a static DNA template, was the driving force behind genetic change, in Oparin's understanding of life.

Third, the "law of the negation of the negation" was for many an assertion of a principle of constant change in the universe. For Oparin and others thinking about the origin of life, this suggested that new qualities emerge as matter evolves, which can sometimes "change the rules" so that the previous quality can no longer appear. An example of this is the assumption, ever since Darwin's time, that once life first appeared on Earth, those existing organisms would consume all newly formed building blocks of life, thus preventing the process of prebiotic chemistry from repeating a new creation of life. If Reich was interested in this third law, it seems clear that he did not interpret it the same way as Oparin, Haldane, and Bernal, since he thought his experiments were demonstrating continuous, ongoing generation of transitional forms between nonliving and living matter (bions).

Now let us back up and look at Reich's initial encounters with Marxist thought, and dialectical materialism in particular.

Reich and Dialectical Materialism

Reich said that he initially learned of Marxist ideas at university, but he became much more interested in Marxist writings after witnessing violent riots in Vienna on July 15, 1927, in which he was baffled at how passive the socialist militia was while many workers were shot in the streets by police.[108] He signed up the same day with physicians' organizations of left parties, to make some contribution to helping the workers' movement. His interest was, from the outset, in the context of his previous volunteer service at a free mental- and sex-hygiene clinic, Freud's Ambulatorium, as well as his growing recognition from his clinical work that neurosis occurs on a mass scale; therefore, *prevention* via improved sex-education, living conditions, and other means was crucial. However, on a theoretical level, Reich also brought from his psychoanalytic work certain central intellectual questions to his study of Marx, Engels, and Lenin.[109]

Reich points out that the key question that motivated his research was always, what is psychic activity in terms of quantitative, scientifically measurable, underlying biological processes? From his 1923 article "On the Energy of Drives," this was Reich's central research question, particularly in regard to sexual drives: physically, what are they and what is their energy source? Thus, as with all the other scientific issues of his time we have already touched upon, Reich approached dialectical materialism from the point of view of this central question. This is evident in his first major discussion of dialectical materialism in 1929, in the essay where he explores what common ground it has with psychoanalysis.[110] Having been interested in Haeckel and Semon, and often citing Friedrich Albert Lange's *Geschichte des Materialismus* as a major intellectual influence, Reich thereby took on intact the older problem complex in which the biological roots of consciousness and sensation were seen as enmeshed with the basic question of "units of life" and with the origin-of-life problem.[111] Haeckel's "plastidules" and Semon's "mnemes" and "engrams" conceptualized a material basis for sensation and for memory, related to heredity. In Semon's theory, an *"engram* refers to the enduring change in the nervous system (the 'memory trace') that conserves the effects of experience across time; and *ecphory* is the process of activating or retrieving a memory."[112] By 1910–1915, Max Verworn and other mainstream physiologists had criticized such explanatory schemes as "imaginative, but of no use to the modern laboratory physiologist."[113] But Reich, coming by way of medical training, then psychoanalytic training, and at a time when many, even Erwin Schrödinger, still took Semon quite seriously, did not accept in an unqualified way Verworn's dismissive attitude.[114]

Reich had a commitment to social-political activism on behalf of workers, a social activist agenda greater than for most of the British scientific socialists.[115] Yet, despite this, Reich also felt strongly that dialectical materialist philosophy—as a method of thought—was far more important than any particular Marxist doctrine or party platform. As he expressed it in 1934,

> Marx and Engels always emphasized that every new discovery in the natural sciences would change and develop the dialectical-

> materialist view of the world. When narrow-minded Marxists oppose the acceptance of new sciences, as they so often do, they ... commit the serious error of confusing the general dialectical-materialist worldview and method with Marxist theory on specific facts. The former is much more comprehensive and durable than the latter, which, like any theoretical construct concerning matters of fact, is subject to change. For example, a theory concerning the middle classes established in 1849 cannot possibly be completely valid for the middle classes of 1934; yet the method whereby we arrive at correct conclusions about the middle classes then and now remains the same. The method of investigation is always more important than any particular theory.[116]

Reich retained this attitude long after he concluded (by 1939) his method was different from anything the Communist and Socialist Parties would ever accept or understand; he left the political parties behind when he left Europe but kept his method and called it instead "energetic functionalism" or "orgonomic functionalism."[117] (It is clear from Reich's later works that he did not distance himself from Marx, whom he held until the end of his life to be a great thinker and humanist, but from the Communist and Socialist Parties of Europe and the Soviet Union.)[118] Having flagged out those important preliminaries, let us look closely at what Reich said about dialectical materialism in his 1929 essay and its 1934 second edition.

Reich begins by declaring that the kind of materialism widely held by many Marxists is not a correct materialist conception; it is crude "mechanistic materialism first put forward by the French eighteenth century materialists [e.g., LaMettrie] and [Ludwig] Büchner and kept alive in the vulgarized Marxism of our own day." In such a mechanistic conception, the psyche itself is considered an illusory phenomenon; psychic processes are merely the manifestation of mechanical ones. Reich cites Engels's critique of Feuerbach, by contrast, arguing that the biology of the late nineteenth century had now reached enough maturity to realize that "the laws of mechanics are indeed, valid [in living organisms], but are pushed into the background by other, higher laws."[119] Many Marxists, said Reich, were overreacting to the extreme, often religious idealism of the past by declaring "there is no psyche, only a body." Unless it can be weighed and measured, it does not exist. But although Marx was deeply critical of ideas of a "soul" or any kind of "subjective" consciousness, said Reich, nonetheless, there is no question that Marx recognized the material reality of psychological activity. Thus, "in principle, the possibility of a materialistic psychology must be admitted." However, if this immediately lapses back into a mechanistic model, he argues, then class consciousness will be "explained" by some chemical formula or some mere Pavlovian reflex, and "our understanding of what pleasure (or sorrow, or class consciousness) actually is will not have advanced a jot." Instead, psychological activity must be viewed "as a secondary function bound up with and developing out of the organic world." And in a 1934 footnote he explains that his article "The Basic Antithesis of Vegetative Life" and the new science of "sex-economy" make it possible to better sum up this point:

> Freud had always assumed that the psyche is based on the organic, but he did not deduce the laws of the psyche from those of the organic. Sexual economy, if it wants to become a proper scientific discipline, must study the sexual process in all its functions, psychical as well as physiological, biological as well as social, and must equally investigate all the functions of the basic law of sexuality; thus it is faced with the difficult task of deducing sexual-psychical functions from sexual-biological functions. In this task it is assisted by the dialectical method which it consciously employs. We may put forward the following principle: it is certainly true that the psychical is a product of the organic and must consequently follow the same laws as the organic; but at the same time, it is the opposite of the organic, and in that function, it develops a set of laws which are its own and peculiar to itself. Only the study of these latter laws has been the task of psychoanalysis; and in the main, this task has been completed. Sexual economy may be expected to solve the problem of the relationship between physical and psychological functions.[120]

This passage is quoted at length because it shows how Reich's use of dialectical materialism was embedded in his developing ideas about the mind-body problem—one can see clearly his movement toward designing the bioelectric experiments—and because it shows the emergence of his concept of "simultaneous identity and antithesis." The psychic as contrasted with the organic yet simultaneously rooted in it illustrates the concept. It is another feature of Reich's concept of dialectical materialism, explained in detail in his 1934 article "The Basic Antithesis of Vegetative Life," that was not shared (or perhaps even understood) by Bernal, Haldane, Oparin, and others.

"Vegetative" life refers to the "vegetative nervous system." In English this is called the autonomic nervous system, that which governs basic, involuntary life functions. It has nothing to do with plant life, despite the awkward resemblance to "vegetables" in English.

Another facet of Reich's line of thought can be understood from a parallel set of comments. In 1929 he wrote, "The basic structure of psychoanalytic theory is the theory of instincts. Of this, the most solidly founded part is the theory of the libido—the doctrine of the dynamics of the sexual instinct.... By the term 'libido' Freud understands the energy of the sexual instinct."[121] For Reich, "the regulator of instinctual life is the 'pleasure-unpleasure principle.' Everything instinctual is a reaching out for pleasure and an attempt to avoid unpleasure." Everything that removes instinctual tension produces satisfaction that is pleasurable.[122] Freud guessed that libido was driven by a chemical process, especially concentrated in the genitals and erogenous zones. Elsewhere, as noted above, Freud spoke of libido as an energy, "something capable of increase, decrease, and which extends itself over the memory traces of an idea like an electric field over the surface of the body." Reich said Freud hoped psychoanalysis "would one day be placed firmly on its organic foundations; the concept of sexual chemistry plays an important part... in his theory of libido. However, psychoanalysis cannot methodically deal with concrete processes in the organic sphere, that being the proper concern of physiology." By 1934 Reich had changed his mind about the latter point, now thinking that a scientific, laboratory-based "sex-economy," rather than the older psychology-only version of psychoanalysis, was the logical next developmental direction for pursuing the libido theory. In October 1934, Reich stressed that, rather than chemical processes, "it is now thought that electrophysiological processes of charge and discharge in the organism are at play."[123]

The "simultaneous unity/antithesis" concept mentioned above seems to have crystallized in Reich's mind in 1933 or early 1934 (though it was inherent in his earlier opposition to psycho-physical parallelism), at least partly in the context of his developing ideas about character analytic technique and of a conflict he had with Freud over the etiology of masochism.[124] Freud had originally put the spotlight on the pleasure principle, which to Reich meant a materialistic psychology of drives. Freud also showed the dialectically embedded relationship between "development of libido and the actual life of the child," that is, social and economic conditions of the child's family, for instance. But no sooner had this view been put out there than "crude materialistic science," crude dialectical materialism, and even Freud himself wished to set up as a dialectic pair Eros and Thanatos (a death instinct). Reich accepted this as a true dialectical pair in the 1929 essay: "This corresponds to a wholly dialectical view of development."[125]

Even then, however, Reich seemed suspicious of the death instinct idea, probing for a possible fallacy, since it conflicts with his alternative damming up of pleasure as the root of destructive impulses theory from his 1927 book *Die Funktion des Orgasmus*.[126] By the time of his 1932 article "The Masochistic Character," Reich went much further and directly challenged his teacher, saying he saw Freud's pairing as a superficial, false analogy.[127] In the 1929 essay one can see that Reich has already formulated the kernel of the reaching out/contraction "Basic Antithesis" approach, which will be discussed shortly.[128] From all of the former, it is clear that very early on in the use of "dialectical materialist thinking" and terminology among those interested in biological and psychological problems, Reich's notion of dialectical materialism differed not only from idealism and "crude mechanical materialism" but also from the dialectical materialism of others, specifically because he viewed it through the primary lens of the pleasure principle and the orgasm function.

To review: by 1932 Freud had begun to invoke a *"Todestrieb"* (death instinct) to account for masochism; Reich disagreed and thought it was a desperate response of an organism so trapped by guilt and muscular rigidity that it could feel pleasure only from a feeling of being cut or pricked open.[129] Freud was putting the ontological cart before the horse, Reich argued, in trying to "psychologize" what was biological: "Freud had attempted to apply to the sources of life the psychological concepts derived from psychoanalytic investigation. This led inevitably to a personification of the biological processes and to the re-instatement of such metaphysical concepts [e.g., a death instinct] as had previously been eliminated from psychology."[130]

The concept of identity and antithesis meant that for Reich, psychic events such as drives and emotions were not merely "caused by" or "parallel with" chemical or physiological events. They were rooted in the biological core, but could manifest as diametrically opposed functions, such as love and hate. Similarly, biological functions themselves such as vegetative excitation manifested in antithetical forms (expansion/swelling and contraction/dehydration) while remaining linked as expressions of the underlying unitary phenomenon of "tissue excitation." Reich realized this was a dramatic theoretical break between himself and the Freudians, including Freud himself.

He also realized that their aversion for his Marxist politics represented a major rift over strategy: he urged social reforms to prevent neuroses, while

Freud grew steadily more nervous with the rise of fascism, about any linkage between psychoanalysis and left politics. Both rifts culminated in the machinations by which Reich was underhandedly excluded from the International Psychoanalytic Association (IPA) at its convention in Lucerne in August 1934.[131] (It is not clear if Reich realized at that moment just how far the psychoanalysts had already gone, especially the German branch of the IPA, in deliberately expelling all Jewish and most politically left members, cooperating with the Nazi *Gleichschaltung*, in order to avoid being specifically targeted for shutdown by the Nazi regime.)[132] Reich felt, in some sense, his exclusion from official psychoanalysis represented a factual recognition by his detractors, that his "sex-economy" really had moved beyond psychology and was therefore not any kind of psychoanalysis any longer. Highly symbolic of this break, he gave his latest paper moving in the direction of the organic, "Psychic Contact and Vegetative Streaming," as a guest of the Lucerne Congress since he was now formally excluded as a member.[133]

This experience helped clarify Reich's understanding of what ideas were his own, not "just extensions of what Freud said." Reich saw the IPA and former friend and Marxist psychoanalyst Otto Fenichel, still a member, as trying to paper over the major rift by putting forward a politically neutered and theoretically distorted version of his orgasm theory, and he tried to warn colleagues not to confuse this with his own work.[134] Regarding theory as well as the need to experimentally verify it, Reich also noted that in the IPA, "I know that the concept of psychophysical functional identity is gaining more and more ground. However, I postulate a different concept: *identity and, simultaneously, antithesis*, which is a problem for dialectical materialism and will have to be developed from concrete facts. . . . I think it is important not only to assert that both psychophysical parallelism and the mechanistic interaction theory are wrong while the unitary (plus antithesis) concept seems to point in the right direction. Above and beyond this we must prove experimentally what this unity *demonstrably* consists of."[135]

Reich is restating here his research goal: to identify in a laboratory the underlying unifying biological function behind apparently antithetical states, especially pleasure and anxiety. As is discussed in Chapter 2, Reich felt the bioelectrical experiments on skin potential in differing emotional states accomplished exactly that.

In his 1933 *Massenpsychologie des Faschismus*, Reich asserts from the political left what Anne Harrington describes as a holism that pushes back forcefully against the Nazis' attempt to take over holistic thinking and draft it into the service of their own political aims.[136]

When Reich moved to Norway in 1934 to begin experimental work, not surprisingly, many of the Norwegians who took an interest in Reich's work were socialists or communists (including many members of the left-wing group Mot Dag), such as novelist Sigurd Hoel (see Figure 1.1) and his wife, psychoanalyst Caroline "Nic" Hoel, as well as poet Arnulf Øverland (later poet laureate of Norway) (see Figure 1.2).[137] They all studied with Reich, in character analysis and in his new vegetotherapy.[138] Yet a diary entry from New Year's Day 1937 shows Reich already realizing how different his ideas are from most socialist or communist thinkers, trying to sort out what was known before him (e.g., "dialectical materialism") from what he himself contributed. In the latter category he includes "Functional identity of psychic and vegetative. Origin of inner contradiction—dialectic law of development," as well

Figure 1.1: Sigurd Hoel, 1937. Courtesy of Wilhelm Reich Infant Trust.

Figure 1.2: Arnulf Øverland, 1937. Courtesy of National Library of Norway.

as "the experimental proof of 'sex = life,'" and "Unconscious = vegetative; Stasis = electrical tension."[139]

In some ways, gradually working himself free of categorizing his work as Marxist or dialectical materialist resembles his earlier struggle that led to working himself free of thinking of himself and his work as "psychoanalysis." The psychoanalytic struggle he described in a letter to his younger analyst colleague Lotte Liebeck: "I've done myself a grave injustice by working for so many years under the impression that my theory of genitality was rooted in Freud. This was merely due to my father fixation. Someday I hope to make a clean break."[140] As I argue in Chapter 4, by 1939 when he called his work "energetic functionalism," no longer "dialectical materialism," Reich had made a similar break with Marxist parties. In both cases, he still greatly valued much of the work created by Freud and Marx. But he realized it made no sense to continue calling himself a psychoanalyst if the vast majority of them rejected his central contributions; similarly it was confusing rather than clarifying to call himself a dialectical materialist once he realized that most socialists and communists also rejected his central contributions.

Movement toward the Laboratory

Reich's movement into the biological realm was driven both by theory and by continued development of *clinical observations* in his patients. The more he pursued their muscular armoring, freeing of respiration and, later, of pelvic movement, the more powerful biological effects broke through; for example, the patient suddenly experienced racing heartbeat, or became totally *white*, or *blue*, that is, had powerful shifts in body fluids and energy, toward or away from the periphery. Thus, Reich's thinking was reinforced that what lay underneath character armor was muscular armor, and beneath that, powerful biological energy processes.[141]

Always (throughout the time of this story, and up to about 1950), Reich tells us that his biological thinking was *tightly linked* to his clinical observations. As noted previously, he was even more universally recognized for his gifts as a clinician—in technique and in acute ability to *observe* small but significant things others overlooked—than he was lauded for his theoretical acuity. "Der beste Kopf" of the young generation, Freud had called him.[142] Reich could *see, notice* what others overlooked. And Reich could tease out the *meaning* behind such seemingly unimportant details. Often, Reich could see the elephant in the room that everyone else's psychic defenses prevented them from seeing. I consider these traits in much more depth in Chapter 4.

Kraus's Fluid Theory

In 1926 Reich reviewed a new work by famed Berlin cardiologist and EKG creator Friedrich Kraus. In 1921–1922 Kraus and chemist Samuel Zondek had published widely regarded work on the effects of potassium ions (hydration, swelling) and calcium ions (dehydration, detumescence) on tissues. Now, in this new book, *Allgemeine und Spezielle Pathologie der Person,* Kraus argued that all living phenomena were best understood by seeing the organism as an interconnected system of membranes that partially separate compartments of ionic, electrolytic fluids. Physical laws described how "electrical tensions would develop at the borders between conducting fluids and membranes." Differences of electrical potential (voltage) in differing compartments and surfaces would provide potential energy to drive bodily processes. Movement of fluids, charges, and charged ions would result, via diffusion from areas of higher potential to areas of lower potential.

Not many of Reich's contemporaries responded as positively to Kraus's "fluid theory" as did Reich. An American specialist in psychosomatic medicine (who went on to study with Reich and train in his therapy methods), Theodore P. Wolfe, wrote in 1941, "The findings and concepts of Kraus, at the time of their publication, were revolutionary, i.e., at variance with the usual mechanistic thinking in medicine. Consequently, they met with little understanding. G. R. Heyer, in one of his books on psychosomatic medicine, states frankly that the book was too difficult for him to understand. Most critics, however, simply declare that Kraus is 'all wrong,' without, however, going to the trouble of really studying his works or of proving or disproving his findings."[143] In Reich's 1927 book *Die Funktion des Orgasmus* and in two important articles he wrote in 1934, however, it was evident that Reich's thinking about the orgasm function was stimulated by his reading of Kraus. He posited that orgasm always resulted from a four-beat process: (1) swelling of the tissues of the genitals by mechanical congestion with fluid; (2) build-up of electrical charge in the genitals; (3) a bioelectrical discharge at the height of sexual excitation; and, immediately following, (4) mechanical relaxation, detumescence of the genital tissues. Reich referred to this as the "orgasm formula."

The Basic Antithesis of Vegetative Life (1934 Papers)

In a 1934 essay, "The Orgasm as an Electrophysiological Discharge," Reich described what was known about the phenomena of orgasm from physiology and from his clinical experience of patients describing their sex lives.[144] Numerous lines of evidence, he argued, pointed to genital excitation resulting not from friction and tumescence alone. For example, he pointed out that when vaginal secretions were more viscous and oily, much greater pleasure and higher excitation was built up by both partners than if those secretions were thin and watery. This suggested that genital friction led to greater charge given an acidic electrolyte solution as electrical conductor between the membrane of the penis and the vaginal mucous membrane; thus, Reich thought it likely that the excitation was a bioelectric charging of the membranes.[145] (Hartmann's "relative sexuality" experiments on protists also suggested to Reich that the key factor in determining sexual charge—even gender—in these simple organisms was electrical and not chemical in nature.)[146] This confirmed his hypothesis that the process of sexual excitation followed a regular series of steps: first, mechanical tension (swelling of the genitals with fluid); then bioelectric charging of the tissues, increased through slow, gentle friction; followed by bioelectric discharge at the climax via contractions of the body musculature; and finally, gradual mechanical relaxation as the excess fluid left the turgid tissues.

Reich felt the "orgasm formula" offered new insights: for example, he thought it likely that the bioelectric discharge was as great or greater a source of the pleasure and release of tension experienced in orgasm than was the mechanical discharge of fluid by itself (e.g., male ejaculation). This made sense of the fact that many of his patients from therapy could ejaculate during sex (they had "ejaculatory potency") but did not experience much pleasure or release of sexual tension. Reich defined "orgastic potency" (in contrast to ejaculatory potency alone) as the ability to fully discharge the dammed-up sexual tension through involuntary contractions of the musculature of the entire body, not just the localized muscles of the genitals.[147] Over time, as he pondered how widespread his "orgasm formula" was in the living world, Reich said he began thinking about bioelectric experiments on human subjects but also about studying worms, starfish, and jellyfish—in which their pulsatory movements could be clearly seen—and even protozoa.[148]

In parallel with this thinking, Reich was inspired by Kraus's work, he said, to think deeply and read widely about the role and effects of potassium and calcium ions, and of numerous biochemical substances, on living tissues, for example, sex hormones, lecithin, cholesterin, adrenaline, and choline. Reich discussed this in depth in a 1934 article titled "The Basic Antithesis of Vegetative Life."[149]

A word is in order here about the German term "vegetative." It means "autonomic," having to do with involuntary functions of life, that is, those under the regulation of the autonomic nervous system (ANS). This could be literally translated "vegetative nervous system" in English, but already by 1940 Reich and his English translator Theodore P. Wolfe were aware that rendered thus, it was very technical; to all but specialists it suggested the word "vegetables," so they sought a better English rendering.[150]

Adrenaline had been known to stimulate the sympathetic response of the ANS—elevated heart rate, feelings of anxiety, and so on, the so-called fight-or-flight reaction. Similarly, in 1921, Otto Löwi had found that acetylcholine stimulated the vagus (parasympathetic reaction) and slowed the heart rate (work for which he won the 1936 Nobel Prize). Following up on this, two medical colleagues of Reich, Drs. Walter and Käthe Misch, had found that choline tablets could also offset anxiety symptoms.[151] Reich had long thought of sexuality and anxiety as opposite *physiological* states, rooted in stimulation of the parasympathetic and sympathetic branches of the ANS, respectively. Now he began to think about pairs of biochemical substances that exhibited the same antithetical relationship in their biological effects. Choline seemed to be parallel in its effects to the parasympathetic, adrenaline to the sympathetic. Potassium ion solutions caused swelling and turgor in tissues, parallel with sexual excitation and thus with the activity of the parasympathetic. Calcium ion solutions had the opposite effect: dehydrating tissues and causing detumescence.[152] Reich considered the idea that states of mind or emotions were parallel to but separate from physiological changes like stimulation of one or the other aspect of the ANS. He had, however, previously found fruitful an alternative conception—simultaneous identity and antithesis—so he again deployed that concept here, rejecting psycho-physical parallelism because it overlooked the deeper, underlying unitary function, of which both apparent opposites were manifestations. Seeing the underlying unity of sexuality and anxiety, Reich realized that both represented excitation in the ANS. But in sexuality that excitation appeared to be moving outward, "toward the world" (an erection of the penis eminently expressed this); in anxiety the excitation moved inward to the sympathetic ganglia at the core of the organism (causing the characteristic tight feeling in the chest with faster heartbeat). It was this approach that helped Reich understand sexual stasis and stasis anxiety, as physiological—not merely psychological—tension. This view was the logical extension of Freud's original libido theory: libido was a real, physical "something," not merely a "psychological urge." Reich was not the only psychoanalyst to take this idea seriously. Siegfried Bernfeld, for example, in collaboration with a physicist, Sergei Feitelberg, published a series of articles in 1929–1930 exploring "psychische energie" and libido, which Reich read.[153] This approach had also helped Reich formulate his solution to understanding masochism (which is discussed further in Chapter 4). Reich also concluded that "real anxiety" and "stasis anxiety" could both be understood as blocked excitation.[154] In all these cases, the payoff of this theoretical approach in therapy, he argued, was to make clear to the clinician that he must help the patient find a way to remove the blockages and reestablish the regular buildup and unfettered discharge of sexual excitation in order to cure the patient of his neurotic symptoms.

Reich thought it significant that all organisms, when in an anxiety state, roll up into a spherical form and draw in all extended parts that they can. Conversely, in expansion "toward the world," the snail extends its antennae, the amoeba its pseudopods, and so on.[155] Reich, in fact, had been struck as early as 1922 or 1923 by the mental image that an amoeba's extending of a pseudopod was in a deep essential sense functionally identical with the erection of the penis in sexual reaching out "toward the world."[156] The "crawling into oneself" felt in anxiety, Reich felt, represented the opposite side of the same coin. The autonomic nervous system was itself contractile, he observed (in transparent mealworms, for example).[157] Reich concluded that the ANS was "the protozoan, still evolutionarily present" within the metazoan.[158]

In the 1934 article, Reich describes the functioning of pseudopods in amoebae based on the cell model experiments of Bütschli, Quincke, and others: in terms of surface tension and osmotic phenomena.[159] He also cited the work of Berlin protozoologist Max Hartmann, stating that the movement of pseudopods is the result of a constant alternation of expansion and contraction. Reich reasoned that mechanical tension alone is not enough to account for what *makes* the plasma begin to flow or to expand. Vitalistic notions of "a force acting from within" must be avoided, he argued; he hypothesized instead that buildup of some kind of physical energy might be the force that initiated motion in living matter.[160] Reich pointed to Otto Bütschli's model experiments, which used oil droplets, potassium carbonate, and dilute glycerin, then squeezing the resultant droplets under a cover glass. This resulted in formation of pseudopod-like processes in which the same type and direction of fluid flow was evident as in the pseudopodia of a real amoeba. The droplets even approached and engulfed solid food particles just like an amoeba. This was interpreted as resulting from changes in surface tension caused by the potassium solution.[161]

Max Hartmann suggested that surface energy is what drives the motion of artificial pseudopodia but that—notwithstanding numerous hypotheses—no convincing explanation had yet been given of why the surface tension increased or decreased. In spite of the lack of a clear account of the cause, Reich argued that the two opposed directions of plasma flow in a one-celled organism were definitely "the prototype of the two psychic currents we have postulated; namely, the 'sexual' toward the world and the 'anxious' away from the world." (See Table 1.1.)[162]

Suddenly, this way of thinking helped clarify what had always been seen as a rather random assortment of effects by the parasympathetic versus sympathetic on various organs of the body. There was no pattern, it seemed; generations of medical students simply had to learn these effects by rote memorization.[163] In one case the parasympathetic stimulated gland functioning

Table 1.1 Antithetical biochemical pairs and corresponding functions[a]

Vegetative group	General effect on tissues	Central effect	Peripheral effect
Sympathetic			
Calcium (group)	Reduction of surface tension	Systolic heart musculature stimulated	Vasoconstriction
Adrenaline	Dehydration		
Cholesterin	Striated musculature: flaccid or spastic		
H-ions[b]	Reduction of electrical excitability, increase of oxygen consumption, increase of blood pressure		
Parasympathetic			
Potassium (group)	Increase of surface tension	Diastolic heart musculature relaxed	Vasodilation
Choline	Hydration		
Lecithin	Muscles: increased tonicity		
OH-ions	Increase of electrical excitability, decrease of oxygen consumption, decrease of blood pressure		

a. Taken from *BISA*, p. 66.
b. Regarding H+ versus OH- ion, Reich later decided the positions of these in the schema were reversed, based on new evidence from bion experiments. This was reflected in the new version of the table in the 1942 ms. of *Function of the Orgasm*. (original ms. for 1942 *Function* is in Mss. Box 8, RA). All of Reich's published works from 1942 onward reflect the change. The table I present here is that later ordering; that is, OH was originally with the sympathetic group through September 1937 when *Die Bione* went off to press. The editors of the standard Farrar, Straus and Giroux editions of Reich's works made this correction retroactively in the 1979 English translation of *The Bion Experiments* as well as in *BISA*. The translators of the 1976–1979 serialization of *Die Bione* in the *J. of Orgonomy* (and their earlier translations of "Basic Antithesis" and *Bioelectric Experiments*) did not. The *J. of Orgonomy* serialized English translation, by B. Koopman and I. Bertelsen, begins in vol. 10: 5–56 (1976) and concludes in vol. 13: 5–35 (1979).

(genital glands); in another case it was the sympathetic (sweat glands). In one case the parasympathetic stimulated muscle to contract (iris of the eye; bronchial muscles, e.g., in asthma); in another instance muscle contraction was stimulated by the sympathetic (speeding up heart rate; hair follicle muscles, producing "goosebumps").

Now, Reich could see that there *was* actually a pattern—but a pattern that made sense only at the level of biological expansion or contraction of the whole organism, not of the individual organs considered separately.[164] Reich later referred to such an approach as "thinking functionally" about the problem in question.

It was not just muscle contraction per se that gave a particular autonomic stimulus meaning that made sense for the whole organism, Reich realized. It was whether *in a total organism contraction response* such as a sharp anxiety reaction, a particular kind of response made sense for each individual gland, muscle, or organ: for example, speeded up heart rate but inhibited micturition and intestinal peristalsis; "cold sweat," meaning more activated sweat glands but constricted peripheral blood vessels. Or in a total organism response of pleasurable expansion, the iris might constrict, narrowing the pupils, while peripheral blood vessels expanded. In this view of individual functions, Reich found the core and the periphery of the organism had opposite functions. For instance, a constriction occurred in anxiety of peripheral blood vessels, while at the same time, vessels in the center of the organism dilated. Thus, there was a net shifting of blood away from the periphery and toward the core, which made evolutionary sense at the level of the whole organism if it were in a fight-or-flight situation in which injury to peripheral blood vessels was more likely. This clearly belongs in the context of holistic views in medicine and biology of the 1920s to 1930s, yet Reich contributed something new here. Reich began to think of diseases as potentially the result of excess or long-term chronic activity by the sympathetic—sympatheticotonia, as in cardiovascular hypertension or angina pectoris—or, conversely, by the parasympathetic—parasympatheticotonia, as in asthma.[165]

Reich tells us that he began to conceive of experiments that could test whether these wide-ranging physiological changes during emotional states correlated with measurements of bioelectrical charge at the skin surface.

2

REICH'S MOVE TOWARD LABORATORY SCIENCE

The Bioelectric Experiments

In preparation for the experiments he was contemplating, Reich conducted a review of earlier work on electrical measurements of the skin. Carl Ludwig and Emil DuBois-Reymond had distinguished between the skin's ability to conduct electricity (its change in resistance) and its ability to be a generator of current. Phillip Keller discovered that the potentials of the skin are constantly changing, while C. P. Richter studied how skin resistance varies with time of day and with time of year. Keller noted a striking negativity of the palm of the hand compared to the rest of the body, and found that even gentle touching of the skin causes a positive potential in the zones that were stimulated. This potential was reversible.[1]

None of these investigations in any way considered the erogenous function of the skin or any connection between this and skin sensitivity. Nor did they relate the emotions of sexuality and anxiety to skin sensitivity. In other words, they did not investigate any systematic comparison between erogenous zones and the rest of the body's skin surface. Keller had attempted to explain the negative potential of the palm during fright as a result of sweat gland activity; Reich said this is correct as a correlation. But he pointed out that this should not be mistaken for causality, since that assumption "conceals the fundamental fact that in a state of fright the whole organism reacts and in this case the palm is only one detail of the whole function." From the previous research, Reich emphasized that three main facts had been established: first, that the electromotive function of the skin was clearly established; second, that the skin possesses the characteristics of a membrane; and third, that skin potentials could not be standardized based on age, sex, or any other variable.[2]

Ivan Tarchanoff and Otto Veraguth had previously found that the skin potential changes in response to psychic stimuli, what Tarchanoff called the "psychogalvanic phenomenon."[3] Reich observed, however, that "not only the electrical reaction of the skin to affective stimuli, but also the nature of the functional relationship between the type of affect and the type of electrical reaction, is significant for psychopathological clinical experience." He noted that the previous literature contained no recognition of functional identity and simultaneous antithesis. That is, "either the physiological phenomena are taken as 'accompanying phenomena' of the affect, or the affect is considered the 'consequence' of a vegetative excitation." Reich said that in the first case, the affect is being treated as though it has no material basis; in the second case, the model is simple mechanism, reminiscent of old claims that ideas were purely a "secretion of the brain."[4] Reich, however, emphasized that the psychic and the somatic were both inseparable parts of a unitary function.

Volunteer subjects in Reich's experiments included numerous associates and students of his. Willy Brandt, in exile at the time in Norway, knew most of Reich's supporters like Sigurd Hoel, Arnulf Øverland, and Christian August Lange through socialist and social Democratic circles. His wife at the time, Gertrud Gaasland, was Reich's secretary, and Brandt was a subject in the bioelectric experiments.[5] Reich consulted with physiologist Wilhelm Hoffmann on the technical details of the experiment. Hoffmann collaborated on the experiments for a time; in addition, Reich hired Hans Löwenbach, a young Berlin physiologist, to assist with the work.[6] (Later, differences of opinion, which will be discussed further in Chapter 6, ended their collaboration on the experiments.) In addition, Dr. Rolf Gjessing, medical director of the Dikemark mental hospital outside Oslo, gave permission for Reich to take measurements on the skin of mental patients, including catatonics, for comparison with normal subjects.

The main findings of the bioelectric experiments were as follows: First, the electrical potential (voltage) of the skin over most of its surface area changes relatively little; however, the erogenous zones show marked increases and decreases in different emotional states and, even in a passive state, the potential "wanders" gently up and down rather than remaining steady as does the rest of the skin. The areas exhibiting this responsiveness were the lips, anus, nipples, penis, vaginal mucous membrane, ear lobes, tongue, palms of the hands, and—to Reich's surprise—the forehead. In "vegetatively free individuals" (those who did not suffer from chronically restricted breathing, emotional expression, or other characteristics), Reich reported, the same erogenous zones might show variations of up to 50 mV (millivolts) or more.[7] (He later used the changing volume of sound from a radio, in place of the oscillograph, as a more audible representation of the same changing potentials).[8] On all the electrical potential curves produced were superimposed tiny cardiac peaks in synchrony with the subject's heartbeat, but these were 1 to 2 mV in size, compared to the swings of 30 to 100 mV caused by varying emotional states.

The stimuli used included gentle tickling or touching (e.g., of the palm or nipple) with a cotton ball, feather, or electrode, which tended to cause increase in potential; pressure and sudden loud noises, which caused decrease of potential; sugar and salt solutions applied to the tongue; and other means. When a sugar solution was placed on the tongue of a subject who did not know sugar was coming, the potential jumped sharply higher. When the sugar was reapplied repeatedly, the amount of increase gradually dropped a bit more each time, as the tongue became "habituated." If sugar was first applied and then unexpectedly the next solution was salt, an initial sharp increase in response to the sugar was followed by an equally sharp drop in response to the "surprise" of salt. After such surprises, the tongue's subsequent response to sugar was not as great, as though the tongue had become "cautious" after its "disappointment." When the charge was measured at mucous membranes such as the anus or the vaginal opening, the charge wandered upward when the subject was in a state of pleasurable excitement; however, it declined to below zero in annoyance or, for example, when the same female subject was in a premenstrual depression.[9]

In another set of experiments, Reich had two persons of opposite sexes act as joint subjects, touching each other, and to complete the circuit, each of whom had one finger in a dish of electrolyte solution, in which one of the electrodes was also immersed (see Figure 2.1). Reich observed the tracing on the oscillograph when the two subjects pressed hands together in handshakes of varying pressure, held hands, kissed, and made other kinds of physical contact. In one curve, the kiss led to a sharp upward deflection, relative to the same couple touching hands but not kissing. But when the same couple kissed repeatedly, the positive deflection was less each time. Finally, there was even a marked downturn, which corresponded with the woman reporting that she was becoming annoyed (though the subjects in one room could not see the oscillograph tracing, since the instrument was in a separate, adja-

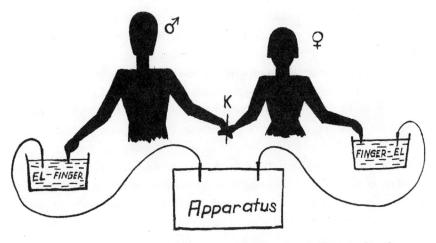

Figure 2.1: Bioelectric setup with human couple in circuit (figure taken from Reich, *Experimentelle Ergebnisse über die elektrische Funktion von Sexualität und Angst* [1937], p. 30)

cent room).[10] In a culminating experiment in this series, a naked couple embraced, with the man kissing the woman's breast. In this case, the curve started off immediately elevated to +100 mV, with peaks and valleys of about 10 mV each, corresponding to the frictional fluctuations of the kiss.[11]

At the erogenous zones, Reich reported, the curve never increased without the subject (who was in a separate room from the recording instrument and could not see it) reporting a simultaneous pleasurable feeling in that region of the body. A dip in the curve always accompanied perceived feelings of unpleasure: annoyance, fright, depression, or anxiety. (Control experiments on nonliving materials showed none of these phenomena.) Conversely, mere turgescence of a nipple did not always cause an upward deflection of the curve. And in a control experiment in which the penis was held firmly at the base to cause swelling with fluid, no increase in potential occurred on the surface of its skin. The potential of an organ rose only if the congestion with fluid was accompanied by a "pleasurable feeling of current" (streaming sensations) in that organ.[12] The findings proved that congestion with fluid, mechanical erection alone, was not the source of the pleasure sensation. The source was the increase in bioelectric charge, the movement of energy into that area. The psychic intensity of the pleasure (or unpleasure) sensation was directly proportional to the measured quantity of increase (or decrease) in bioelectric potential. Strikingly, "in several experiments, the subject was able, on the basis of his or her sensations, to tell what the apparatus in the adjoining room was showing."[13]

Reich observed that "these experiments confirm some well-known aspects of erotic relationships." For example, "In the same organ, the same stimulus can produce lively and rapid responses in one affect state, and in another affect state slow and indistinct responses, as if the organ were 'sluggish.'"[14] From this, Reich concluded that the biophysical reaction of sexual organs does not correspond to the stimulus alone, but to an even greater degree on the state of readiness of the organism. Blood flow alone, for instance as in an erection, does not produce pleasure sensations unless it is also accompanied by an increase in electrical charge. Thus, sexual excitation cannot be commanded or produced on demand. Put differently, the reaction to a sexual partner is not always, or only partially, directly correlated with the attractiveness of that person. It is essentially dependent on the state of sexual readiness. If one's partner is not in that state, Reich suggested, it is misguided to take that personally.

As in the sugar- or salt-on-the-tongue experiments, Reich noted that a "disappointed" organ reacts sluggishly and "cautiously" to new stimuli, even if one might have expected (as with sugar) a strong positive response previously. Similarly, when an organ becomes "accustomed" to a previous stimulus, the positive or negative potential change decreases, and the positive and negative reactions are nearer to the zero line.[15] This suggested to Reich that boredom or dulling in monogamous sexual relationships over time was a biological, not "just a psychological," phenomenon. Again, the extent to which a partner took this personally would probably be counterproductive to seeking ways to improve the matter.

In his earlier paper, "The Orgasm as an Electrophysiological Discharge," Reich had noted, "The fact that sexual compatibility exists between certain men and certain women is a very remarkable phenomenon which until now has remained completely unexplained and has merely been glorified in mystical terms."[16] After the bioelectric experiments, he felt the results could go some distance toward demystifying and giving a scientific explanation for such facts. The scientific understanding, he thought, would serve as a far better basis for trying to help patients solve their sexual difficulties.

Reich doubted it could ever be possible to directly measure electrical excitation of the genitals during coitus, though that result would be informative from a scientific point of view. There were numerous practical difficulties. "The manipulation [of the electrodes] alone would suppress any excitation," Reich observed drily. Also, direct measurement was subject to the possibility of broken contact if movements were not very gentle.[17] Modern sex research may still not have found a solution for this problem.[18] Nonetheless, the overall trend of the findings seemed to Reich to strongly validate his "orgasm formula," the theory that only mechanical swelling of the tissues, accompanied by electrical charging, would lead to pleasure sensations and that bioelectrical discharge was necessary, along with detumescence, to produce the decrease of pent-up libido needed for a fully gratifying orgasm (thus diminishing the energy to fuel irritability or neurotic symptoms).

Reich describes in detail control experiments that involved rubbing the electrodes on cotton, on a metal flashlight, on other metallic materials, and on other substances.[19] In some cases, one could get potential to vary, up or down, but the changes were not rhythmical; in only one case did the changes ever have the rhythm of organic wandering that was shown by living skin. That was an experiment in which the electrodes were separated from each other but both touched a piece of cloth soaked in a KCl (potassium chloride) or NaCl (sodium chloride) solution. No manipulation of the cloth with a piece of nonliving insulating material caused any change in potential. However, if one rubbed or pressed the wet cloth with a finger, "the typical wanderings, etc. appear at once." This did not invalidate the main point, Reich thought. (Hoffmann disagreed, as will be discussed in Chapter 6.) It only showed one had to use care in control experiments not to inadvertently introduce one's finger (or any other part of the body) into the circuit between the two electrodes, particularly not if they both touched an electrolyte solution.[20] Errors could also be produced by poorly insulated electrodes, Reich found. Attention to proper insulation prevented such problems. Electrodes were reversed, to ensure they had comparable reactions. For indirect measurements, silver electrodes were used. For direct measurements, nonpolarizable electrodes filled with 0.1 N KCl solution were used. Later, 0.9 percent NaCl solution was substituted when measurements were made on a mucosal surface, as it was found that a drop of KCl electrolyte solution produced discomfort if it accidentally came in contact with a sensitive mucous membrane.[21]

Reich found it was crucial that the bioelectrical measurements be conducted within a Faraday cage. Otherwise all light (and other electrical) circuits in the building must be turned off, or disruptive oscillations were recorded on the apparatus from electromagnetic fields created by current flowing in the building electrical wires.[22] This may have been the first time Reich found the use of a Faraday cage in his experiments important; it was not to be the last.

Reich concluded also that the vegetative ganglia of the autonomic nervous system (ANS) appeared to be the "vegetative center" from which the bioelectrical energy expands outward and to which it retreats in anxiety. Among these autonomic nerve plexuses, he cited especially the plexus coeliacus (the solar plexus), the plexus hypogastricus, and Frankental's genital plexus.[23] "Expressed cautiously," he ventured, "the cerebrum, according to this hypothesis, would be merely a specially designed apparatus for implementing and inhibiting the general vegetative bodily functions."[24] He believed these ANS plexuses are, by contrast, "the producer of bioelectrical energy in the human body."[25]

Further, Reich argued the ANS itself must be contractile. It must, that is, stretch and expand during pleasure states and contract during states of anxiety, from a middle position he called "vegetative equilibrium."[26] He observed this kind of pulsation in the ganglia of living transparent mealworms, viewed under a low power microscope. Vagotonia and vagotonic pathological states (such as asthma), he argued, would thus result when the ANS became stuck in a fixed state of expansion; chronic sympatheticotonia would correspond to a fixed state of contraction. Muscular armor, he thought, was thus a pathological response as the organism tried to avoid either anxious contraction or (out of fear) the pleasure of expansion, particularly in involuntary orgastic convulsion. "*The vegetative nervous system is thus a contractile plasma system, a contractile organ running throughout the entire organism. It represents the 'amoeba in the multicellular organism.' This is the explanation and the basis of the uniformity of the total body function.*"[27]

Since modern brain science has diverged so far from this view, it could be of great interest to replicate these experiments and learn in what way the results can or cannot be accommodated to modern neurobiology, neurotransmitters, and other areas. Masters and Johnson's more recent experiments from the 1960s did not replicate Reich's most central work: measurement of changing bioelectric skin potential in differing states of emotion or sexual arousal. Instead they measured blood pressure, heart rate, and other non-electrical variables.[28] Indeed, a 1971 review article on studies of sexual arousal points out that Reich was the first to study changes in skin potential that occur during sexual excitation. The review notes that no other researchers made electrical measurements of the erogenous zones until 1968; even then, skin resistance was the primary object of study (rather than voltage) and no distinction was made between positive and negative charge.[29] None of these researchers, Myron Sharaf has observed, "ever approached electrodermal functions as Reich did—as aspects of a *unitary pleasure function* in the body."[30]

Hoffmann had participated at first in the experiments and was quite interested in seeing whether Reich's hypothesis was right, that catatonics would have a lower electrical potential of the skin than normal subjects. He stated later that he and Löwenbach had measured the skin potentials of catatonics and found they were not any lower than normal people; he claimed he also could not detect in them the differences Reich found between erogenous and nonerogenous zones, with the exception that he did find a wandering potential on the palms of the hands. Hoffmann raised a number of technical objections to Reich's methods and to his interpretation of the data. For example, he claimed it was well known that physical stimuli (such as touching or tickling with cotton) usually caused an increase in potential, while mental influences, regardless of the kind, usually produced a decrease in skin potential. Thus it was likely, he argued, that Reich would get such results even if sexuality had nothing to do with it.[31] Hoffmann also criticized Reich's controls as inadequate, particularly in the experiments with two subjects touching one another (though the details of this were only in a private letter).[32] Hoffmann also saw Reich's confidence as "cocksureness," which he found personally off-putting. He criticized what he saw as "propaganda-like certainty" in Reich's writings, not least that Reich would publish in his own journal, rather than more mainstream scientific journals. Both of these things undermined his confidence in Reich's scientific credibility.[33]

Reich's own later version of the story differed considerably from Hoffmann's. For one thing, he claimed he considered Hoffmann an assistant or consultant from the outset. Reich said he had "already observed the increase of the potential with pleasurable vegetative excitation" by late May 1935. He then said he asked Hoffmann and Löwenbach to carry out further experiments while he "went abroad for a lecture course." When he returned six weeks later, Reich said, "My assistant told me that 'nothing had shown' at the erogenous zones, that there was no increase of potential with pleasure, and that therefore my hypothesis was erroneous." However, when Reich asked for a demonstration of the experimental technique, he saw that the assistant (it is not clear whether he means Hoffmann or Löwenbach in this passage; probably Löwenbach) "for six weeks, with the greatest precision, . . . had fastened glass cups [containing KCl solution] over the subject's nipples with adhesive tape. The glass cups were filled with electrolyte fluid and supplied with electrodes.... Mechanically speaking, the arrangement was faultless, absolutely correct in every detail." But in the context of the question the experiments had been designed to answer, "only one fact had been overlooked, and that was the decisive one: *no living organ gives a pleasure reaction if ones ties glass cups to it with adhesive tape.*"[34] Reich said he felt this was rather missing the forest for the trees. He summed up the main point thus: "my assistant had excellent training in mechanistic concepts and methods, but he did not know what to do with the biological concept of 'emotion' and had no realization of the faultiness of his procedure." In the midst of their first disagreements in September 1935, Reich wrote to Harald Schjelderup, asking him to witness the experiments, because he felt their attitudes (especially Löwenbach's) were so out of touch with the main hypothesis to be tested, their objectivity was in doubt. Since June, Reich told Schjelderup, "the experiments have been continued on catatonics in Dikemark but from the standpoint of a theory that is totally irrelevant to the problem we are investigating. The workers there measured only the resting potentials. What we are interested in are the potentials that exist at the peak of sexual excitation. It is clear right from the start that an electrode firmly attached to the nipple, no matter how correct it may be, tells us nothing about the specific orgastic sensation occurring in the glans penis, even if it obtains positive results."[35] For Reich, then, if one did not understand that human emotions were part and parcel of the setup, one of the most fundamental elements to grapple with using—among other tools—our common sense, then the observer could change the very phenomenon under study, even make it disappear, by his intrusive presence or manipulations. For Hoffmann and Löwenbach, it was unsound for Reich to enter this field when he was not thoroughly trained in it. For them mechanistically fastidious controls were foundational, without which they felt unable to interpret any of the results the experiment produced. I discuss these methodological issues further in Chapter 4.

Next Steps

During the time Reich carried out the bioelectric experiments, his therapeutic techniques evolved in tandem with his evolving understanding that emotions were real, physical processes, involving changes in the organism all the way to the involuntary depths of the autonomic (vegetative) nervous system. In his successive papers on masochism (1932), "Psychischer Kontakt und Vegetative Strömung" (1935), and "Orgasmusreflex, Muskelhaltung, und Körperausdruck" (1937), Reich reported that when his patients' emotional blockages were removed suddenly, sometimes by his imitating their facial or bodily expression to them, sometimes when a stiff neck suddenly gave way and relaxed, or in similar circumstances, the emotions released powerful biological forces—resulting in sudden pallor, sudden flushing, or alternation between the two. The character armor appeared to be rooted biologically in muscular armoring, and that muscular armoring seemed to interfere with deep, involuntary functions. Reich tells us that he began to work in therapy on trying to help the patients dissolve the physical tensions in their body, since that seemed more effective than character analysis alone. Reich referred to this new, combined physical and character therapy as "vegetotherapy," to indicate that it was working directly with functions of the vegetative, or autonomic, nervous system.[36]

From the bioelectric experiments, Reich further concluded that "we are justified in assuming that erogenous excitation represents the specifically *productive* process of the living organism. It ought to be possible to verify this in other biological processes, such as cell division, where concurrent with the biologically productive activity . . . the cell ought also to develop a higher surface charge. *The sexual process, then, is the biological-productive energy process per se.*"[37] This line of reasoning led Reich to begin thinking of electrical measurements on simpler organisms, unicellular organisms, and perhaps individual cells in a multicellular organism.

By late 1935, Reich said he wondered whether the vegetative currents—movement both of fluid and of bioelectrical charge from the core of the organism to the periphery (expansion, pleasure) and back again (contraction, anxiety)—were more basic life functions. Could they be found not only in humans or vertebrates but also in less complex organisms? He wondered, he said, whether the four-beat orgasm formula—validated in the bioelectric experiments—might in fact be a more fundamental phenomenon common to all living organisms.

Another fascinating finding of the experiments resulted when Reich measured the bioelectric charge on the skin of the abdomen. The electrical potential of the skin decreased upon inhalation (which places pressure on the abdomen) and returned to normal after exhalation. Reich knew a number of the experimental subjects as patients in vegetotherapy. So he was familiar with their emotional and vegetative reactions, which allowed him to notice a pattern: "In patients who [because of what Reich called 'muscular armoring'] have a rigid diaphragm and are unable to exhale fully, the fluctuation in potential when they inhale and exhale is not as clear or as extensive as that in subjects who are able to breathe freely."[38] Reich interpreted this to mean that tensing the diaphragm and other muscles must decrease the bioelectric charge because it puts pressure on the underlying autonomic plexuses. But it also pointed the way to the possibility of scientifically documenting the effects of armoring—though armoring differed in different individuals—in a quantitative way. And it suggested long-term health consequences, if Reich was correct about diseases that could result from chronic vagotonia or sympatheticotonia.

Reich began to study these phenomena in octopi and other marine mollusks, in starfish, jellyfish, and worms.[39] In transparent mealworms, as stated above, he tells us he observed that the autonomic nerves themselves lengthened during expansion of the organism and shortened when the organism contracted. In other words, the entire ANS was contractile; its expansive and contractive movements mirrored the motion of fluid and bioelectric charge in different emotional states. Reich concluded that the feeling of "sharp contraction inside" during a sudden fright response was literally a contraction of the ANS, not merely metaphorical.[40] Reich also decided, by early 1936, to study these phenomena microscopically in protozoa such as the amoeba.

To pursue his hypothesis that the bioelectrical expansion and contraction functions were also present in simpler organisms, Reich studied the literature on what was known about bioelectrical effects on amoebae. He found reports by Ludwig Rhumbler of experiments in which currents of fluid within amoebae were documented. Reich thought these might be parallel to the "vegetative currents" his patients felt in therapy and which he had objectively measured in the bioelectric experiments. After initial characterization, the amoebae in Rhumbler's experiments were subjected to tiny electrical currents through the solution they swam in on the microscope slide.[41] When a current of c. 0.5 mA was passed through the preparation, the plasmatic streaming within the amoebae sped up. The amoebae also moved faster in the field of view. Reich found these observations easy to confirm when he replicated the experiments. He created grass and moss infusions to obtain protozoa; thus his mixed cultures also contained *Paramecium* species. While the amoebae became livelier in response to mild electrical currents, the *Paramecia* by contrast "seemed to become highly disorganized." Their rapid nonstop swimming ceased and they began to go in circles; "it appeared as if each application of current had a sudden 'traumatic' effect," reported Reich.[42] If the current was turned up, by about 1.5 mA the amoebae stopped moving as well and curled up into a spherical shape.

Reich eventually concluded that "the *particular combination of mechanical and electrical functions* [i.e., mechanical swelling → bioelectric charge → bioelectric discharge → mechanical relaxation] *was the specific characteristic of living functioning.*" He felt this made a major contribution to the age-old mechanism-vitalism controversy. "The formula of tension and charge showed both schools to be right, though not in the way they had thought," he asserted. Living matter, Reich said, "does function on the basis of the same physical laws as non-living matter." This validated the mechanistic position. "It is, at the same time, fundamentally different from non-living matter, as is contended by the vitalists."[43] His contribution as he saw it was recognizing that what made living matter different was the combining together in a specific sequence of the "functions of mechanics (tension-relaxation) and those of electricity . . . *in a specific manner which does not occur in non-living matter.*" This fit well into Reich's simultaneous identity and antithesis concept: "*the living is in its function* at one and the same time *identical with the non-living and different from it.*"[44]

Reich anticipated that "the vitalists and spiritualists will argue against this statement by pointing out that the phenomena of *consciousness* and *self-perception* are still unexplained." This did not, he argued, justify the assumption of a metaphysical life principle; in addition, he felt the bioelectric experiments had brought science "within sight of final clarification" of the relationship between matter, living matter, and self-perception. "The electrical experiments have demonstrated the fact that the biological excitation of pleasure and anxiety is functionally identical with its perception. The assumption is justified, therefore, that even the most primitive organisms possess the ability to perceive pleasure and anxiety."[45]

The Bion Experiments

Amoeba cultures particularly interested Reich for several reasons: as stated above, because German biologists Max Hartmann and Ludwig Rhumbler had previously described plasma currents or streamings in them, and such currents were important for his theory of vegetative (autonomic) functioning.[46] The reader will also recall Freud and Reich's mental imagery of an amoeba "putting forth a pseudopod" as analogous with a person reaching out toward the world, emotionally or otherwise. But Freud and Reich were hardly alone in this: the amoeba had widely been taken as a "model organism," the simplest living system in which one could study behavior, by zoologists such as Herbert Spencer Jennings and by psychologists.[47] It is interesting to note that the *origin* of amoebae or other protozoa does not seem to have been a prominent issue on Reich's mind when he began replicating these experiments.

In addition to bioelectrical measurements, Reich paid attention to the antithetical pairing of chemicals in living things that cause expansion or contraction, swelling or dehydration of the tissues, as he described in "The Basic Antithesis of Vegetative Life" in 1934. Potassium ion and lecithin, for example, cause swelling of tissues, whereas calcium ion and cholesterin cause the opposite: dehydration and detumescence. Reich simultaneously sought to investigate whether combinations of those paired substances—for example, potassium/calcium and lecithin/cholesterin—might be able in the right combination to simulate some of the tension-charge and/or discharge-relaxation phenomena. I discuss the amoeba experiments first and then describe these synthetic experiments, which in some ways fall within the tradition of "cell model experiments," also sometimes referred to as "simulacra." Cell model experiments had for decades explored what characteristics of life could be investigated through model systems using only physical or chemical forces such as osmosis and surface tension.[48] For example, a classic work by Otto Bütschli is referred to by Reich in his article "The Basic Antithesis" and in *Die Bione*.[49] Potassium solutions were also used by M. Traube and S. LeDuc, like Bütschli, to produce their various lifelike "osmotic" structures by similar simple models that simulated movement, feeding, and division of cells.[50]

Alfonso Herrera in Mexico City had also worked on creating synthetic cell models—"plasmogeny," as he called it—since the 1910s.[51] Left-leaning and anticlerical, Herrera was as forceful a supporter of Darwinian evolution in Mexican science as was Huxley in England or Haeckel in Germany.[52] In 1939, Herrera and Reich began correspondence. In this same "cell model" tradition, famed surgeon George W. Crile Jr. in 1932 also created "autosynthetic cells," followed by much distortion in the popular press.[53] In these experiments, Crile produced many cell-like structures exhibiting amoeboid motility, respiration, and bioelectric charge. The work is extraordinarily similar to Reich's in many ways, except in its use of only complex organic substances to produce the "cells." A sensationalist reporting of the results by a newspaper man ("Artificial life created in laboratory!") caused controversy and avoidance of the work by most of the scientific community in a manner very much like that which awaited Reich's work.

Reich distinguished in his preparations between purely mechanical surface tension phenomena—for example, lecithin tubes—and more lifelike movements connected with pulsational changes in surface tension. In many ways these kinds of experiments are foundational for what science today refers to as being able to distinguish "genuine biosignatures" from products of purely nonliving processes, a discussion that has intensified since the 1996 controversy over whether structures, crystals, and chemical deposits in Martian meteorite ALH84001 truly implied the existence of past Martian microbial life—or might actually be fossilized Martian "nannobacteria."[54] For the most part, I use Reich's terminology in describing his experiments (rather than modern language about "biosignatures") since it is so conceptually central to understanding both his choice of experimental approach and his interpretation of the results. This is not tantamount to endorsing Reich's terms and concepts (as I make clear in the Epilogue), but it is helpful in giving an account that takes Reich's laboratory science claims seriously—precisely what I argue has rarely been done until now by either scientists or historians of biology.

Because Reich wished to observe the plasmatic streaming in amoebae and replicate Hartmann and Rhumbler's work on how that streaming was affected by electrical current, Reich needed to create an apparatus to send tiny, variable electrical currents through a preparation on a microscope slide (see Figure 2.2). This he was able to do with a Pantostat, purchased from Siemens in Berlin, capable of "exact measurements and metering of current down to 0.2 mA [milliamperes]." The leads from this device were connected to a copper wire, thence to tiny platinum wires attached to either side of the fluid well on a microscope well slide.[55] The leads could be reversed to reverse the direction of the current across the microscope slide. Reich began the experiments in February and March 1936, eventually setting up a laboratory in Oslo on Ensjövejen (later moved to his home at Drammensvejen 110).[56] As described, Reich was able to quickly replicate the earlier observations. Reich thought these phenomena important, though how they connected up to his tension-charge hypothesis was not yet clear to him.[57]

More important, as it turned out, in order to obtain protozoa for the electrical experiments, Reich ordered prepared cultures. These had very few amoebae in them, however, so he went to a technician in the Botany Department at Oslo University and asked how to prepare fresh cultures. He was told to just soak dried hay or moss in water and in ten to fourteen days the amoebae would be numerous. Though he had extensive biology training in

Figure 2.2: Apparatus for sending electrical current across microscope slide, to test electrical charge of bions. Courtesy of Wilhelm Reich Infant Trust.

medical school some eighteen years previously, Reich inquired, "Where do the amoebae come from?" Startled, the technician replied that they developed from spores or cysts that were ubiquitously present in nature and needed only water and food to hatch and begin to grow. Reich wrote that he found himself skeptical of this explanation, observing, "Obviously I had deliberately, though unconsciously 'forgotten' the 'germ theory'" as an explanation for protozoa appearing where none had been.[58] He was very specifically interested in the *swelling* of tissues because of his tension-charge theory (swelling produces mechanical tension), this seeming to him a possible alternate way in to checking the validity of the theory. So he decided to prepare the cultures as the technician recommended but to watch the entire process over many days and nights, to see the swelling of the grass or moss in water and thus to observe directly if the amoebae germinated from spores or cysts in the swelling plant matter. He watched closely at the margins of a piece of moss or grass, where it was thinnest and where swelling proceeded most rapidly.

Over two days the fibers lost their green color, and many chloroplasts were seen floating loose in the water. But in addition, as the plant tissues swelled and disintegrated, they showed many tiny, bacteria-sized vesicles and clumps of these vesicles within them. (Reich later called these vesicles "bions.") The previously striated structure of cells in rows gave way over time to fibers that were more and more completely vesicular in appearance. Reich watched these changes over four continuous hours and summarized what he saw as follows: "(1) Swelling plant fibers disintegrate into vesicles. (2) Dried-up amoebae have a vesicular structure like the clusters of vesicles. (3) Vesicular formations with definite borders detach themselves from the disintegrating plant. (4) The distinct border that forms gives the clump of vesicles a definite shape and fullness. (5) Some rapidly moving cells have a sharply defined border and a uniform vesicular structure."[59] Reich said he was struck by the similarity between the amoeba and the clumps of vesicles along the margin of a disintegrating plant fiber. He wondered whether the similarity might be more than appearance only: "Could it be possible that an amoeba or other protozoan with a similar vesicular structure is nothing more than a cluster of vesicles enclosed and shaped by a membrane?"[60]

Meanwhile, Reich observed protists that look like *Vorticella* forming as well. These swelled up, sharply contracting on a thin stalk, then swelling again in a cycle that Reich found confirmed his hypothesis that the tension → charge → discharge → relaxation cycle existed in the simplest life forms, not only in humans. Because they acted out his "orgasm formula" so visibly, Reich referred to them as "Org-protozoa."[61]

When watching the swelling and breakdown of moss in water over many hours and days (switching off between himself and an assistant), Reich said he saw the plant tissue gradually become more and more vesicular in structure. It became filled with tiny vesicles (bions) and clumps of these bions. Once those clumps of bions formed, they then gradually developed more and more motility, a clump eventually after several hours breaking away from the margin of the moss and swimming off looking indistinguishable from free-living amoebae. Reich described this process in detail in his 1938 book *Die Bione* (The Bions), adding, "Amoebae are formed continually in this way. Sometimes two or three form simultaneously within a short period of time; sometimes they form one after the other at certain intervals. The amoebae, once detached, undergo division. They can often be observed grouped together . . . and are difficult to distinguish from the margins of moss fibers that have undergone swelling."[62] Reich saw immediately that his "observations and the resulting hypotheses clashed severely with the 'germ theory'" and said "I deliberately avoided considering the contemporary views on 'the origin of life from life germs.' Following these initial observations, it was more than ever necessary to approach my work in an unbiased manner."[63] Chastened by what he felt was less-than-objective reception of much of his earlier work in psychoanalysis and in the bioelectric experiments, Reich had been concerned from the beginning about his ability to see microbial activity without excessive theoretical blinders from mainstream biology. As a result, he had a research assistant, Arthur Hahn, write up a thorough literature review of previous work on spontaneous generation, origin of life, and other experiments and theories. But Reich refrained from reading it, he said, until well into the experiments.[64]

As a kind of control, Reich made similar nonsterile water preparations of other plant material (a tulip leaf, a rose petal, grass) and of simple earth from his garden. The tulip and rose preparations showed no development of protozoa, even after three days. The grass showed motile rods, vesicles, and a few protozoa.[65] Since they were nonsterile, they only served by way of comparison with the moss preparations, but this showed that not all plant material yielded protozoa as one might expect if the germs of protozoa really were ubiquitous in nature.

The earth soaked in water showed surprising things, Reich said. There were a few common rod-shaped soil bacteria and inorganic mineral crystals with occasional vesicular formations in the liquid between them. By the third day of soaking in water, the crystals had begun to show internal vesicles. The preparation was "full of motile angular structures moving in exactly the same manner as the rods and vesicles. . . . The vesicular inclusions in the structured crystals were indistinguishable from the vesicular formations floating free and tremulous in the liquid."[66] By day seven, the crystals were still further swollen and vesicular in structure, with clumps of vesicles on their edges beginning to stretch and contract. Reich called these motile clumps of vesicles "plasmoids." Their bending and stretching increased (as did the quivering of individual vesicles within the clump), he observed, when he passed a current of 1–2 mA through the preparation. They moved toward the cathode, revealing that they had a positive electrical charge. As a control, Reich reversed the electrical leads; the direction of movement of the objects promptly reversed, confirming that the charge was the reason for their movement. Turning up the current increased the speed of their movement.

Reich said he felt that the moss preparations and the earth preparations were very different phenomena. Nonetheless, the vesicles formed in both preparations showed such similar movement, charge, and other characteristics that Reich hypothesized the vesicles produced by swelling in both cases were identical.[67] Since he believed the swelling of matter was producing the bions, Reich tried replacing the water with 0.1 N KCl solution, knowing that potassium promoted swelling of tissues. In both the grass and the soil preparations, this led to more rapid, more consistent breakdown into vesicles and clumps of vesicles. This result was reproducible again and again, he found; therefore, he used 0.1 N KCl solution as the standard solute for all the bion experiments thereafter.[68]

Reich had hypothesized as early as late 1935 that an amoeba might be no more than an agglomeration of the bions (vesicles) he studied;[69] however, the studies on earth bions had led him away from pursuing this question. When he returned to it, he decided that time-lapse photography through the microscope might perhaps resolve the question by showing continuously the entire process of bionous disintegration (breaking down into bions) of grass or moss, up through the formation of complete amoebae. Filming the process over days or weeks presented quite a technical challenge, though, and "technical difficulties prevented the problem from being solved for another year."[70] Standard equipment for time-lapse microcinematography was not yet easily available for sale;[71] Reich had to improvise a photo tube, relay-timer, and motor to turn the Kodak movie camera and expose frames at the desired intervals (see Figure 2.3).[72] In addition, as Reich put it,

> Time-lapse cinematography requires a completely still preparation with no flow whatsoever in the liquid. It was soon found that preparations of grass embedded in paraffin die off because of lack of oxygen. But the filming of vesicles being enclosed within a membrane and the subsequent development of amoebae could be one of the main proofs of my theory. The technical problem was resolved in the following way: Two or three carefully separated pieces of moss or grass were placed on a hanging-drop slide and the liquid for the preparation was added. A cover glass was provided with four small wax feet, one at each corner, and placed over the slide in such a way as to prevent formation of air bubbles. The two long sides of the slide were sealed off with paraffin. It should be noted that about one-quarter of the concavity of the slide was not covered by the cover glass. Water was then deposited along the two unsealed sides to form reservoirs from which the evaporating liquid could be continuously replaced. These two

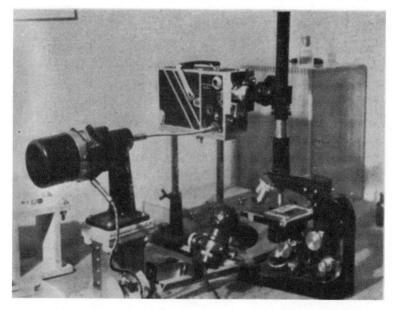

Figure 2.3: Time-lapse apparatus for microcinematography. Courtesy of Wilhelm Reich Infant Trust.

reservoirs of water had to be replenished about once every two hours. Thus, the technical problem was solved and time-lapse filming of the preparation could be carried out over a period of several days.[73]

In addition, according to Reich's photography assistant, Kari Berggrav, sunlight was at first the only source of light bright enough for the filming of living preparations, though this had the drawback of the occasional dimming of the picture when a cloud passed in front of the sun.[74]

By July 1937, Reich had completed at least three time-lapse films, one of earth bions, one of protozoa developing from disintegrating moss, and another simply labeled "bion film."[75] These films, preserved and transferred to DVD format, can be viewed with a request to the Reich Archives. Excerpted versions are on display at the Wilhelm Reich Museum. The protozoa film shows what appears to be dried (autumnal) moss or grass breaking down into bions and clumps of bions.[76] Berggrav says that the accidental discovery—that dead or dying autumnal moss, but *not* fresh moss collected in spring or summer, led to what Reich called the "natural organization of protozoa"—was extremely important. It was a clue that led Reich to imagine, by analogy, that cancer cells might form in a similar way from dying animal tissues disintegrating into bions, then the bions clumping together as amoeboid cancer cells. Reich's exploration of this hunch will be discussed in much more depth in Chapter 5.[77] Any attempt to replicate this experiment needs to take into account that only autumnal moss or grass gives unequivocal results.

Before he solved all the problems to produce time-lapse films, Reich thought he could test his hypothesis that an amoeba was just a clump of bion vesicles in another way, more in the vein of "cell model experiments." He wanted to make the "vesicles formed during disintegration of various substances coalesce artificially by using some suitable agent."[78] So he added some very diluted red gelatin solution to the earth-KCl mixture. Indeed, vesicle clumps did form which had not been there before. They showed a crawling, amoeboid movement, though the movement was a bit jerky; the movements and internal streaming of vesicles within the clump were not as smooth or fluid as in living amoebae. Reich referred to these artificial creations as "pseudo-amoebae."[79] *Could* these be a result of germs of organisms in one or more of the ingredients since these preparations were uncovered and unboiled, Reich wondered? As a control, he boiled the preparations for fifteen to thirty minutes in closed glass containers.[80] The result was totally unexpected. In the boiled preparations, there were far more numerous and more active bion vesicles, already seen immediately after boiling, than in the unboiled preparations after days of swelling. The liquid in the boiled preparations became colloidally opaque. The vesicles were still positively charged (they migrated toward the cathode). And when dilute sterile gelatin was added, the same "pseudo-amoebae" formed. Thus it was clear the "pseudo-amoebae" did not come from germs. The appearance of actively moving vesicles in large numbers immediately after boiling also tended to suggest they could not have come from spores or cysts, which would have required at least a few minutes to germinate.[81] Six days after boiling, a sealed preparation of earth and KCl solution showed many of the familiar movements, with most of the mineral crystals now having a largely vesicular structure. Compared to unboiled preparations, the vesicular disintegration advanced far more quickly; the contractile tubes and marginal swellings seen before were far more abundant as well. When these were observed at 2300 to 3000× magnification using a water immersion objective lens (which has high magnification and a close working distance), Reich "was able to clearly observe *pulsation* at individual points in the contractile formations."[82] "Thus," said Reich, "it became increasingly clear to me that *the more vesicular the structure of such a crystal, the more motile it is.*"[83]

In the earth "pseudo-amoeba" preparations, Reich felt he had to assume "the vesicular particles were *swollen, electrically highly charged units of matter*, held together by a gel-like substance. These particles move around inside the amoeboid structures just as they would move individually as anucleate vesicles in a free, unrestricted space."[84]

Repeat experiments with boiled earth always yielded the same results: "motile vesicles, nucleated and anucleate formations," and pseudo-amoebae.[85] Could these phenomena possibly be caused by errors in the preparation? Once again, Reich made control preparations of unboiled earth, leaves from trees, tulips, and ferns and left those preparations unsealed.[86] Again, there was much less vesicle formation, which took a much longer time: *"It was much more difficult for motile vesicles to form or for the marginal layers to break down into vesicles than was the case in the boiled preparations,"* said Reich.[87]

Reich sought to discover if these earth-KCl-gelatin formations had other lifelike properties in addition to motility, expansion, and contraction. Watching the same pseudo-amoebae under the microscope steadily for hours from the moment the boiled preparation was made, Reich observed that after several hours they could be seen undergoing divisions. He came to think that "these were probably living forms which were, so to speak, incomplete, for example, the movements of the pseudo-amoebae were abrupt, slow, tremulous without any inner flow, i.e., 'mechanical.' The objects must therefore be in *preliminary stages of life.*"[88]

At this stage, Reich did not want, he said, to attack the problem of whether the rod-shaped structures, spherical ones, and others were true bacilli or cocci, or whether "other types of structures [only superficially resembling the morphology of bacteria] *had formed in the boiling process*."[89] A trained bacteriologist might have been inclined to pursue that question first, but Reich's far different theoretical background led him to first emphasize other questions. If it had been possible to create pseudo-amoebae with many lifelike characteristics from KCl, boiled earth, and gelatin, then, he reasoned, it should be possible to add other substances or use a different formula, to "eliminate the 'unfinished' character of the formations."[90]

Inexplicably, boiling seemed to actually accelerate the process of bionous disintegration of matter from many different sources, resulting in more nu-

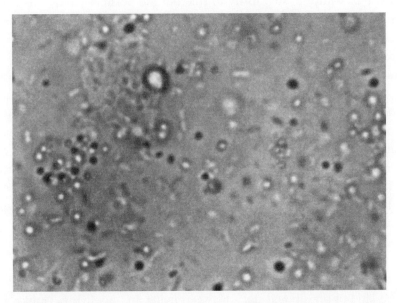

Figure 2.4: Earth bions: from boiled earth, water filtered off, and then sediment examined. The microscope was a Zeiss Universal frame, Achromat 1.4 condenser, 40× Nikon CF Planapo 0.95 objective, camera Canon Powershot G9. Magnification is 1500×. Photo by Stephen S. Nagy, MD.

merous structures that had more lifelike properties. Though this seemed to argue against the likelihood that the observed structures were produced via contamination by germs, Reich said he was nonetheless baffled: "At this stage of the work, it was most important to refine and control the experiments. *I was continually faced with the question of how it was possible for boiled substances to contain more life than unboiled, non-sterile substances.* I originally thought that the rapid quivering motion of the boiled vesicles was a heat phenomenon, but several experiments in which electrical current was applied demonstrated that this motion was electrical in character, because the quivering increased when the current was applied."[91] Reich later switched to autoclaving the bion preparations at 120°C and 15–20 psi of steam pressure. This accelerated the process still more: very few unstructured crystals remained in earth-KCl preparations, and the solution contained almost entirely very motile bions and clusters of bions.[92] Later, Reich established that this also produced more highly electrically charged vesicles. A year later

Reich took a further step to try to rule out the possibility of germs being contained in the original solid matter used. He took pulverized coal or soot, heated that to incandescence on a metal spatula in a Bunsen burner flame at red heat for thirty to sixty seconds, and then plunged it via sterile transfer into previously autoclaved KCl-nutrient solution. The result was yet again greater numbers of bions, more highly motile, with greater electrical charge in a solution that was at first black with colloidal turbidity.[93] Over the course of many hundreds of experiments with a large number of different preparations, Reich found colloidal turbidity of the initial solution to be strongly correlated with the lifelike properties of the bions, including eventually the ability to grow one type of bions in pure culture, transfer it to sterile media, and have the same pure culture grow up again through successive generations (see Figure 2.4). I provide more on this in Chapter 3.

Preparation 6 Bions

As stated in the preceding text, Reich thought he was creating structures in some kind of transitional stage between living and nonliving. Unlike previous "cell model experiments," his were based on his tension-charge theory and on his synthesis of the biochemical antithesis of vegetative life. Many substances caused swelling (expansion), many others caused dehydration (contraction), and still other chemical systems showed electrical charging and discharge. In Reich's view, the specific combination characteristic of life was the sequence of mechanical tension (swelling) → electrical charge → electrical discharge → mechanical relaxation. Thus when he sought to eliminate the "unfinished" or not fully lifelike character of preparations such as pseudo-amoebae, Reich believed one needed to combine substances causing both expansion and contraction. Thus he added lecithin (swelling) and cholesterin to his earth bion preparations. Lecithin in water or KCl solution by itself produced various membrane-enclosed sacs and tubes which grew, sprouted, and budded as they took up fluid. Reich did separate controls, adding each new item singly to the original earth mixture, to determine the effect each contributed.[94] Reich next added egg white (albumen) as a nutrient substance to the mix. The result was dramatic and unexpected: round, cell-like structures formed immediately, which had dark central objects like nuclei and which "underwent frequent division in rapid succession."[95] Only the mixture with all of these ingredients produced the cell-like structures. The mixed egg white preparations, like those before, showed far more numerous, far more motile structures when boiled than when unboiled. Boiling produced particles with more electrical charge, which therefore stayed in colloidal suspension for longer, even up to six to eight weeks. The lifelike motility and electrical charge of the particles disappeared when they settled out of suspension.[96]

Next, Reich added the same dilute red gelatin solution used in the pseudo-amoebae preparations. While the previous pseudo-amoebae had shown little or no internal streaming of vesicles, or flowing, bending, and stretching motility, the new egg white-cholesterin-lecithin-etc. mixture showed all these movements more fully and appeared much more lifelike in character. They divided and took up vesicles from the surrounding solution ("eating") much more actively as well. "In addition, after a few days it was possible to observe formations that possessed homogeneous plasma and a flowing pseudopodal movement. The questions of *metabolism, culturability, stainability, and the bacteriological characteristics* of the structures that had arisen still remained unanswered."[97] As it turned out, the bions did take up biological stains such as methylene blue, or Gram stain.

Reich tried additional nutrient ingredients: Ringer's solution, meat broth, milk, and egg yolk. Since Reich reasoned that carbon was a crucial element for life, he also tried adding different forms of it, for example, "finely pulverized, dry-sterilized coal; coal dust autoclaved in KCl; coal dust heated to incandescence, dry or in broth; . . . [or] dry-sterilized soot or soot heated to incandescence. The best results were obtained by adding soot: very small, but extremely motile, finely structured amoeboid objects formed."[98] Reich referred to this as Bion Preparation 6. By September 1937, Reich was studying a number of variants of Preparation 6 and had developed a coding system to refer to them: "(6a) Non-sterile mixture. (6ab and 6ac) The non-sterile mixture is boiled (100°C, ½ hour) or autoclaved (120°C, ½–1 hour). (6c) The substances are pre-sterilized. (6cb) The mixture is prepared under sterile conditions and then boiled. (6cc) The sterile mixture is autoclaved once more. (6ccc) The sterile mixture is autoclaved twice." (See Figure 2.5.)[99]

Reich foresaw that some would construe his description to be a claim of creating artificial life. But he argued for a more nuanced claim: "If, in metaphysical terms, life was a realm completely separate from non-life, I would be entitled to say that I was creating 'artificial life.' But my experiments showed that there are *developmental stages* in the progression from lifelessness and immobility to life and that, in nature, life is being created out of inorganic matter by the hour and by the minute. . . . *I merely succeeded in revealing, experimentally, the developmental process of life*. It was possible that a new

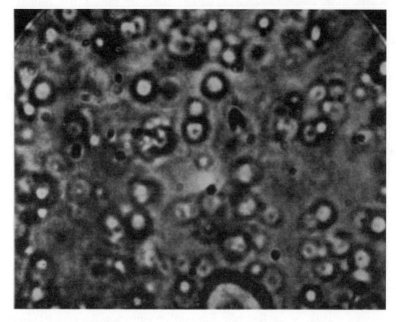

Figure 2.5: Preparation 6c bions, 2000×. Courtesy of Wilhelm Reich Infant Trust.

form of organism was artificially generated in the process."[100] So startling was even such a qualified claim and so contrary to widely believed ideas of biology that Reich repeated all of these experiments many times over, for several months through October 1936, to be certain of his conclusions. At that time, Reich began attempts to culture the Preparation 6 bions through successive generations on sterile growth media.[101] Two tables of numerous bion types Reich produced and their characteristics are reproduced in Tables 2.1 and 2.2 from his 1938 book *Die Bione*.

(In private moments, Reich allowed himself in a private diary to be more speculative and exuberant, for example, when on December 7, 1936 he jotted down, "'Life must have a father and mother'—there must be law and order!! Science! I'm going to plant a bomb under its ass! Abiogenesis does exist!!")[102] Despite all these precautions, Reich's foreboding was correct: many critics read accounts of his experiments fairly superficially and immediately accused him of trying to revive outdated claims of "spontaneous generation" from the nineteenth century, "long ago laid to rest once and for all by Pasteur and Tyndall." These reactions are discussed in Chapter 3.

Table 2.1 Culturability of bions from ordinary earth, cultivated earth, coke, and soot

Substance	Sterilization type	Swelling type	Electrical reaction	Movement Immediately	Movement 2–6 weeks	Culture in 2–6 days	Culture in 3–10 weeks	Electrical reaction of culture	Animal experiment
Ordinary earth	a	none	none	none	+	none	none	none	?
"	b	KCl b	++	++	+++	++	++	++	?
Cultured earth	b	KCl c	++	++	+++	++	?	++	?
Ordinary earth} Cultured earth}	c	KCl c	+++	+++	+++	+	?	++	—
"	d	none	none	none	+	none	none	none	?
"	d	KCl c	++	+++	++	++	++	++	—
Powdered coke	a	none	none	none	+	none	none	none	?
"	b	KCl b	++	+	++	+	?	+	?
"	c	KCl c	++	++	++	+	+	++	?
"	d	none	none	none	none	none	none	none	—
"	d	KCl c	++	+	++	++	++	++	?
Soot	a	none	+	+	none (+?)	none	none	none	?
"	c	KCl c	+++	+++	+++	++	?	++	?
"	d	none	none	none	none	none	none	none	?
"	e	KCL+bouillon	+++	+++	+++	+++	?	+++	—

The designations have the following meanings:

a — nonsterile
b — boiled for one-half hour at 100°C
c — autoclaved for thirty to sixty minutes at 120°C
d — dry-sterilized for one to two hours at 180°C
e — heated to incandescence for thirty to sixty seconds in a benzene gas flame
+ weak movement, little movement, occasional positive reaction
++ strong movement, frequently positive, clearly positive reaction
+++ very vigorous, mainly positive result
— negative, animal injection experiment without any visible effect

Table 2.2 Electrical charge of various bion types and their cultures

Type of bion	Charge	Culture	Charge	Remarks
Mixture Prep. 6 bions	electrically negative	yes	negative	Exception: 6cI = electrically positive
Earth bions	positive	yes	positive	
Coke bions	positive	yes	positive	
Soot bions	positive	yes	positive	stains Gram +
Muscle tissue bions	negative	so far, no	—	
Liver bions	negative	yes	negative	Culture findings still need further checking, interpretation
Lung bions	negative	yes	negative	
Yolk bions	negative	no		
Moss bions	negative	possible	negative	
Grass bions	negative	?	?	
Cow's milk bions	negative	possible, after autoclaving at 120°C	negative	

Reich was well aware of the experiments of Pasteur and Tyndall. They are described in detail in Hahn's literature review, which was published in 1938 as part of Reich's book *Die Bione*. But Reich—like many nineteenth-century opponents of Pasteur and Tyndall, such as Pouchet and Bastian—did not find those experiments conclusive. While the French Academy of Sciences claimed Pasteur's "swan-necked flask" experiments conclusively disproved the possibility of spontaneous generation of microbes in boiled infusions, for instance, Pouchet pointed out that (a) Pasteur never used hay in his infusions, the most successful ingredient for generation of microbes in his own infusions; and (b) by showing no microbial growth occurred when dust was kept out, all Pasteur had really proven was that dust must contain some ingredient essential for spontaneous generation. Prominent observers like Richard Owen, Gilbert Child, and Jeffries Wyman agreed with Pouchet that on this point Pasteur's experiments were not conclusive. Similarly, Tyndall and Pasteur showed that boiled infusions exposed to only pure air high atop an Alpine glacier never developed any microbial growth. They concluded that when growth of microbes occurs in boiled infusions, it can only be because of outside contamination from impure air. Reich responded that this conclusion was unjustified and suggested that from these results one could just as well conclude that the living organisms were formed in the infusions—just like in his bion preparations—but then those organisms were killed off by glacier air.[103] No single experiment in either case ever conclusively answered these claims.

Pleomorphism

Reich often describes bions that appear as rods on one type of culture medium, but cocci or a mixture of the two on a different medium. Similarly, the morphology often changed in aging cultures, though they had been kept sterile throughout. For example, in a February 19, 1937 postscript in a letter to Roger du Teil, Reich notes, "I have so far propagated four strains of various bion mixtures in broth but mainly on agar (strains IV–VII). Strains I–IV were grown on egg-white nutrient medium, but this type of medium was discontinued because of its unreliability and the similarity of the substances. I would therefore be very grateful if you would write and let me know the results of your control tests. Judging by the reliability of the findings, there is nothing to contradict the statement that bions can be cultivated in broth. So far, I have only noticed that the rod form predominates in broth while mainly cocci develop on agar. I do not know what significance this has. It is possible that the nutrient medium exerts some influence on the selection of the various types."[104]

Many of Reich's observations on bions and their extremely changeable morphology—changing from rods to cocci, even to what look like fungal mycelia—make sense in the context of the doctrine of microbial pleomorphism, revived as "bacterial cyclogeny" in the 1920s and 1930s. Many bacteriologists during this time reported similar wide-ranging changes in morphology under varying environmental conditions or as different phases of bacterial life-cycles.[105] Like revived claims of spontaneous generation (and often seen to be linked to those since some observers claimed the microbes could even sometimes originate endogenously, i.e., from the culture medium itself undergoing transformation), these pleomorphist ideas were under attack in some quarters in bacteriology by the mid-1930s.[106] Thus, in wading into this territory, Reich was up against yet another intellectual barrier. Criticizing the pleomorphist ideas of Carl Nägeli, nineteenth-century mycologist Oskar Brefeld had famously stated that "if one doesn't work with pure cultures, one gets only molds and nonsense." Reich was not uninformed or naïve about Brefeld's dogma; he did work with sterile media. However, he was formulating a view of life coming into being via formation of vesicles, which under different environmental conditions could develop into different forms, reviving many of the pleomorphist ideas.

In this Reich was not alone: for twenty years previous to Reich's bion experiments, a significant number of biologists were again making the pleomorphist case, including Felix Löhnis, Gunther Enderlein, Philip Hadley, and Arthur I. Kendall, among many others, often describing the phenomenon as "bacterial cyclogeny."[107] A few even wondered—though this was not widely accepted—whether viruses might also be one of the forms that bacteria could produce under the right conditions. (Reich thought, at least provisionally, that a "virus is nothing more than a vesicle [i.e., some kind of bion], probably very much smaller, that is produced under certain conditions [physiologically].")[108] Theobald Smith and Ludwik Fleck decried how Koch's "monomorphist dogma" had prevented recognition of the extreme degree of bacterial variability; indeed, Frederick Griffith's investigation of variability among *Diplococcus pneumonia* forms was a crucial opening that led eventually in 1944 to Avery and colleagues' famous recognition that DNA was the transforming principle.[109] Reich was solidly within this tradition; he was defending and exploring a fairly strong pleomorphist approach at a time when such ideas were at a peak. There was significant resistance from mainstream bacteriology, however. Bacteriologists such as Hans Zinsser and virologist Thomas Rivers fought against this trend; they pointed out that it would lead to a revival of spontaneous generation ideas. Indeed, several of the pleomorphists, for example, Kendall, were making that argument, and Zinsser publicly attacked Kendall's position at a major microbiology meeting.[110] The debate within bacteriology was seemingly resolved only at the expense of the pleomorphists by the mid- to late 1940s.[111]

Claims of a symbiotic origin of organelles such as mitochondria are also connected in some ways to pleomorphist ideas. The "small living particles" described in the 1890s by Shôsaburô Watase, for example, sound similar in conception to how Reich understood bions.[112] As Jan Sapp has pointed out, symbiogenesis ideas were "kept on the margins of 'polite biological society'" in the 1930s.[113] This was another association that would have made most biologists in the 1930s skeptical about bions.

Packet Amoebae

Reich observed one common, actively mobile bion type that looked like a tetrad-shaped cluster of four bions, sometimes more, the entire clump crawling actively along together. He called these "packet amoebae" because of their appearance and motility. They were first seen in Preparation 6, but later from several different kinds of bion preparations (see Figure 2.6). In an August 31, 1937 letter, du Teil says to Reich that these look like well-known bacterial forms called *Tetragenus* or *Sarcinae*. The yellow color of packet amoebae resembled species called *Sarcina lutea* (today called *Micrococcus luteus*) and/or *Sarcina ventriculi*, a cause of gastrointestinal and lung illness in humans.[114] Du Teil stated that they were different in their motility, however, since the forms familiar to bacteriologists were immobile; further, he said, even if they were what bacteriologists were used to seeing, it was a major discovery that they could be produced by sterile means from bion preparations. "Clearly nothing can change the fact that we have seen them develop from our mixtures; but if one wants to categorize them, it appears to me preferable to

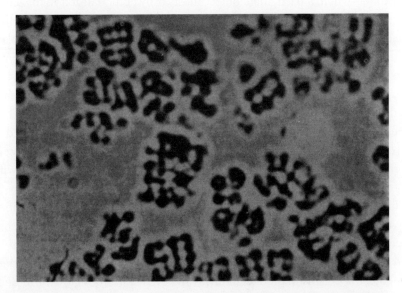

Figure 2.6: Packet amoebae 6cI bions, 3000×. Courtesy of Wilhelm Reich Infant Trust.

use the established classifications and notations."[115] Du Teil felt that using the bacteriologists' nomenclature, rather than inventing new names such as "packet amoebae"—which Reich preferred—would give them a double message: (1) we are not unaware of your reports of these forms, and (2) we will meet you halfway at least when it comes to terminology. Reich, however, was resistant to this approach, for reasons I discuss shortly.

There is some cause to wonder about the possibility of the packet amoebae being *Sarcinae* from air contamination (sometimes also called *Tetragenus*). When Reich first obtained them, it was from a Preparation 6cI culture. The first culture of the preparation, produced in nutrient broth, contained rods, cocci, and amoeboid forms. But only when this initial culture was inoculated onto an agar nutrient medium did there appear at the inoculation site, twenty-four hours later, a thick, yellow growth that when viewed microscopically revealed a pure culture of packet amoebae. As of early September 1937, fifty-three successive generations of these had been propagated on agar from the first one. But never again did they appear in any Preparation 6 mixture experiments.[116] The one-off appearance is suggestive of contamination during an inoculation procedure, although a contaminant would usually grow near the edge of a flat dish containing agar—or near the mouth if a test tube—where a particulate object most easily could fall if it managed to get under the lid while it was briefly lifted during inoculation. It is plain that Reich was aware of the distinction: if growth occurred on the area streaked with inoculum, near the center of the plate or tube, it was almost certainly the result of the inoculum. He made this clear via figures 52 through 55 in *Die Bione* (see Figure 2.7).[117] Reich also pointed out that the packet amoebae had a strong positive electrical charge, while most other bion types that had been successfully cultured had been negatively charged.[118] He emphasized in a report to a Paris scientific academy that while he was not yet 100 percent certain the rod forms were from the bion mixture itself, he had clarified with complete certainty that the packet amoebae were.[119]

On August 24, 1937 Reich reported to du Teil that he had recently obtained a pathogenic bacillus from metastasis of liver sarcoma. When packet amoebae were placed on a slide with this bacillus, they paralyzed and killed it; they did the same to tuberculosis bacilli.[120] The "S-bacillus," as Reich first called the sarcoma organism, produced illness and even cancerous growths in mice injected with it, whereas mice injected with packet amoebae suffered no ill effects afterward. These phenomena—in addition to their appearance in a sterilized culture on the inoculation streak—suggested to Reich that the

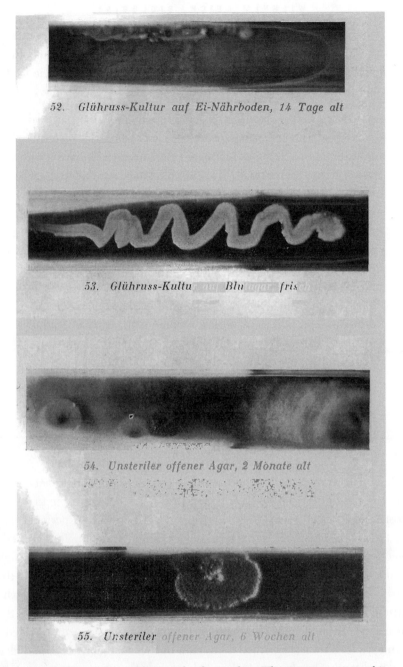

Figure 2.7: Culture tubes with growth of microbes. The top two are soot bion cultures; the bottom two are unsterile tubes with air contaminants. In the top two the soot bions grow up where the inoculum was streaked. Courtesy of Wilhelm Reich Infant Trust.

packet amoebae had much greater significance than and an identity distinct from a commonly known bacterium that might happen to look similar. Reich summarized to du Teil his objections to using the name *Sarcinae* for the packet amoebae in a letter of September 7:

> As to the question of the legitimacy of referring to "amoebas" and "amoeboid," the following: the terms I choose are based on terms from formal bacteriology, although they necessarily have a totally different connotation. Two examples: the cocci that we grew from autoclaved lung tissue were diagnosed as white staphylococci by bacteriologists here, but the diagnosis was based solely on the form, while our diagnosis especially included the electrical reaction. It is possible for the form to be the same, yet the electrical reaction different. Our packet amoebas can not be identical with the tetragenus bacteria, first because they have totally lost their original shape due to being inoculated onto egg medium, and present as extremely mobile contractiles instead of square structures at a 3000× magnification; second, because they do not cause any pathological phenomenon in organisms.
>
> I would also like to state that one should attempt to correlate obtained structures to other already familiar forms of bacteria, but one should not force the issue because there are too many factors that extremely change the image in bacteria diagnostics.[121]

Reich felt strongly that if the bacteriologists were not able to accept the origin of the forms from sterile mixtures, then they were refusing to see the most crucial aspect of what the bions signified. His previous experience with the Freudians' resistance to his terminology, and that it actually signified a rejection of his concepts, had formed in him a strategic preference to fight for the distinct terminology as key to defending a conceptual structure that clashed with the dominant paradigm. Reich might well have asked, what sense does it make to call them *Sarcinae*, when to the bacteriologists that means they *cannot* come into being by the means we have demonstrated?

In 1938, Reich was confirmed in his view by the fact that he obtained identical packet amoebae from autoclaved blood and other sources, not only from the original Preparation 6 source.[122] When gram-stained, they lost their packet appearance and artifactually looked only like simple cocci (see Figure 2.8). Later when Leiv Kreyberg and Theodor Thjøtta looked only at stained prep-

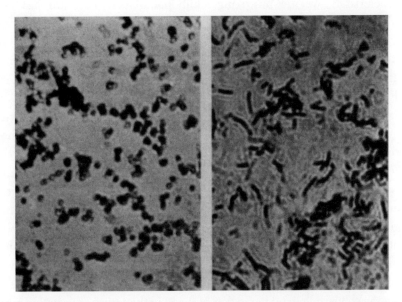

Figure 2.8: Packet amoebae, stained (both 1000×). Courtesy of Wilhelm Reich Infant Trust.

arations, they were thus misled to think that Preparation 6 bions were "merely staphylococci from air infection."[123] Reich had studied staphylococci and streptococci in some detail. He noted that most bions were electrically negative while staph bacteria were electrically positive in charge; most bions were gram negative while staph germs were gram-positive. In this case, the packet amoebae were an exception; they actually were gram-positive. However, Reich noted that all types of bions are generally larger than staphylococci.[124] Reich's "distinction based on size, for example, between bions and staphylococci, is ambiguous. Reich measured PA ["packet"] bions to have a diameter of 2 to 6 μm (micrometers), while the individual bacterial cocci he studied were only 1 μm in diameter. Bacterial cocci, however, cover a wider range of sizes, from 0.4 to 3 μm, which overlaps with the 2 μm size Reich commonly cites for PA bions."[125] The electrical charge distinction remains, however.

Du Teil responded that he was fully satisfied with Reich's explanation about the term "amoeba" or "amoeboid." He added, "The question of terminology, i.e. the word 'amoeboid,' has a specific meaning to me. It is correct that when we use an expression that usually identifies a specific thing in present day bacteriology and apply it to something with completely different characteristics, we are provoking additional objections. I am just now working on this in order to possibly eliminate dumb objections, so that only the intelligent—useful—and, yes, even indispensable objections remain. Objections that target words waste time, that is all. That was the exact purpose of the objections I made to you—which were, by the way, just as dumb."[126]

Reich was repeatedly asked, even by sympathetic scientists, "Don't you know about *Sarcinae?*" Over time he grew exasperated by this, since he felt they were assuming the knowledge deficit was his (when in fact he did know about them)—giving the benefit of the doubt to "establishment bacteriology" when it had failed to grasp any of the breakthroughs bions implied—and missing the point about conceding terminological ground to hostile, irrational opponents who would never accept his concepts in return. This same issue was behind Reich's dispute about "Brownian movement" and about Reich's developing theory of how bions were related to the origin of cancer tumors, both of which are discussed at length in Chapter 3. When physicist W. F. Bon collaborated with Reich on experiments on a new bion type of the same yellow color and tetrad-packet shape in the spring of 1939, Bon asked a similar question.[127] This episode is discussed in Chapter 7.

3

REICH AND DU TEIL: CONTROL EXPERIMENTS BEGIN

In February 1936, a month before Reich appears to have met with Trotsky, a group of Reich's students in France brought him to hear a lecture by Roger du Teil of the Mediterranean University Center in Nice, France (see Figure 3.1). Du Teil was a philosopher; his primary interest was in biology, especially the mechanism-vitalism controversy.[1] After his lecture, du Teil spoke with Reich about Reich's bioelectric experiments on the nature of sexuality and anxiety, which the French professor found fascinating. He told Reich he wanted to present a summary report about those experiments to the French Academy of Sciences. In addition, by the summer of 1936, du Teil took over from two students of Reich's in France the task of publishing a French edition of Reich's monograph on the bioelectric experiments.[2] Du Teil was very well connected in French scientific and humanities circles. He was a member of the Academie des Sciences Morales, as well as a Consultant member of the Centres d'Etudes des Problems Humain, "where he sat along with Alexis Carrel, Aldous Huxley, Pierre LeComtc du Noüy, Andre Siegfried, Teilhard de Chardin, and other professors from the College de France."[3] Over time it became clear that du Teil was inclined toward vitalism far more than Reich.

Reich apprised du Teil in December 1936 of the development of the bion experiments, just as he was tackling the problem of whether the bions could be cultured through successive generations on sterile media. Reich had just days earlier demonstrated his bion Preparation 6 to Albert Fischer, head of the Rockefeller Institute in Copenhagen and a specialist in tissue culture, who had previously taken an interest in Reich's use of time-lapse microcinematography.[4]

When Reich visited Fischer's lab to demonstrate the bions, he said Fischer at first took on a half-joking attitude, asking, "What kind of paste are you

Figure 3.1: Reich (left) and du Teil in Oslo, August 1937. Courtesy of Wilhelm Reich Infant Trust.

preparing to brew?" When Fischer saw the bions, however, Reich said Fischer became more serious. Fischer was unable to provide magnification greater than 1500× with the standard microscopes in his lab, to verify Reich's claim of pulsatory movement within the vesicles. One of Fischer's lab assistants suggested trying to stain the bions with a biological stain, Giemsa stain, usually only taken up by cells and thus used to render cells, for example, bacteria, more visible. The reaction was positive: the bions readily took up the stain. Reich left feeling that despite Fischer's initial skepticism, the demonstration had been a success.

A month later, Reich's friend Dr. J. H. Leunbach in Copenhagen visited Fischer and was shocked to hear him recounting a very different version of the events. Fischer was now trying to claim that Reich had made ridiculous, unreasonable demands during his visit. As Leunbach expressed it in a letter to Reich, " 'You may never again introduce us to such a horrible person!' 'We didn't immediately throw him out only because we are so nice and polite.' 'Such insolence to conceal the material from us.' 'We have never experienced anything as uncritical and sensational in our life!' 'And then he demands a ridiculously high magnification, when if anything were there to see, it could also be seen using customary magnification.' 'No, we can really not devote our time to such things.' 'Please do not be angry with us, we do not wish to offend you, it was not your fault—but that person!' 'And this combination of biology and psychology—and what else!' " Leunbach summarized this stunning about-face by Fischer thus: "Mr. Fischer and Rubanow [were] quite different when we were there in the institute. But perhaps that was only due to courtesy? If that is the case, I think that one can also be too polite! Anyway, there is no point to bother those gentlemen any further." Leunbach said that Fischer claimed Reich demanded absurdly high magnifications, for which there could be no use and which might not even be possible. He said the movement of the vesicles had seemed to him clearly caused only by fluid movement on the slide. The vesicles themselves were simply "cholesterin- and lecithin-structures," said Fischer. Furthermore, he accused Reich of being "uncritical" and "a fantast," saying all Reich's claims that bions had living properties were just "old fairy tales" [Märchen] from before Pasteur's time.[5]

Reich refuted the charges one by one in a long letter to Fischer, to set the record straight as he saw it. For example, in response to Fischer's claim that all the movement seen was caused by fluid flow, Reich replied "that the preparations were all produced by you or your assistants, and the goal when doing so was to eliminate all fluid flow. In fact, none occurred; instead, you and your assistants were forced to admit that *instead of all the structures moving in one direction,* they were moving in relation to one another."[6] One could easily resolve this question, Reich asserted, "by producing a paraffinated, hermetically sealed preparation" and observing it between 2000 and 3000× magnification. When he did this, said Reich, the movements could be seen very clearly, "namely locomotion, expansion and contraction, as well as movement of the contents."[7] The unmistakable difference between a simple lecithin structure and the more complex, motile bions had been confirmed by Reich seven to eight months previously when he observed the individual ingredients of Preparation 6 one by one in solution, *before* mixing any of them together.[8] Regarding the idea that it was improper "to mix psychology and biology," Reich told Fischer that he was "familiar with the principal works in these fields" and that "in epistemological terms scientific biology does not disallow the intermeshing of biological and psychological problems any more than scientific psychology would. I am very eager to make it clear right now that it was pure chance [*Zufall*] that led me from my work in the field of clinical psychotherapy to the experimental study of physiology and biology."[9]

Regarding the issue of whether Reich's demand for magnification of 2000× and greater was "excessive" or unreasonable, biologists usually draw an important distinction between magnification and resolution. Resolution means the ability to distinguish two objects very close together clearly as two separate points. If one increases magnification, for example, of a photograph, without any increase in resolution, the image merely becomes blurrier and blurrier. After a certain point (1000× to at most 1500× with standard optical lab microscopes), the extra magnification is useless because the image becomes so blurry one cannot make out anything clearly.[10] Reich, however, had obtained research microscopes with lenses of a far higher quality (and correspondingly more expensive) than standard "workhorse" lab microscopes. So his microscopes could genuinely give more information at magnifications of 2000× and above. Furthermore, Reich stressed that what high magnifications were critical for was not distinguishing fine details of structure, but rather for being able to see details of *movement* clearly enough to distinguish, for example, internal pulsatory movements from external Brownian movement. At higher magnifications, this ability is degraded more slowly than the ability to distinguish fine details of structure, Reich argued. Electron microscopy can provide much higher magnification without loss of resolution. But in the 1930s and for decades after, it could look only at dead specimens, because the fixing, staining, and even viewing procedures killed the specimen. Thus, for answering questions that required seeing a live object in motion, electron microscopy during Reich's time was useless. This subject will be discussed further below. For the moment it is sufficient to say that Fischer was wrong in thinking Reich unreasonable for calling for magnification of 2000× and above.

Notwithstanding his appearance of firm confidence in his letter to Fischer, Reich was shaken by (as he saw it) Fischer's deeply irrational reaction, and proactively trying to protect his experiments from similar reactions from the scientific world, Reich began to document his experiments with a heightened attention to detail.[11]

In that context, Reich agreed to du Teil's offer to carry out replications of key bion experiments in a laboratory in Nice and to give a report on them in March 1937 to the Natural Philosophy Society in Nice. More than any other scientist—indeed, almost alone among all other scientists—du Teil took the time, trouble, and expense required to learn and carry out key details of Reich's technique. A close look at du Teil's role in the bion experiments, then, is warranted here. Du Teil and Reich exchanged dozens of letters—most between February 1937 and October 1938 just after the Munich crisis and Chamberlain's appeasement of Hitler. Thus, a fairly detailed record exists in which many crucial details of experimental technique are discussed.

In his letter of December 12, 1936, Reich told du Teil his first culture experiments, begun in October, over the past two weeks had succeeded. In November 1936 Reich ordered new microscopes from Reichert in Vienna, their latest Model Z, which was capable of magnification up to 4000× in darkfield (using a 100× objective; see Figure 3.2).[12] (Reich later obtained a special 150× fluorite objective, allowing even greater magnification in bright field).[13] On January 8, 1937, Reich sent du Teil two sterile bion preparations and a preliminary report on how they were created; it was Reich's Preparation 6b bions.[14] On January 26, du Teil was able to observe the bions microscopically, using, as Reich directed, a binocular microscope capable of 2500 to 3000× magnification. Du Teil did not have such equipment but was helped and instructed in bacteriological technique by "Dr. Ronchese and Dr. Saraille, who run a superbly equipped analytical laboratory in Nice."[15] Du Teil sampled the bion preparations using sterile technique, placed them on a sterile microscope slide, and immediately placed a second slide atop the first, sealing the edges to prevent evaporation. Then the bions were immediately examined

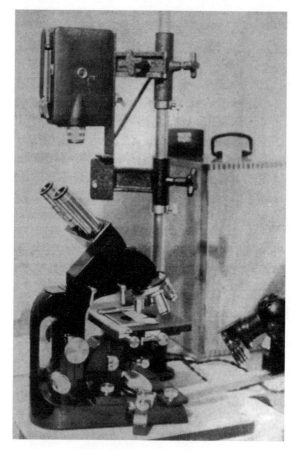

Figure 3.2: Reichert 1937 model "Z" research microscope. Courtesy of Wilhelm Reich Infant Trust.

under the microscope. Du Teil observed four principal forms, immediately after opening and sampling the cultures: (1) Rods that moved vigorously longitudinally, interspersed with undulating, crawling movements. These underwent division into two similar forms as du Teil watched. (2) Unicellular, mushroom-shaped bions that had "a luminescent, constantly quivering nucleus." (3) Large forms that "looked like mycelium with spores on the end of each branch." These were much larger than the other forms and they constantly expanded and contracted while remaining stationary." (4) Cells with no nucleus, "which clearly move in a much more mechanical way [than the other 3 types], as if moved by an outside force." Du Teil added that his "overall impression is that the preparation contains living microorganisms," despite the "precisely stated fact that the preparation was obtained by boiling and is thus absolutely sterile."[16] The mycelium form with spores sounds like a fungus or actinomycete, which might suggest to some that the preparation must have been contaminated with mold spores. Yet du Teil emphasized that Dr. Ronchese, a trained bacteriologist, "although he reserves his interpretation of the observed phenomena, he never for one moment doubted the accuracy of the experiment which produced the bions."[17] Another bacteriologist, M. Deel from Cannes, agreed: "he too had little if any doubt about the appearance of microorganisms in the culture which had been kept for two hours in the sterilizer at 180°C. He even added that the cultures looked very much like a 'pure' culture, which would exclude the hypothesis that these are so-called air germs."[18]

By February 3, du Teil wrote to say that he could confirm in all respects Reich's observations and descriptions of a sample of one of the cultures Reich had sent, asking if Reich wanted him to report this to the French Academy of Sciences or rather to report it there himself. Du Teil advised that the report should note the experiments of Stefan LeDuc, which thirty years previously had produced results in some ways analogous in form and movement.[19]

By this time Reich was pondering in his lab notebook, "*Doubts*—many bion strains can no longer be cultured.... At what point does metabolic activity begin? Where does the jump from physical movement to life occur?... Everything is confusing!"[20] Reich wrote back to du Teil on February 8, informing him that despite these problems he had methodically cultivated the bions and that he shared du Teil's opinion that the success of these cultures was of extreme importance for the interpretation of his discovery. Reich simultaneously wrote and sent bion cultures to the French Academy of Sciences in Paris and to physicist Emil Walter in Zurich, who had some familiarity with Reich's work.[21] By late February, he had sent samples of cultures to the Rockefeller Foundation in Paris as well. Reich gratefully accepted du Teil's offer to verify some of his bion experiments independently at a local lab in Nice.[22]

Odd Havrevold was a young physician who worked as an assistant in Reich's lab group; he had completed his MD in 1926 at Oslo University Medical School and thus knew most of the professors there (see Figure 3.3).[23] Reich valued Havrevold's expertise in neurology and endocrinology. Havrevold was interested in psychoanalysis; he was a patient of Reich's in vegetotherapy. In

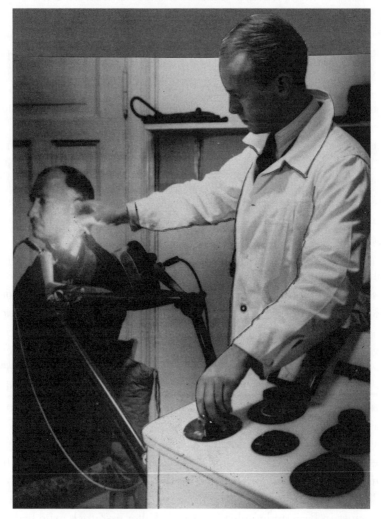

Figure 3.3: Odd Havrevold (right). Courtesy of National Library of Norway.

March 1937, Havrevold, under pressure from Oslo University Professor Kristian Schreiner (who was skeptical of Reich's bions; he became dean of the Medical Faculty in December 1937), urged Reich to send some of his bion cultures to Oslo University bacteriology professor Theodor Thjøtta, for identification by standard bacteriological techniques (see Figure 3.4).[24] Reich had his doubts, recalling the irrational behavior of Albert Fischer, but eventu-

Figure 3.4: Theodor Thjøtta. Courtesy of National Library of Norway.

ally he assented. By March 8, Thjøtta had received the cultures, which Reich described as "an uncooked bion preparation 6a, in which rods had appeared a few minutes after the mixture had been made."[25] Although Reich heard from him about the cultures only by phone, Thjøtta told Havrevold that same day "that it contained simple rods, i.e. [Bacillus] subtilis and [B.] proteus as found in decaying matter."[26] Havrevold grew extremely anxious with a senior bacteriologist and Oslo University professor telling him Reich's findings were meaningless.[27] "Havrevold declares that he is vacillating," Reich wrote in the lab notebook on March 16; "He cannot break free from the traditional theories. He tries to do my work by proceeding from alien theories. That is impossible. I cannot adjust myself to the compulsive manner of bacteriologists. . . . Will it all work out?"[28]

Reich had applied for permission to conduct bion injection experiments on mice and guinea pigs, to test whether the bions had any beneficial or harmful effects. Harald Schjelderup inquired of Thjøtta for approval, since the bacteriologist was "the official expert for the Medical Directorate." Schjelderup relayed Reich's request to Thjøtta by phone. By May 20, 1937, Thjøtta told Schjelderup he was not going to give permission for Reich to carry out animal tests with bion injections. Reich wrote in his laboratory notebook, suspecting anti-Semitism as at least a partial motive: "He says that everything in that field has already been thoroughly studied. He asks Schjelderup whether I am an immigrant."[29] Reich stated, "The answer was that I had sought employment outside of Germany following Hitler's rise to power. 'So he is Jewish,' was the bacteriologist's logical conclusion. Thus I was denied permission to test the effects of the bion cultures on mice."[30] Thjøtta's response, Havrevold's panicky reaction, and Reich's suspicion of anti-Semitism are discussed further in Chapter 6. For now, suffice it to say that if Fischer's hostile and visceral reaction had made Reich cautious about dealing with "formal authorities" in biology, Thjøtta's reaction only deepened Reich's sense that he could expect quite a bit of irrationality from the world of official science.

Havrevold knew the head of the Medical Directorate personally; he intervened and quietly obtained permission for Reich's animal experiments. However, during the press campaign that boiled over after the publication of the bion experiments beginning in March 1938, Anaton Schreiner in the government Social Services Department formally filed charges against Reich, claiming Reich had inoculated animals despite having been forbidden from doing so.[31] This led, later in 1938, to a formal filing by the Social Services Department, advising that Reich's residency permit in Norway should not be renewed.[32]

Reich moved ahead, but felt increasingly cautious; only du Teil so far seemed to Reich to approach the subject with a rational and open-minded attitude. When du Teil asked, for example, about the possibility that staphylococci or heat-resistant spores might have been present in the gelatin used in his Preparation 6 bions, Reich replied.[33] "I already heard this argument from the bacteriologist, Thjøtta," said Reich. "I cannot comment from personal experience, but wish only to share that we sterilize the gelatin three different ways. We either store it dry in a sterilizer at 160–180°, or second by putting it in potassium chloride [KCl] and the test tube in a sterilizer at the same temperature, or, third and last, by autoclaving it in potassium chloride at 120° under pressure. I have personally seen similar cocci-shapes at high magnifications, but my experiences up to now show them to be immobile, and they thus have nothing to do with the very mobile and electrically charged bion-cocci."[34] Du Teil accepted this as an adequate answer to the question, while Thjøtta had simply refused to discuss the matter further after his initial peremptory diagnosis.

By March 25, Reich recorded a breakthrough in his lab notebook: "Eureka! Discovered the real thing! [*Ei des Columbus!*] It is claimed that spores are killed off at 130°C. Good. We heat earth, coal, gelatin up to 180°C in a *dry* sterilizer. Then we boil the earth and coal, which has been heated to 180°C, in KCl at 100°C. It breaks down, disintegrates into vesicles. Immediately there is amoeboid movement. If we prepare cultures, growth occurs from autoclaved coal."[35] The following day, he elaborated on how he understood the paradox of germ "contamination": "*The germ problem is solved!* Germs are vesicles that form in matter as a result of gradual swelling and become living organisms on culture media. They can be 'killed off' at 100°–180°C, i.e., they are dried out to such an extent that they do not swell back again in broth. Heating to 180°C and, from anew, boiling at 100°C is merely a radical acceleration of the normal drying and reswelling process of the so-called spores. *The living arises from the nonliving!!!!!!*"[36]

Brownian Movement and Bions

As discussed in Chapter 2, based on his theoretical approach, that is, combining substances that would create swelling (mechanical tension), electrical charge and discharge, and dehydration (mechanical relaxation), Reich believed he would be creating structures with many of the properties of life. He made chemical mixtures of paired substances associated with expansion or swelling and contraction/dehydration in biological systems. In attempts to sterilize these mixtures by autoclaving, he found that they became still more colloidally turbid, and—to his surprise—the microscopic structures produced were more likely to be electrically charged, and more strongly charged, than in unsterilized mixtures. The unsterilized structures after swelling exhibited the Brownian movement typical of all tiny, bacteria-sized objects suspended in solution—random bumping around from place to place. The sterilized ones, in addition to Brownian movement, showed some other highly interesting movements. Reich thought these were true living motility, especially when he found he could successfully *culture* the sterilized bions by transferring them through numerous generations in pure culture on sterile growth media. From the beginning, biologists skeptical of Reich's results said that what he was seeing was nothing but Brownian movement and that, because he was not a trained microbiologist, he was unaware of it, very much as T. H. Huxley had made the same claim against H. C. Bastian in their 1870 debate over Bastian's experiments purportedly demonstrating spontaneous generation, about which I have written in *Sparks of Life*.[37]

By January 18, 1937 it is clear that Reich was aware of and paying attention in his experiments for the distinction between Brownian movement and living motility.[38] He distinguished four different states of movement within his bion preparations:

1. The vesicles *do not move* and therefore do *not* change their position relative to each other (vesicles in gelatin, dust, unheated soot in water, blood charcoal in sodium chloride [solution], etc.).
2. The vesicles *move back and forth on the spot* in an almost rhythmic manner. They do not move from place to place, nor do they influence each other (e.g., in unboiled milk).
3. Movement from place to place, e.g., in the case of streptococci and staphylococci. In this type of motion, the various vesicles exert a force on each other. If an electrical current is applied to the staphylococci long enough to give them a negative charge and render them immotile, then with time they also lose their cataphoretic ability. Thus, motion and electrical charge must in some way be related.
4. *Movement within the organic system;* i.e., relative displacement of the parts of the organism. This mainly takes the form of expansion and contraction.[39]

Even Brownian movement was absent in (1), when the matter had not yet undergone sufficient swelling, a fact Reich noted was in contradiction to the standard physical interpretation that Brownian movement should *always* be present.[40] His condition (2), looking at normal pasteurized milk under the microscope, is a demonstration of what counts as "textbook" Brownian movement commonly used by high school and undergraduate biology teachers. Reich was paying very careful attention to subtle differences in movement usually overlooked by biologists who act as though only one of two states is possible: Brownian movement or true living motility.

The key distinction Reich observed was whether the bions exhibited internal pulsation, as opposed to merely the random motion from place to place. Often, the fine pulsatory movements within tiny bions could only be clearly distinguished at above 2000×, magnifications that Reich's high-quality

Reichert and Leitz research microscopes could easily attain but that most standard laboratory microscopes could not. By a month later (if not before) Reich had decided that the concept was very often being used by scientists as a "fetish term," that is, a label to falsely explain away truly vital movements by conflating them with purely physical-mechanical movements caused by uneven, random collisions with fluid molecules surrounding the object in question.[41]

Above and beyond the pulsatile movement of the bions, the culturability of the bions proved that they were true living organisms, Reich argued. Reich was arguing, much as Fleck did in 1935, that microscopists would refuse to see things that "could not exist" within their closed epistemological system—indeed most of his opponents refused to even look into his microscope—or, failing that, they would assign the anomalous phenomena to "epistemological wastebasket" categories (see, e.g., "involution forms" to explain away bacterial variability during the heyday of Koch's monomorphist dogma).[42] Even today, when I have shown biologists what Reich called "bions," they commonly say, "Oh, those are just spherosomes," "just fat globules," "just mitochondria moving within the cell (or in the fluid from a ruptured cell)," and so on. These statements are made without any kind of testing (e.g., staining with Janus Green or DNA sequencing if the objects contain DNA, either of which could support the view that they might be mitochondria). Such an attitude, particularly the word "just," is a red flag that this behavior could be what Fleck describes.[43] Reich claimed that Robert Brown himself had at first thought at least some of the movements he saw (in particles of soot, for example) were vital movements; only later did physicists successfully take over this piece of territory from biology.[44]

In a more recent critique, one writer attempts to dismiss bions as purely optical illusions. Their motion is *not* Brownian movement, he agrees (with Reich). Rather than "just Sarcinae," he argues they are "just the blue field entoptic phenomenon or Scheerer's phenomenon," which appears to our eye as "small vesicles moving in a juddery fashion."[45] This is a phenomenon by which the outlines of white blood cells passing through the capillaries of the retina can be glimpsed under certain lighting conditions (the effect is strongest when gazing at a smooth blue surface, such as a clear blue sky). The effect can be used as a helpful test of circulation in the capillaries of the retina, important in diseases like diabetes that can cause retinopathy.

This seems to illustrate rather well Reich's point that many scientists tend to rather reflexively dismiss his findings about bions as "just" this or that phenomenon, "already well-known," without having read his experiments with any real care. In this case, if the writer had actually read in detail any of the publications about the bion experiments, he would immediately be faced with some facts that make his confident assertion evaporate. Why would optical illusions migrate in an electrical field, for example, or reverse their direction of migration when the current direction was reversed? Why would optical illusions accept biological stains? How could they be made to grow in pure culture through multiple successive generations? Unfortunately, since very few have read the original reports about the bion experiments, such assertions—if made with sufficient confidence—tend to stick. Seen from Reich's point of view, this might seem like a constantly nonlevel playing field.

Historical Background on Brownian Movement

It is useful here to step back from the story, to briefly review the history of the "discovery" of Brownian movement and the process by which the phenomenon came to be "black-boxed" into the conceptual tool that Reich found deployed—deceptively, as he saw it—against him. In the summer of 1827, Scottish botanist Robert Brown was observing some plant pollen suspended in a drop of water under a microscope. As the particles swelled and broke open, he saw that they spilled out tiny, bacteria-sized particles that moved actively around in the field of view. Brown thought the movement was strikingly lifelike and called the tiny particles "active molecules." Attempting to ascertain their nature, he tried many different substances and found active molecules were produced by almost any matter he put in water and observed under his microscope, from dead plant matter a hundred years old to bits of granite and even a fragment taken from the Egyptian Sphinx. He fastidiously controlled for numerous factors such as heating of the solution on the slide, and concluded that none of these was the cause of the "molecules'" movement.[46] (In Brown's original notes, it is clear that some pollen particles produced had movement much more lively than others, "in some cases as brisk as that of the analogous infusory animalcules," as Brown expressed it in his notes.)[47] Brown made the discovery just as achromatic lenses were first pioneered and before their general availability; thus, even to clearly see the "molecules" and their motion required the best-quality single lens microscope of the time, one that not every lab could afford. Brown went on two tours of Europe, taking his microscope with him, to show the phenomenon firsthand to all the prominent scientists on the Continent and in Britain.[48] His scientific reputation was impeccable; nonetheless, many scientists could not see the molecules until Brown showed them the details of the method using his own high-quality microscope.

Seventy-five years earlier, Comte de Buffon and John Turberville Needham had made similar observations and concluded that the motile particles were the self-active "organic molecules" Buffon had predicted as the basic constituents of all living matter, into which it would dissolve again after death and decay. Brown was aware of these earlier observations and of the result: that naturalists considered this claim of self-active matter to be inseparably connected with Buffon and Needham's claims to have observed spontaneous generation of protists via the clumping together of "organic molecules" released from decomposed plant and animal matter in infusions, boiled and unboiled.[49]

Perhaps, then, Brown was not completely surprised, despite his rather cautious language in his 1828 paper, that immediately after it appeared naturalists from all over Europe immediately assumed Brown was asserting (or at least implying) a claim for spontaneous generation. This doctrine had overwhelmingly charged implications—rendering a Creator God obsolete for creating primitive life—and so the vast majority of those naturalists strongly objected to Brown's perceived claim.[50] His term "active molecules" seemed to imply to most that Brown believed in a doctrine of self-active matter quite similar to Buffon's. (In Brown's time, after the French Revolution and Napoleonic wars had rocked Europe, the *political* implications of such a doctrine seemed every bit as disturbing—at least in Britain—as the theological overtones. As Desmond and Moore have put it, this was a question of "democratic, bottom-up organizing matter."[51]) Brown was widely criticized for seeming to imply self-active matter; in an 1829 follow-up paper, Brown went to great lengths to argue that he never meant to imply the particles were self-active or the motion vital motion. Even among those critics, some, like Berlin microscopist Christian Ehrenberg, saw the particles as possessing *"lebendigen Bewegungen"* (living movements), though Ehrenberg, too, was anxious in his discussion to show that this did *not* imply the possibility of spontaneous generation, a doctrine he opposed.[52]

Even as late as 1868 and 1870, scientists were still posthumously chiding Brown for succumbing to the temptation to "see what he wanted to see," that is, "the ultimate atoms of life," "Leibnizian monads," or other such life particles. When physicist Emil DuBois-Reymond, in his talk on July 1, 1870, made this argument, he might have been responding to new claims of spontaneous generation, such as those of Henry Charlton Bastian.[53] Indeed, the talk bears striking resemblances in its discussion to Huxley's Liverpool British Association for Advancement of Science Presidential address being written at just that time, in which Huxley was planning to attack Bastian. At that meeting in September 1870, Huxley accused Bastian of failing to recognize the difference between true living motility and nonliving Brownian motion and, thus, of mistaking dead objects for living microbes.[54]

It is in many ways striking how the pattern of these earlier controversies played out again in the story of Reich's 1930s bion experiments. Indeed, despite their twentieth-century context, Reich's bions must be understood squarely in the tradition of what I have elsewhere called "histological molecules," such as the "molecules" of Hughes Bennett or the "microzymas" of Béchamp in the nineteenth century, because of so many common features.[55] Reich claimed that what Brown saw looking at soot "molecules," for example, were the same objects that Reich was now calling carbon or soot bions. In these, the pulsatory movements were biological, argued Reich, so he was claiming that at least *some* of what was lumped into the category of "Brownian movement" was biological movement, not physical movement caused by uneven molecular bombardment.[56] Reich actually saw the category of "Brownian movement" as it applied to bions and other structures similarly to some nineteenth-century biologists. As far back as the 1870s and before, some biologists argued that what was called "Brownian movement" encompassed a spectrum of different kinds of movement, "a kind of gradual transition from inorganic to organic motility." They saw that within this spectrum was a range "from non-lifelike movements of mineral particles to the much more lifelike movements of pollen granules."[57] This fit well Reich's view of bions as transitional forms between nonliving and living matter, showing varying degrees of motility depending upon their electrical charge.

However, numerous scientists did not even look at Reich's bions because they were so certain he must be making the same mistake of which Robert Brown had been accused and of which Huxley accused Bastian.

Recall that Roger du Teil became interested in Reich's experiments and ran independent controls of them at a laboratory in Nice, eventually setting up his own lab at the university there. When du Teil first repeated the experiments, with the help of two bacteriologists, Mr. Deel and Dr. Ronchese, at their lab, both men stated that they saw the same forms and their lively movement in a bion preparation viewed microscopically immediately after its preparation. Ronchese, however, said the movement must be Brownian and not truly living. Reich's main rejoinder was that the bions could be grown through successive sterile transfers in pure culture, which would in no way be seen with nonliving objects.[58] But later, with very high magnification, Reich stated one could clearly distinguish internal pulsation of the bions, not mere place-to-place movement that might be Brownian. Only after he visited Reich's lab in Oslo in the summer of 1937 to learn important details of technique could du Teil reproduce many of the phenomena and verify the observations at magnifications in excess of 2000×. Afterward, du Teil took one of Reich's large Reichert research microscopes back to France on loan in early August 1937. Du Teil was well connected in Paris scientific circles, and he set up demonstrations of sterile bion preparations using the Reichert microscope, as well as showing time-lapse films Reich had made of several bion preparations—to groups of scientists at the French Academy of Sciences and from the Faculty of Medicine.

Here we may consider for a moment an alternative interpretation by modern biophysicist Steven Vogel, relevant to bions formed from dead, disintegrating plant tissue. Though not directly referring to Reich's experiments, in a general discussion of Brownian motion Vogel puts forward a useful hypothesis for the topic under discussion here. He points out that *within* cells, the internal structure of the cell "seems to slow macromolecular movement disproportionately. Indeed, the loss of much of this structure on the death of the cell is marked by an increase in Brownian motion that is quite conspicuous in time-lapse movies."[59] Vogel's hypothesis would suggest that when Reich observed disintegrating plant cells and noted the formation of bions and the steady increase of their motility, the motion they exhibit is actually Brownian motion, because of the removal of physical barriers to molecular movement as the cell's internal structures disintegrate. In this case, one can actually collect evidence to evaluate whether Reich's interpretation or Vogel's is more likely to be correct, or whether both kinds of motion are going on simultaneously. As Reich pointed out, one could examine the movement of the objects using magnification of 2000× and above, to distinguish whether their motion is internal pulsation, random place-to-place motion, or a combination of the two. One could also do longer-term time-lapse observations on (and/or attempt to culture) the bions, to determine whether they develop into protists or objects with other lifelike properties. If that were so, then even if their initial motion was entirely Brownian, most of Reich's interpretation of their significance—as transitional stages between nonliving and living—might well still be valid.

The "Brownian movement, not true living motility" argument—identical to when Ronchese put it forward—was advanced again when du Teil showed the bions and films of their movement in Paris to the abovementioned group of French scientists on August 10, 1937. However, most of those who looked at the bions through the Reichert Z microscope at high magnification were impressed and inclined to be more open-minded. As du Teil described it to Reich, eleven members of the Academy of Medicine were present, one of whom was Dr. René Allendy, who was interested in psychoanalysis and had visited Reich in Oslo. Also attending were "Dr. Barbe, Professor [Raoul-Charles] Monod from the Academy of Surgery, who had brought with him Dr. [Henri] Bonnet, head of the Laboratory of the Academy of Medicine; Dr. Symbi, biologist, known for his work on nutrition; Dr. Vincent, biologist; Dr. Nordberg, Consul General of Finland at Paris and an international journalist; Mr. Janvy; attaché at the Consul's residence, etc."

Also present was a physicist, M. Bourdet. Du Teil emphasized how impressed the observers were by the high quality of the large Reichert microscope: "The atmosphere was favorable from the beginning. I had also set up the microscope, which produced an impression all by itself. Everyone was filled with admiration for such an instrument, and Dr. Bonnet, head of the Laboratory of the Academy of Medicine, worked with it all evening and did not want to stop!" Several of these doctors and scientists enthusiastically volunteered to repeat the experiments at their own laboratories, according to du Teil's report:

1. In the end, Dr. Monod, who did not hide his interest, told me: "I ask you from the time of the rentree in October to send me all that is necessary. The Laboratory of the Academy of Medicine is going to follow methodical procedures on the bions and their cultures under my and Dr. Bonnet's direction. . . ."
2. On the other hand, Dr. Allendy asked me to officially present the film at the Sorbonne in October or November . . . to the Group of Philosophical and Scientific Societies which he directs, and he promised me that all the great French biologists of the Museum and the Pasteur Institute would be there, including Tissot, who has already stated that the BK [tuberculosis bacilli] are endogenous and not exogenous.

3. Dr. Martiny, who is on vacation, writes: "Give M. du Teil my great esteem for his person and his research. The laboratory of the Leopold Bellan Hospital is at his complete disposal."[60]

The "holistic" trend in French medical thought at this time meant this audience was particularly receptive. Allendy and Martiny were interested in neo-Hippocratic medicine and "had been seeking to create a scientifically respectable form of homeopathy more closely integrated with mainstream medicine."[61] Allendy was a psychoanalyst and homeopath who had written an influential 1922 book trying to restore the importance in medical doctrine of the "temperament" of the patient, what in other places was called the patient's "constitution."[62] This was also occurring at the time a young Rene Dubos was arguing for an "ecological" understanding of the relationship between germ and patient, in explaining variations in susceptibility to disease—the older "seed versus soil" debate, but now cast in twentieth-century scientific concepts.[63] Reich's later concept of "biopathies" was very much rooted in this tradition of trying to understand the large role of the patient's constitution in modern scientific terms.

According to du Teil, "The meeting began in an atmosphere of diverse attitudes; sympathy on the part of those already acquainted with the work; interested curiosity on the part of those who had not yet heard of it; occasionally also skepticism as, e.g.," from the physicist Bourdet, "who offered the argument of the Brownian movement and tried to prove at all cost that the proteins were the carriers of unknown and naturally invisible germs."[64] However, many of those present were not so sure the bions might not be considerably more important. Bonnet in particular, whose opinion among all those present, according to du Teil, "was by far the most important, showed the understandable surprise of a laboratory man who daily carries out successful sterilizations. Nevertheless, he demonstrated great interest and seemed determined to approach the matter scientifically and to acknowledge the relevant arguments. In brief, the atmosphere was highly favorable and was made more so by my position in the C.E.P.H. [Center for the Study of Human Problems] where I am well known, and also by the microscope which immediately forced one to take seriously the owner of such a magnificent instrument."[65] Thus, several of those present agreed with du Teil to undertake a series of experiments in the Paris lab of bacteriologist Robert Debré, where Bonnet worked, and at other labs, to attempt replication of the key claims, particularly the culturability of the bions.[66] Du Teil, responding tactically to the objections, began by October 1, 1937 to diverge from Reich on the best strategic argument to make regarding Brownian movement.[67] The experiments in Debré's lab appear to have produced mixed results, which Reich never got the opportunity to clarify. This episode, and du Teil's tactical differences with Reich, will be discussed in Chapter 6.

Reich and du Teil had an exchange in September 1937, in which they differed about the meaning of the bacteriologists' label *Bacillus subtilis*, which Thjøtta had used earlier in dismissing Reich's cultures. This is a common bacterium, particularly in hay and in cheese, which produces heat-resistant spores that can survive even boiling for hours, only to germinate into live bacteria again once the temperature has cooled. In the spontaneous generation debate between Tyndall and Bastian in the 1870s over Bastian's claim to have produced microbes from hay infusions and turnip-cheese infusions, the discovery of these heat-resistant spores in 1876 by German botanist Ferdinand Cohn allowed Tyndall to claim these spores were responsible for all growth in Bastian's boiled infusions. Because of these spores, Tyndall introduced a cycle of repeated boiling, cooling, then reboiling, to finally achieve sterilization. Later to simplify sterilization the autoclave was invented in 1882, once it was realized that boiling alone was insufficient to sterilize; autoclaving samples keeps them at 120°C and 15 psi (pounds per square inch) of steam pressure for at least thirty minutes. This kills the spores.

In one of Reich's cultures of soot bions, easily recognizeable by its consistent bluish-gray color when grown upon blood agar, du Teil found that *B. subtilis* as—he thought—a contaminant had been introduced at a certain point in reinoculations, so that it was present in cultures inoculated from that tube thereafter but not prior. It was easily recognized on solid media by its different appearance; microscopically, du Teil described these organisms as "real sausages."[68] Reich replied, "I cannot, emotionally, warm up to the subtili question for the time being. When a film appears [i.e., a pellicle on top of a liquid broth culture], it might not necessarily indicate subtili, for it could also be that the bacteria consolidated into a membrane. Besides that, we had agreed here that fundamentally the subtili-bacilli themselves cannot be anything else than some of the forms we obtained through artificial means."[69] Du Teil may have had second thoughts since leaving Oslo a few weeks earlier, about whether he agreed with this idea of Reich's that even when *B. subtilis* appeared in infusions, it might be of endogenous origin. Once one accepts the idea that life, transformable into what look like known bacteria, can arise in sterilized media, the question of how to know for certain that a given organism originated endogenously versus from contamination becomes more difficult. If the growth appears only near the edge of an agar plate and not at all on the inoculation streak, that is clear presumptive evidence it is a contaminant. But no such clarity is available in liquid broth cultures, which is, of course, why Robert Koch developed solid media in the first place around 1880–1882—to be able to isolate colonies that came from a mixed culture.

But Reich very much valued du Teil's independence and willingness to disagree with him. They discussed publishing their correspondence, since it contained so much detailed back-and-forth about technical issues in the bion work. And Reich included du Teil's September 1937 essay, "Life and Matter," in *Die Bione*, along with some of their early correspondence. Reich imagined their working relationship visually: "I understand very well that you will proceed along a different track than I. I only wish to add this: both tracks are branches of one railway line, which will, at a later point in time, unite to form a single railroad track. I believe this analogy characterizes my full understanding of our two positions that exist out of necessity."[70] Responding to du Teil's intelligent, immanent criticisms helped Reich clarify his own thinking. Perhaps not surprisingly, since some of Reich's laboratory assistants in Oslo (e.g., Havrevold and Wenseland) were also in therapy with Reich, they were overawed by him enough that they were less forthright about criticizing him. Also, they felt conflicted since they had future careers in Oslo at stake if they made enemies of Oslo scientific and medical authorities. Perhaps because Reich understood this, he valued du Teil's open-minded but critical collaboration very highly indeed.

Reich continued the discussion of *B. subtilis* in his next letter, on September 9. He had asked du Teil to mail him some of the contaminated tubes and was awaiting them to examine, so he could decide for himself if the "real sausage"–shaped microbes du Teil found were contaminants. He told du Teil, "My gut feeling is to reject your analysis." He explained that he had previously "observed here several times that the surface of a creamy-type growth begins to change after several weeks. Examining the different layers under a microscope, one sees formations that are different. It looks like a mixed infection, but I must inevitably bear in mind that forms change under circumstances that are not yet understood. And the formations, which you described as 'real sausages,' I saw in absolutely pure cultures a long time ago."[71] What a standard bacteriologist would call a "mixed culture caused by contamination," Reich was more inclined to interpret as containing different stages of a microbe undergoing pleomorphic transformation over time.

Du Teil was no "yes man;" he replied to Reich's September 7 letter (albeit having not yet seen the missive of September 9), saying, "We cannot deny that there are a large number of microorganisms in the atmosphere. We, ourselves, have collected and cultivated such. And, furthermore, we cannot hope to guarantee that never, ever did one of these microbes fall into our bouillon during an experiment. Now, with the impossibility of completely preventing germs in the atmosphere from contaminating our experiments, I believe it would be better for us to stand together and all agree that here and there a germ could be present. For this purpose, it is vital that we understand how to recognize and identify them."[72] Further, du Teil told Reich, "Take note that when I determine *B. subtilis* in a culture over the initial, authentic, pure culture, I thus verify the purity of the last one made," that is, the last culture before the contamination occurred.[73] If one could prove the bions produced under sterile conditions were a pure culture, this was was strong evidence against the presumption of contaminated ingredients, since that would more likely result in a mix of different organisms. But du Teil thought this situation so clear-cut that it would all but alienate any bacteriologist to hold out, as Reich was doing, for a pleomorphist interpretation. He argued that the sausage-shaped organisms were present in any culture inoculated from his August 27 culture on agar, along with cocci, which were the original bions. Other cultures inoculated from tubes produced before August 27—or even on August 27 into bouillon rather than the agar plate—contained only the cocci. So du Teil found it obvious that the sausage-shaped rods were accidentally introduced during creation of the August 27 agar culture: "The agar, after it had shown a creamy, gray cocci culture the morning following the inoculation, is suddenly infested with a dry culture, which is rippled like a wilted leaf—characteristic of *subtilis*—and which covers the other, thoroughly aerobic culture. . . . While the bouillon, which had been inoculated at the same moment, presented only with cocci, no pellicle. These circumstances make it evident that the new culture, which shows up on the strain and on the agar, has been caused by an event that does not occur on the bouillon."[74] Caution over the possibility of contamination by such a common organism as *B. subtilis* inspired du Teil to design and build a completely closed system of connected glass flasks and tubes, in which sterilized bion ingredients could be mixed without ever opening the system to outside air after sterilization. Du Teil's "Sy-Clos" apparatus will be discussed later in this chapter.

Reich felt that du Teil's idea of presenting the contamination explanation to bacteriologists was a good strategy, though he still thought the evidence argued more strongly for these organisms' having come from endogenous infection. The assertion of du Teil's point of view "could, after all, be refuted at some point." Writing on September 16, Reich in the meantime had time to examine carefully the "contaminated" tube du Teil had sent him. Reich had also made fourteen fresh incandescent soot preparations for comparison with du Teil's sample. Three of those produced no growth; seven produced a pure culture of the blue-gray soot bions that were the most common outcome of this preparation, which appeared microscopically as cocci. On another occasion Reich made four soot preparations all at once. Each of them produced a brownish growth consisting of rods, which were identical to the rods in the culture du Teil considered contaminated. As Reich summed it up,

> This growth appears in all four of my culture mediums that were made on the same day, which all contain cocci that definitely originated from the soot and rod-shaped organisms that one does not normally find in soot cultures. I am requesting that you carefully re-examine the two culture mediums that I am returning because I found cocci as well as the *Bacillus subtilis* rods you diagnosed in both of them.... Please report to me if the rods did not show very rapid movements when they were inoculated directly with airborne germs. The rods in your preparation showed almost no motion. There is no difference between the rods in your preparation and those in mine. Comparing the homogenous cocci, your rods show no structure, rather resemble small sausage links. Mine, on the other hand, show either a streptoarrangement of vesicles or are cudgel shaped when viewed at 3000× magnification. I obtained similar rods when I autoclaved moss, as I once demonstrated to you.
>
> I consider the question of airborne infection extremely important, not so much from a practical standpoint, rather from a theoretical one. Recall that we experimented with eggs in order to determine whether the putrefactive bacteria in rotten eggs were a result of the self-accelerating egg white decomposition or a result of airborne bacteria. Yesterday we opened the first egg that had been coated with tincture of bronze. It was completely rotten. We made a culture of it and will send you a sample. Based on

my experiences, the decomposition must be endogenous. This does not, however, totally exclude the possibility that air contributes to the decomposition of eggs.[75]

Reich was already becoming absorbed in cancer research and bion injections in experimental animals by September 1937; soon it became his most time-consuming line of work. In spring 1938, a campaign of vilification broke out in the Norwegian press over his bion work, taking up even more of his attention and eventually resulting in du Teil's temporarily losing his university job. Thus, this remained an unresolved question between du Teil and himself. Du Teil, if he was right, was pointing to the exception that proved Reich's rule: only in an occasional culture did a clearly recognizable contaminant organism get into the cultures. That made it even clearer that the bion cultures—whatever they were—were not contaminants from the air.

Professor Lapicque and the French Academy of Sciences

The "only Brownian movement" argument was trotted out again, completely unchanged, by Professor Louis Lapicque, writing Reich as the official representative of the Academie des Sciences in January and February 1938. Reich had appealed to the academy directly, sending sealed preparations of freshly prepared bions and films of them, in January 1937, and asking that if his observations were verified by the academy, the results be published immediately in their *Comptes Rendus*. (Had Reich known of the past history of the academy's intensely biased treatment of Pouchet and Bastian's experiments on spontaneous generation, he might have had second thoughts about this strategy. His own interactions with the academy show striking parallels with those earlier stories.)

Lapicque was "the acknowledged *doyen* of French neurophysiology."[76] Du Teil told Reich he was an older man, quite set in his ways. He at first agreed with all of Reich's observations, even noting that it was striking how motile the formations remained a year after they had originally been sent.[77] However, he rejected Reich's tension-charge theory and its implication that bions represented a transition between nonliving and living matter. He offered to publish his verification of the observations of bions and their motility, but only if Reich would accept an added section in which Lapicque explained bionous motility as caused by entirely physical-chemical causes, that is, Brownian movement as understood by physicists. Reich argued that "Brownian movement" of his soot bions, Preparation 6 bions, and other bion types was actually biological in character. (Lapicque either did not know of attempts to replicate the culturability of bions in Paris or he chose to ignore them as well as to ignore du Teil's successful replications in Nice.) Despite Reich's request, Lapicque refused to include the successful results on growing bions through successive generations in culture, a decisive matter.[78] Privately, Reich recorded in his diary, "Lapicque of the French Academy is attempting to downplay the cultures and pass them off as Brownian movement."[79] Reich consulted with du Teil about how best to negotiate with a French professor.[80]

Throughout two weeks of negotiations, Lapicque used his power as the official representative of the French Academy to stonewall Reich's efforts to get his own interpretation of the movement in his bions published in the *Comptes Rendus*. He gave Reich what he intended as a final offer:

> It is certain that the agreement between you and me is very limited. It is a question only of the reality of the "lifelike" movements announced in your communication of 8 January 1937. I am ready to confirm the reality, but in interpreting it, under my responsibility, physico-chemically and not biologically, by a simple hypothesis, which attributes it to the Brownian movement. Do you want to take advantage of this offer?
>
> I am not at all disposed to go further and to discuss the questions of culturing or artificial culture of the bions, questions which would not be within my technical competence. Moreover, it would not interest me to research the determinism and the physicochemical mechanism through your metaphysics.
>
> If you have not made known your acceptance of my offer before two weeks' time, in agreement with your first affirmative reply, in accordance with my preceding letter, I shall present a negative report at the Academy, and your communication will be purely and simply classified.[81]

Why Lapicque felt that ruling out lifelike movement was within his competence but culture experiments were not—both, by his standard, would require bacteriological training—is not clear. There is a hint in his wording that he may be uncomfortable with what he considers excessive materialism in Reich's interpretation of life arising from nonliving matter. Reich replied,

> After extensive consultation with Prof. du Teil, who controls the culture experiments and was the first in France to do so, I came

to the conclusion, to ask you to either publish my first statement *in extenso*, including the central point of the whole question of the successful cultures, or not to publish at all. To explain, I must add that I had not reported the production of Bion mixtures alone without the cultures to the Academy of Sciences. . . . Successful culture of the mixture of the colloid Preparation 6 changes the nature of interpretation. I understand that you don't accept the novel Professor du Teil interpretation that is imposed by the cultures. There is no absolute justification for a chemical-physical meaning in contrast to the biological. It does not coincide with that of Prof. du Teil.

If we were to publish only the purely physical as opposed to "biologically understood" Brownian motion, the established fact of culturability of bions would not be taken into account. Thus, there is a risk that my studies would be completely misunderstood.

I therefore see no other option than: publication with the inclusion of culture results or no publication.[82]

Lapicque took Reich's position as a straightforward "no," sticking with the terms of his earlier ultimatum. Reich apparently felt it best to reject the offer of publication, when he was being forced to link mere notice of the facts of his discovery to what he considered a wrong (not to mention heavy-handed) interpretation, which would imply his assent to that interpretation.

To du Teil, Reich wrote, "That does not matter. The public interest is growing not only in France but also here [in Oslo]. The authorities are using, generically speaking, a lot more cautious language [in criticizing the bions]. . . . We have until 22 February. Time. Lapicque may still allow the general publication, if he sees that the matter is not easily dismissed."[83]

Du Teil replied to Reich, "Concerning Lapicque, don't worry: I sent him a copy of your telegram and . . . I must tell you that Lapicque is a member of both the Academy of Medicine and of the Academy of Sciences, and is now very aged, which explains why he's rather set in his opinions." Du Teil was optimistic he could convince Lapicque as a fellow French academic, that his work on the nervous system shared a fundamental point in common with theirs, because "He kindly wrote me a long letter and sent me a brochure of his works, in which I found interesting connections with the peripheral electrization, which we think basic. I pointed this out to him, moreover, and I await his reply."[84]

Du Teil also hoped intervention by another prominent French biologist might help. On February 8 he had been contacted by Dr. Guy Van Den Areud, president of the Psycho-Biology group at the Sorbonne in Paris. Van den Areud had heard from Allendy about du Teil's talk and demonstration of the bions the previous August, and he was extremely interested to learn more.[85] Du Teil thought he would take an interest primarily in the philosophical implications of the experiments (e.g., for the mechanism-vitalism controversy); however, Van den Areud answered that it was the technical details of the experiments themselves that most captured his interest. He asked du Teil, "Do you believe that the continuous dynamics [*le dynamisme aeterne*] of charge and discharge shown by bions are enough to put them side by side with life forms such as amoeba and bacteria? If you don't mind, I will speak to Herr Andre Mayer concerning your discoveries: I believe that his opinion about the organization of lower life forms cannot be ignored [*n'est pas a negliger*]."[86] Du Teil sent him bion cultures to examine. On March 8, Van den Areud asked for another bion culture and stated that he had spoken to Professor Robert Debré—in whose lab under Henri Bonnet du Teil had conducted some of the experiments in October 1937—about proposed bion research. Du Teil reported to Reich, "I sent a sample of it [bion Preparation 12], at his request, to M. van den Areud, President of the Psychology and Biology Group at the Sorbonne. He passed it on to Prof. Debre of the Faculty of Medicine, who just the day before yesterday again asked me for more of them. In this way I could give you the identification that you requested from me, and this shows that little by little interest is growing around your works. It's just that much patience is required in this respect."[87] On April 30, du Teil wrote about coming to Oslo again in July 1938 and added further news: "Perhaps Debré would like to come also, and this would be significant. But it's necessary to wait until he has finished the verification, or rather the study he is conducting at this very moment at the demand of Mr. Van den Areud. It is necessary to avoid speaking of any result whatever, positive or negative, as long as the verifications are not finished. And there are none yet save mine, which really are—that is to say, none which has been made during the number of weeks and even months necessary to assure conviction and commit to a responsibility."[88] Du Teil's hope was that Van den Areud's intervention (and authority) would tip the balance in French labs (including Debré's) toward taking the bions more seriously. Again, however, du Teil's optimism was not borne out. No more was heard from Van den Areud.[89]

In the event Lapicque was not persuaded, and he exercised his power to completely ignore the culturability of the bions and to consign Reich's manuscript to the oblivion of the French Academy's archives, where few, surely, have ever read it.

In the summer of 1938, Reich acquired an apparatus that could send pulses of voltage through a preparation while one was observing it under the microscope. In early September, he thought this had produced a new way to distinguish between two fundamentally antagonistic bion types: PA bions and t-bacilli (more on this distinction in Chapter 5) and—incidentally—to clarify still further the difference between living pulsation and motility from place to place versus Brownian movement *in place*. When subjected to these voltage pulses, Reich noted in his lab notebook, "Bions are grayer, remain alive perhaps. T-rods die, retain Brownian motion *on the spot*."[90] T-bacilli—a much smaller bion type originally cultivated from sarcoma tissue—did exhibit their own active, darting motility when alive. But the Brownian motion was all the movement that remained once they were dead.

A year after he last heard from Lapicque, Reich was contacted again by the French Academy, offering again to publish his findings in the form Lapicque had proposed. Reich attempted, yet again, to get across what he considered necessary while simultaneously hoping for publication of his discovery of an important new bion culture, SAPA (sand packet) bions (these are discussed in Chapter 7):

> According to the letter from Professor Lapicque, the fact that the bion preparation produces *cultures* will not be published, and the report will be followed by an interpretation that explains the observed results as a physical phenomenon, namely "Brownian movement." Now, the essential aspect of the bion experiments is that the culturing experiment has shown the concept of "Brownian movement" to be an inadequate explanation for the sterile-produced bions and their *cultures*. Since the culturing experiment is the decisive factor, I can only accept your proposal to publish the report if, at the same time, the culturability of the bions is made known. I would therefore like to make the following suggestion.

Among the bion cultures prepared in the last six weeks, there is one which is extremely interesting [SAPA bions]. I would be glad to send you a sample of this culture with a brief description of its origin and properties for your verification. Then my first, already verified report could appear with this new one. In that way the essential character of the bion experiments would be expressed. Please let me know whether you accept this proposal. I will then send off the culture together with the report.[91]

There is no evidence of any reply from the academy to this offer. In all likelihood, Reich finally gave up on getting what he considered fair treatment there. He immigrated to the United States only six months later.

High Magnification and Other Technical Issues

Through a student of Reich's in Denmark, sociologist Jørgen Neergaard, Reich met his brother, the plant pathologist Paul Neergaard, and showed him some of the bion preparations when Neergaard was in Oslo in February 1937.[92] Neergaard, like du Teil, offered to carry out some control experiments, replicating Reich's procedures, in his Copenhagen lab. Reich sent him some bions prepared in Oslo to get started on his studies and to familiarize himself with their basic characteristics. Neergaard also offered to use standard bacteriological diagnostic methods to tell Reich how a bacteriologist would classify the bions.

Having viewed bions under several different types of microscopes, Reich was aware by December 1936 of the importance of high magnifications for rendering visible the crucial fine details of pulsatory movement within bions. He pointed this out in April 1937 to Paul Neergaard in Copenhagen and to du Teil in Nice, who were attempting to replicate his results but using only standard lab workhorse microscopes, the maximum magnification of which is usually 1000× and occasionally reaches 1500×. Reich emphasized from his first publication on the bion experiments that the particular types of high-quality research microscopes he used were essential to being able to replicate his results.[93] In addition to the magnification, the clarity and three-dimensionality that made the internal pulsation impossible to miss he found only with the outstanding optics of the 1937 Reichert Z microscope's inclined binocular tube. This had a 50 percent magnification increase built in, but even with the same magnification under a straight tube, Reich found the results much less clear.[94] In addition, a Reichert darkfield condenser, allowing observations in darkfield up to 3000×, was necessary in many cases.[95] Often, Reich found, the color of the bions he studied—many of them were a striking blue color—could be accurately assessed only using apochromatic lenses, rather than the much cheaper achromats used in standard laboratory microscopes.[96] Despite his clear statement that using the expensive Reichert Z or a comparable model (e.g., from Leitz or Zeiss) was nonnegotiable to see the crucial phenomena, almost no other researcher, with the exception of du Teil, ever did so. Even Paul Neergaard, sympathetic to Reich's work, said after a couple months of attempted replication,

> Thank you for the bion cultures you sent. I have inoculated part of them onto pure agar and part onto malt agar. The bion structures grow outstandingly on malt agar; however, not on pure agar, which is to be expected.
>
> Your suspicion that I only have a straight and not an angular curved binocular microscope is correct. Unfortunately, I do not even have a binocular, rather only a monocular microscope; however, at times I have worked at magnifications of 2500×. It is, however, a fact that I have not obtained a sharp enough image with my microscope and for this reason cannot as yet be certain whether or not I have really seen bions. After I used the cultures you sent me to complete comparison tests, I am now absolutely convinced that what I have seen since the beginning is really bions.[97]

Rather than state they could not afford such a high-quality research microscope, however, all Reich's opponents simply ignored this statement and many insisted that magnification above 2000× was useless due to loss of resolution. Some, for example, biologist Albert Fischer of the Rockefeller Institute in Copenhagen, even said that Reich requested "ridiculously high magnifications" when demonstrating bions at his lab in December 1936 and that it was impossible to achieve such magnifications with light microscopes. On this rather important point, Fischer was simply factually wrong.[98]

That Fischer was mistaken about contemporary possibilities for high magnification and that researchers might put such magnifications to important uses can also be seen in a 1935 letter from L. C. Dunn to Otto Lous Mohr, in which Dunn said that a special configuration of equipment can allow drawing the banding pattern on chromosomes at 5600× (the same maximum magnification Reich could achieve using a special 150× objective lens). I have examined most of Reich's equipment in the Wilhelm Reich Museum and found that it meets the specifications Reich described. In the bacterial pleomorphism controversies of the 1930s and 1940s, some evidence supporting the pleomorphist position of bacteriologists such as Arthur Kendall also depended upon extremely high magnifications with clear resolution up to 7500× and well beyond, such as the Rife Universal Microscope.[99] While the Rife microscope may remain controversial, the ability to attain greater than 5000× with optical microscopy is not.

Trying to encourage Neergaard, Reich responded to his difficulties, saying the results were not always easy to obtain or obvious immediately to the untrained eye:

> To encourage you, I would like to tell you a little story. When I instructed my current assistant, Miss Berle, in the technique of performing the earth tests, I showed her among other things a contractile earth structure at approximately 3000× magnification. She looked for no less than ten minutes without seeing the totally clear movement. I kept on asking her whether she could see something, and she always replied, "No, I can't see a thing." After about ten minutes of uninterrupted observation she suddenly cried out, "Yes, it's moving"—that is, she suddenly saw something that had been taking place already although she had been unable to observe it. The same thing will also happen to you; it happened to me as well, and that's the way it is with all new discoveries. So please allow the earth to swell up and be sure to carry out long and precise daily observations for at least eight days. Not only look for the movement of the rods and vesicles, about whose biological nature one can argue, but focus your attention in particular on the vesicular disintegration of the crystals of earth and the motility that then sets in. For modern scientific ears this is naturally an unbelievable process. But it is absolutely true; it exists. I have filmed it and that is certainly an objective method. The organisms you saw in my laboratory were already motile amoeboids of earth and coal. Please repeat the experiment over and over again until you really see this. I believe that this is a prior condition for our cooperation, because everything else is based on that observation.[100]

(Neergaard was never able to obtain a microscope of the quality required; however, he contented himself by reporting to Reich the classifications that standard bacteriology methods would produce for a number of bion cultures.)[101] The requirement Reich calls for, to continue looking at the preparation carefully over hours or even days, makes sense in pioneering new research. William Dallinger and William Drysdale required similar continuous microscopic observations for days on end, in experiments in the 1870s trying to resolve the question of whether microbes appeared de novo or from spores.[102] In Reich's anecdote to Neergaard, one is reminded of the card experiment by Bruner and Postman, cited by Thomas Kuhn as evidence from cognitive psychology for why many can look directly at a paradigm-violating phenomenon but not see it.[103] But it is difficult to imagine many others willing to invest such time and care, merely for replication, which rarely gets the author a publication of any significance—even when they claim to have disproven the original claim.[104] And in the story of Reich's bion experiments, that is indeed how things played out.

Bion "Cookbook"

As an aside, it is important to note that in Reich's laboratory, all the media were produced from scratch from fresh ingredients. For example, beef was cooked in water to produce broth; potatoes were peeled, cooked, and hashed to add to a later medium Reich called egg medium IV. This was still true after Reich moved his lab to the United States in the summer of 1939.[105] Fortunately, in the Reich Archives the file of all the recipes used in the laboratory has been preserved. This was published in 2009 by the Wilhelm Reich Museum. All the small (but often key) details of preparing culture media, staining, temperature of incubation and other matters are here.[106] Reich often found that small, key alterations of the culture medium or the incubator temperature made the difference between bions growing in culture and completely failing to do so. Microbiologists are quite familiar with "fastidious" microbes, which have highly particular nutritional requirements. Hence they have developed a wide range of media, such as blood agar, for such organisms. For any serious attempt to replicate the bion experiments, Reich's original recipe file ("Bion cookbook" for short) is an indispensable source of crucial technical instructions for carrying out those experiments. In these detailed recipes may reside answers to why some researchers in recent years have been unable to replicate important parts of Reich's work, most notably the culturing of bions (especially on solid media) and the production of SAPA bions.

Figure 3.5: Reich's lab group, Oslo, August 1937 (left to right, front row: Gertrud Brandt, Kari Berggrav, Vivi Berle, Hauser; back row: E. Kohn, Odd Havrevold, Roger du Teil, Wilhelm Reich, Odd Wenesland). Courtesy of Wilhelm Reich Infant Trust.

After Roger du Teil visited Reich's lab in July and August 1937 and observed at first hand the preparation of media, he was more consistently successful in replicating Reich's results (see Figure 3.5). Very often in the letters between Reich and du Teil one finds mention of small but crucial technical matters like this: Put the inoculated tubes in the incubator flat (i.e., horizontal); otherwise no growth will occur. Or "use only red gelatin in Prep. 6," since du Teil initially tried clear gelatin and found it did not produce bions.[107]

Such details are not the only potential source of difficulties. Since all the culture media in Reich's time were prepared from scratch from fresh ingredients, one can only wonder about the effects of our pesticide- and antibiotic-laced meats and milk and pesticide-laden potatoes today. Or in the case of the Norwegian sea sand used in producing SAPA bions (discussed in Chapter 7), whether oil spills from tankers or from North Sea drilling have contaminated beaches in a way that will influence results. This might well apply to beaches worldwide since the increasingly common oil spills of the past four decades. Reich's findings on the crucial effect on bion experiments of carbon, especially carbon compounds heated to incandescence (see the following section), certainly ought to inspire caution in this regard. Despite the difficulties, the more one knows, the more likely one is to locate and solve the problems with replication.

The only two of these recipes previously available in published form were "Culture Medium e" (Bion cookbook, p. 20) and "Culture medium IV (egg white medium)," which appeared in the 1939 booklet *Bion Experiments on the Cancer Problem*. For example,

> Egg Medium IV. Combine: 38 grams of potatoes. (Thickly peeled and finely hashed)
>
> 75 ccm milk. [Note that milk has been crossed out and bouillon substituted, as of Dec. 1937 when Reich realized the medium gave better results with this substitution.]
>
> 3 grams of potato flour
>
> 5 grams of peptone (Whitte). This mixture is boiled for half an hour, being stirred for the first five minutes until it forms a pulp. It is then cooled to about 50°C. Then to it are added: 10 cc of 0.1 normal [also = 0.1 molar] KCl solution, ½ gram of lecithin, 10 cc Ringers solution [see p. 58 of "Bion cookbook" for formula. Many variants exist.]
>
> 6 cc glycerin
>
> 2–3 eggs (sterile, washed in alcohol)
>
> The mixture is filtered through gauze and poured into sterile tubes. It is left in the dry sterilizer (80–90°C) for two hours until it hardens. The next day it is again left in the hot air [dry] sterilizer for two hours at 90°C, then for 24 hours in the incubator for control. Adjust pH to 7.4 with about 3 drops of NaOH solution.[108]

Even the last, seemingly minor point, written in by hand on the typed recipe, is the kind of tacit craft knowledge from a lab's working experience that often does not get recorded in print yet can make a crucial difference in how experiments turn out. In the experimental spontaneous generation debates of the 1870s between Bastian, Tyndall, Roberts, and Pasteur, the pH of the medium was at times a crucial factor behind different results.[109] Du Teil wrote to Reich at one point, "It is even possible, that with systematic changes in the pH, our cultures could have succeeded under totally different conditions. I was questioned in Pontigny by two doctors about the pH and I had to discuss it with them for 25 minutes."[110]

Reich was not surprised that minor differences in technique can be crucial and that this was also a very significant factor with attempts to replicate the bion experiments. He warned du Teil, "If you use [tubes of] egg-medium, then please only fresh ones. Old egg-medium gets too dry and often yields no results."[111] He had already become familiar with the importance of what might seem small details of method a decade earlier, through the inability of many psychoanalysts to confirm his finding "that every correct analysis finally reaches a stasis neurosis. . . . For many years I could not understand these objections, until it finally became clear that they involved differences in technique."[112]

In this context, it is worth repeating that, once du Teil had learned in detail at Reich's Oslo lab how to repeat the methods, from July 27 to August 7, 1937, and then demonstrated those methods to a gathering of physicians, biologists, and physicists in Paris on August 9–10, those scientists took a much more serious interest in the experiments and put several laboratories at du Teil's disposal for replication of key experiments.[113] Indeed, even the Oslo press took notice of the scientific excitement in Paris over du Teil's presentations there. After his visit to Oslo, they reported, du Teil "took a series of extremely interesting microfilms [to Paris], which, prior to his departure, were previewed by an informed group that included the French ambassador [to Norway], M. Rene Ristelhueber."[114] A detailed report of the experimental protocols, broken down day by day, for the work du Teil did at Reich's lab can be found in the Reich Archive. On the first day alone, for instance, Reich showed du Teil (and had him carry out himself) the preparation of bions from soil, charcoal, grass, and unsterilized moss. Then du Teil observed all those bion types under the Reichert Z microscope at high magnifications.[115] Reich reviewed the preparation of his various media with du Teil, sterile technique, and other details during the unusually sunny, clear summer weather in Oslo.[116]

Incandescent Carbon Experiment

Reich had puzzled for months over why some bion preparations could not successfully be cultured, while others (even from the same recipe) succeeded.

On May 15, 1937, he wrote to du Teil that he had realized the strength of the electrical charge of the bions was a crucial factor: "For the longest time, we could not explain to ourselves why some bion mixtures grew poorly or not at all compared to others. It is not the degree of sterilization. In contrast, we found that bion solutions containing structures with no electrical charge, which behaved cataphoretically neutral, grew no cultures. It is surely not the complete explanation, but a very important part of it. Why some bion mixtures result in neutral and non-charged structures continues to be a puzzle."[117] Reich noted that a bion mixture 9 he had sent to du Teil turned out to be electrically neutral and produced no growth in culture, in contrast to mixtures 10 and 11 he sent in the same shipment. Reich felt the cultures should be tested for electrical charge, and only those with a charge used in culture attempts. Otherwise, the percentage of successful culture attempts would appear artificially low.

Reich continued with "another bit of news": he had told du Teil previously that boiled soil and coal bions become electrically positive (in at least one trial; further trials were underway). Now, with bions from coal or soil that had, instead of boiling, been dry-sterilized at 180°C and then boiled or autoclaved, "It was a pleasant surprise to find with [those bions] . . . that we got a few times were likewise electrically positive in contrast to the rest of the usual cultures from bion mixture 6c/b, which are electrically negative, as were the cultures that we have obtained from tissues. As you see, we are in the midst of a vast amount of questions and necessary new experimental designs."[118]

In addition, Reich continued to seek for a control experiment so conclusive that no reasonable person could believe "spores" from without were the source of his bions. On May 18, 1937 he thought of a new way to sterilize soot, charcoal, or any other carbon source: heat it to incandescence in a Bunsen burner flame for thirty to sixty seconds and only then plunge it, using sterile technique, into previously autoclaved KCl-Bouillon culture medium. When he did this he received another striking surprise. The intense degree of sterilization did not produce fewer bions. It produced far more bions, far more actively motile than any preparation he had yet produced. The soot bions also had a stronger electrical charge and were far more easily cultured than any previous bions. In a letter to du Teil, Reich said he first did this experiment on May 21; the lab notebook makes clear that the date was actually May 18. In every other detail, however, Reich's account to du Teil, just three to five days after the initial discovery, and his subsequent description of the sequence of these experiments in the published version of *Die Bione* are exactly in accord with the description in his laboratory notebook.[119] The laboratory notes and the letter to du Teil are many pages long and offer great enough detail to say conclusively that Reich's published account offered no rationalized reconstruction of the events as they occurred. In his letter of May 24, Reich reported to du Teil that he had already successfully repeated the experiment five times.

Reich's procedure was as follows (though he pointed out, again, that unambiguous observations were possible only with a binocular microscope with inclined tubes, providing at least 2500× magnification, one that could be adjusted and used with immersion): First, take a small amount of coal dust (previously dry sterilized for several hours) on the tip of a thin metal spatula; hold it in the gas flame at 1000 to 1500°C long enough for the entire mass of coal dust to become red-hot. Then immediately introduce the red-hot coal into the autoclaved nutrient-KCl solution. (Reich's lab assistant had previously produced a phosphate-bouillon solution, which she mixed with the same amount of previously autoclaved 0.1 N KCl, then reautoclaving the mixture.) "In the process, I was following the rationale that in this experiment not only nutritive substances but also swelling substances must be present," Reich explained.[120]

The solution immediately became black and colloidally turbid upon addition of the coal; Reich had been unable to obtain such colloidal turbidity in previous, merely boiled bion preparations. After fifteen to twenty minutes the dark black turbidity diminished; it was replaced by a "cloudy, gray-whitish turbidity of the solution" that lasted for hours or days (see Figure 3.6). Reich took a sample and studied it under the microscope immediately after preparation. In addition, he

> examined this solution microscopically after 24, 48, and 72 hours. It showed that immediately after production in dark field, much more movement was present than by the non-incandescent coal combined with potassium chloride. However, I experienced the greatest surprise on the following and next following days, as it became apparent that, in all cases, the solution was filled with robust, moving cocci, different types of rods, and even contractile amoeboid structures. The structures proved to be electrically charged. We will now go on to culture experiments. Although I have a suspicion that these structures can eventually produce

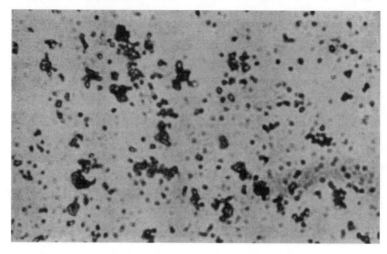

Figure 3.6: Incandescent soot bions, 2000×. Courtesy of Wilhelm Reich Infant Trust.

> growths only on a very complicated and all-purpose culture medium. Thus, I am producing a polyvalent culture medium, the recipe of which I will send you shortly should it turn out to be good.[121]

Reich told du Teil he had to form a working opinion about the phenomena, though it was tentative and totally hypothetical. He elaborated his hypothesis:

> In coke crystals that have already experienced high-grade sterilization and distillation, the individual particles are tightly bound to one another. During cooking, and especially when heated to incandescence, the connection between the particles is destroyed and through this electrical energy is apparently set free, because individual particles could not otherwise be electrically charged, as the larger pieces are not. When potassium chloride is mixed with dry incandescent coal, the particles into which coal decomposes when heated to incandescence voraciously absorb liquid, viewable under a microscope at a magnification of 2000×. Strange electro-magnetic phenomena appear. Individual small coal par-

ticles move vigorously until they come close to a larger one. The manner in which the smallest coal particles rush towards the larger ones is clear evidence of electromagnetic forces at work.[122]

On the other hand, when Reich used a pebble (*ein Kieselstein* [silicium]), ground to fine powder then heated to incandescence, while motile bions were produced, their motility was short-lived and they did not grow successfully in culture.[123]

Du Teil expressed interest in visiting Reich's lab in person for two weeks in July and August and learning all the techniques at first hand, since he had come to realize how many details of method were crucial to guarantee replication of Reich's experiments. When du Teil was in Oslo, Reich made certain he did the incandescent soot experiment and compared it with bions from carbon that had only been dry sterilized at 180°C.[124] Reich felt that heating the material to incandescence was key proof that it could not possibly contain "spores" or "germs"; the fact that it nevertheless produced more, more motile, more culturable bions was the final nail in the coffin for the "spore theory," in Reich's view. Reich later heated other materials to incandescence—even inorganic matter like iron filings and sand—and also found highly motile bions were produced. In the case of the iron bions, the vesicular structures from a disintegrated iron filing continued to be influenced by a magnet.[125] (The bions from incandescent sand are discussed at greater length in Chapter 7.) Reich believed the electrical energy that charged the bions was formerly the energy of cohesion that bound the molecules of a solid tightly together, but which he thought was released by the heating.

Pursuing this line of reasoning, Reich wrote du Teil, "Are you aware that Italian farmers prefer using lava rock [*Lavamasse*] to fertilize the soil? I would imagine that because of the incandescent heat of the material, the cohesion between particles has broken down, and the electrical charges released are directly manifested in the growth of the plants."[126] Almost a year later, Reich had the opportunity to look at lava samples microscopically and confirmed they bionously disintegrated in KCl solution into charged bions.[127]

By August 24, 1937, Reich felt the incandescent soot experiment had finally conclusively answered any possible claims that Brownian movement was all the bions exhibited. As he expressed it in a letter to du Teil (who had gone on from Oslo to Paris) at that time, "It is now no longer difficult to counter the Brownian movement argument. The experiment with incandescent soot, charcoal, ash, etc., the steps of which I described to you in my last letter to you, is irrefutable." Reich thought this new experiment came at an opportune time, since it "is very simple to complete without the many preparations and complications of, for example, preparation 6." Preparation 6 did not apply a high enough temperature, Reich thought, now that he saw how much more effective heating to incandescence was. Reich still felt Preparation 6 "remains . . . one of the major experiments"; however, "the winner in every sense is the incandescent soot experiment. Since your departure, I have completed it roughly eight times in a row and have not had even one cultivation error. Macroscopically, the same type of blue-gray microbe grows again and again. I have demonstrated this several times for friends and on the side. It will ease our overall work very much."[128]

Just as with Reich's previous experiments, however, very few ever took the trouble to replicate the combination of optics Reich said was crucial to recognizing unambiguous pulsation in soot bions. Not surprisingly, then, those very few of Reich's opponents (e.g., Fischer and Ronchese) who actually looked at bions under a microscope gave no credence to his claim that the structures exhibited an internal pulsatory movement easily distinguishable from Brownian movement. (The one significant exception occurred when, as mentioned above, Roger du Teil took one of Reich's Reichert microscopes to Paris and allowed distinguished French biologists, physicians, and a physicist to look at the bions—already prepared for them—through its lenses.)[129] Nonetheless, at least one current microscopy expert who has personally examined and carefully tested Reich's equipment in detail concludes that in most of his claims, Reich was quite correct.[130] We must conclude, therefore, that on one of the most crucial points of experimental technique, Reich's opponents were handicapped to the point of being quite unable to judge his claim: their observations simply do not count as valid replication of his experiments. A proper conclusion would be that we simply do not know the correct interpretation of what Reich saw, but—much as with Philip Sloan's analysis of Buffon and Needham's microscopic work of two centuries earlier—he probably did see exactly what he reported seeing, and his opponents with inferior microscopes could not see it, not because Reich was imagining it, but because they had inferior equipment.[131] A modern replication of Reich's experiments, and an up-to-date analysis of what he saw in light of, for example, a current understanding of the hydrodynamics of bacteria-sized objects' movements in fluids at low Reynolds number, might produce some interesting new insight.[132]

Du Teil's Independent Initiatives

In September 1937, a few weeks after du Teil's return to France from Oslo, he wrote to tell Reich he had, on his own initiative, devised some new control experiments: the "Sy-Clos" apparatus, the "Sy-Clos tube," and the "h-tube" (see diagrams from *Die Bione* depicted in Figures 3.7, 3.8, and 3.9) and an experiment du Teil did on the role of KCl in the bion preparations. The Sy-Clos apparatus was designed so that the separate ingredients for bion Preparation 6 could be autoclaved simultaneously in separate but connected containers. Thus, after the ingredients were sterilized, those in the upper flask could be mixed with those in the lower flask merely by opening a stopcock between the two. The possibility of introducing any external germs during mixing—even though a very small chance because of careful sterile technique—was thus totally similarly eliminated. Du Teil argued that after the bions had been formed by mixing the ingredients, one could eliminate any possibility of introducing outside germs by having the inoculation of the culture medium also take place within this totally closed system. By

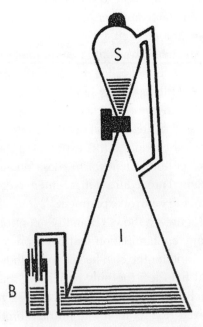

Figure 3.7: Roger du Teil's Sy-Clos system. Courtesy of Wilhelm Reich Infant Trust.

heating the emptied upper flask (S), the expanded air would force a small amount of the mixture from the lower flask (I) into the small tube of autoclaved cuture medium attached at its side (B). Du Teil reasoned that this was superior to having to reautoclave the Preparation 6 mixture after its creation or after inoculating the sterile culture medium to kill any possible external contaminant bacteria, since that procedure "also carries the risk that the bions, too, might be killed off, or at least their vitality weakened, damaging their culturability," whereas, by using the Sy-Clos apparatus, "resterilization is not necessary and the bions can be cultured in exactly the way that they form."[133] The Sy-Clos system would be an excellent approach for any scientist today wishing to attempt replication of Reich's culturing of Preparation 6 bions; resterilizing should be avoided for the reasons du Teil so clearly pointed out.

Du Teil reminded scientists that for many bion preparations this degree of control was overkill since, if one sampled a preparation and examined it under the microscope *immediately* after preparing it from sterilized ingredients, one usually saw a field crowded with enormous numbers of bions. Even if a small number of contaminant organisms had gotten in despite sterile technique, they could not multiply to such large numbers so quickly. Thus, anything present in very large numbers immediately after creating the bions must be from the preparation itself.

Figure 3.8 shows an alternative configuration made by the glassblower at du Teil's university, what du Teil called the "Sy-Clos tube." The basic idea is the same, but mixing of the separately sterilized solutions and then inoculation of a culture medium can all be accomplished merely by tipping the structure enough to spill contents from one tube into the one to its left. In both cases when du Teil conducted this experiment, he obtained Preparation 6 bions in pure culture, identical to those obtained in a nonclosed system in Reich's original procedure. This seemed to du Teil, and to Reich, a remarkably convincing control experiment.[134]

The "h-tube" in Figure 3.9 shows yet another control devised by du Teil. Since purity from any contamination had been proven for the broth (by limpid control tubes) and for the coal heated to incandescence, Du Teil felt this apparatus could help test the only remaining source of any possible contaminants, namely, the KCl used. When du Teil repeatedly used this means to mix sterilized KCl solution into nutrient broth, he found growth of a pure culture even after five successive rounds of sterilization.[135] Du Teil

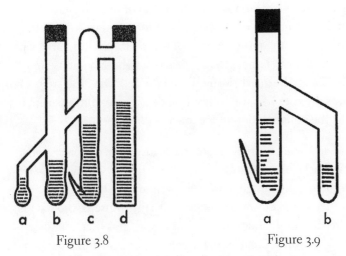

Figure 3.8   Figure 3.9

Figure 3.8: Alternate arrangement of closed system for bion Preparation 6. Courtesy of Wilhelm Reich Infant Trust.

Figure 3.9: Roger du Teil's "h-tube." Courtesy of Wilhelm Reich Infant Trust.

said one could interpret these results in either of two ways. One way was that the KCl was contaminated by microbes that could not be killed off despite five successive rounds of autoclavation, and even then it remained to be explained why they remained completely invisible in the pure KCl only to "appear at the precise moment when the KCl is mixed." The other way was "interpretation in accordance with Reich's theory of synthesis," namely, that the broth can produce bions when it has an excess of potassium added to it. Du Teil appears to have thought the latter explanation by far the less strained of the two.[136]

These independently devised experiments convinced Reich that du Teil was becoming a full collaborator in the experiments, such as Reich had not yet ever had. On October 4, Reich wrote du Teil to say that he saw the Centre Universitaire in Nice as "the primary, decisive control station" for his experimental work; moreover, Reich said, he saw du Teil as a full partner in the work, adding, "I believe it is good to already begin to envision the development of this work relationship."[137]

Du Teil told Reich he had been approached after a recent lecture he gave at Pontigny by "Dr. Philip Decourt—very well known in France for his remarkable works on malaria—and who is the clinical head of the Faculty of Paris, [and who] invited me to come later for two days to redo the experiments at the laboratory of the Claude Bernard Hospital in Paris, which the famous bacteriologist [James] Reilly directs for the laboratory and bacteriology." The Claude Bernard Hospital, said du Teil, was "the largest hospital for infectious diseases in Paris and in France." There, the experiments were redone with the collaboration of Decourt,

> who had already learned about the matter from a representative of the Rockefeller Institute [sic] after a trip to Oslo, and who had become very interested in these experiments. There also, the appearance of organisms was noted by Reilly himself, who having put his eyes to the microscope simply said "It is exact," and continued his works. There also, I left the tubes containing the preparation, for a culture which will be attempted in a few days. There also, finally, Doctor Decourt declared to me that he would redo the experiments as many times as necessary to arrive at a result. Finally, I also left a tube of cultures there, to submit the organisms to their examination. The atmosphere at Claude Bernard [Hospital] is even more sympathetic, if that's possible, than at the Faculty of Medicine.[138]

Du Teil had also told Reich in his October 8 letter that he had been in Paris on Monday, October 4, and had met with Dr. Henri Bonnet, assistant to professor of bacteriology Robert Debré, "one of the greatest French bacteriologists," at Debré's lab. Debré had just come from teaching a course on practical bacteriology. That evening, Bonnet and du Teil began microscopic observation of bion cultures, and inoculated one of du Teil's cultures on agar medium. By Tuesday morning it had grown up visibly on the agar plate. Du Teil reported excitedly,

> The experiments resumed at 2 o'clock Tuesday afternoon at the Faculty of Medicine. With Bonnet I created bions following your formula [Preparation 6], although without milk, such as I had done successfully myself several times. We used the Sy-Clos tube, which stayed over there and which will be used to inoculate bouillon for a culture several days from now. Moreover, Dr. Bonnet decided to redo the experiment several times and this test did not yield anything. Also Professor Debré, whom numerous

professors and assistants, who had heard about the affair and gathered in great numbers around me, all listened with great scientific curiosity and with an attitude that was perhaps astonished but certainly sympathetic. We can be certain that the matter is going to be studied over there with all necessary seriousness and conscientiousness and if, as we must hope, the results are positive, the thing will be officially accepted and known about a few weeks henceforth.[139]

Reich replied on October 14, saying, "If the cultures with Bonnet succeed, then we are relieved of our greatest problems. If they do not succeed, we will have a lot of difficulties." Reich said that in any case, he was "in awe of your ability to confer with specialists. I wish I had the possibility to learn from you."[140] Reich's "one major concern" before du Teil worked with Bonnet was "whether three days are really enough . . . to give a sufficient overview of the many different facts so essential to understanding if one wishes to actually understand the bion experiments and not just use a one-time, or even repeated failure at cultivation as grounds for total rejection."[141] Reich was wary because of his experiences with the Fischers and Thjøttas of the scientific world.

When Reich asked du Teil a few weeks later about the progress of Bonnet's experiments, du Teil informed him "that the work in Paris was still going on and had yielded contradictory results."[142] Reich became involved in experiments on several new bion types and ever more deeply involved with his experiments on the relationship between PA bions, t-bacilli and cancer mice (which I discuss in Chapter 5). Thus, he let the matter with Bonnet go for some months until attacks on his work became very heated in the spring of 1938. That story will be picked up in Chapter 6.

Meanwhile, Reich reported to du Teil that he could now make some general statements to distinguish bions from common decay or rot bacteria. For one thing, most bions that could be cultured had either a positive or negative electrical charge, whereas "rot bacteria and my S [t-bacilli; Reich at first called them S-bacilli] are electrically neutral, *neither* positive nor negative in charge." Reich had a suspicion that the red (gram-negative) or blue (gram-positive) reaction to the gram stain might be connected to the electrical charge. He asked du Teil, "Please compare with us several types of bions and other bacteria for a similarity between positive and negative Gram and the electrical reaction. We have already begun with this."[143]

Du Teil had prepared Preparation 6 bions from scratch and sent a sample to Reich, who reported that in Oslo du Teil's bions "sprouted magnificently in bouillon and today on egg- and blood-medium, too. I can verify that it is a pure type 6 bion culture, electrically negative as well as gram-negative." In the meantime, said Reich, he had found other characteristics that distinguished bions from bacteria. "All types of bions as well as their cultures are generally larger (2000× magnification) than staphylococci. They move very differently than rot [bacteria] or other wriggly rods, more on the same spot, instead of darting through the field. I believe that there are two basically opposite types: bions that carry energy to start life and are created from inorganic matter and rods that develop when living matter decomposes."[144]

BOPA, HAPA, and BLUKO Bions

BOPA (Preparation 12) and HAPA (Preparation 13) bions were derived from autoclaving the blood of a forty-year-old woman and a thirty-six-year-old man for half an hour.[145] This produced blue PA bions between 2 and 6 μm long, slightly less wide. They could be cultured on egg medium IV, though Reich said that cultures succeeded only about 60 to 70 percent of the time in twenty-five preparations he had made.[146] Reich found that if the cultures were not adequately maintained, the PA bions degenerated into cocci similar to "pus-staphylococci of only 1.5–2 μm of blackish instead of bluish color." However, the PA bions could sometimes be regenerated if the cultures were treated with rhythmic electrical pulses and immediate inoculation afterward onto fresh egg medium IV. These PA bions were gram positive, "about 3–4 times as big as ordinary staphylococci, electrically positive, contractile, vibrating," and they were found to "paralyze ordinary *B. subtilis* and short, gram negative rods of the *B. fluorescence* group." As of Reich's publication about their properties in July 1939, mice had shown no pathological reactions in response to injections of these bions.[147]

Reich had also developed bions that looked like small, nucleated cells. These were of two types, which he designated SEKO and BLUKO (from *blut* + *kohle*: blood and charcoal). Reich discovered SEKO I bions in late January 1938 by autoclaving BOPA bions, according to his laboratory notebook.[148] These looked like nucleated cells but were gram negative, as opposed to the gram-positive BOPA bions from which they had come. SEKO II cells originated from the tenth Preparation 12. Macroscopically this bion culture was orange-red and creamy, and it grew on egg-medium IV at room temperature at pH 7.4, usually with a whitish border. Microscopically, Reich reported, it consisted at first of large vesicular bions, which one could see organize together in cell formations under the influence of KCl.[149]

BLUKO bions—produced by autoclaving a HAPA bion culture—are discussed further in Chapter 5, as Reich thought them a model for development of cancer. SEKO I cells injected into a healthy mouse also led to development of tumors.[150] Here, suffice it to say that, observed microscopically at 3000 to 4000×, BLUKO also formed when blood and carbon bions mutually interpenetrated (Reich called this Preparation 10), and the small cell-like BLUKO bions had a bluish glimmer when observed using an apochromatic objective. Thirty-five experiments produced successful cultures 80 percent of the time. The cultured cells grew quickly at room temperature and pH 7.4, and the cultures ranged in color from grayish-white, to an intense salmon red, to yellow-brown. They reacted positive to Gram stain; they also took up Eosin-Haematoxylin stain.[151]

In Chapter 4, we will step back a bit from the experiments to look at Reich's method of scientific work. In what ways are all these unusual experiments and discoveries rooted in a particular style of thought and of experiment and in Reich's initial grounding in medicine, psychiatry, and psychoanalysis?

4

AN INDEPENDENT SCIENTIST: THE BASIC THEORETICAL AND METHODOLOGICAL FEATURES OF SEX-ECONOMIC RESEARCH

Dialectical Materialism and the Bion Experiments

Taking a step back from the experiments to better understand Reich's approach to experiment, let us first of all back up to Reich's use of dialectical materialism, from where the discussion left off in Chapter 1. Here our subject is specifically how Reich applied dialectical materialism during his bion experiments. Around May 1936, after the first month or so of the actual microscopic work, Reich was writing in a notebook:

> *Dialectic of death and life*
> 1. When the metazoan dies, the protozoan appears.
> 2. As the unicellular organisms coalesce, so the multicellular organism is formed.
> 3. Life can form through the gradual swelling of matter.
> 4. It can form through the effect of sunshine. Spring.
> 5. It can form by boiling dried-out material.
> 6. Inorganic material organizes itself through swelling and the build-up of charge into organic material.[1]

The concept of the dialectic was framing, here, how Reich thought of death and life as paired opposites. Similarly, in a lab notebook entry of June 15, 1936, we see Reich's thinking is being *led* by the concept of the dialectic: this is leading him to anticipate and begin searching for antithetical life-promoting and life-destroying microorganisms (bion types), *well before* he ever discovered PA bions or t-bacilli (both in 1937):[2]

3. Since everything is antithetically arranged, there must be *two different types* of single-celled organisms: a) *life-destroying* organisms or organisms that form through organic decay, b) *life-promoting* organisms that form from inorganic material that comes to life.
Tests needed: a. injection of type-a bacteria and type-b bacteria
b. study of the reaction to electric current: (a) bacteria die immediately, (b) bacteria are accelerated
4. *Cancer brings together the vesicles*, organizes them, but disrupts their motion. Cf. the general, combining, paralyzing function of cancer.[3]

This nexus of thinking was also steering him toward the beginnings of his cancer theory—that is, cancer as an endogenous disease—not long after seeing protozoa organize in moss infusions and *before* seeing any live cancer tissue. It also appears to have started him thinking about the idea that bions from many different sources might all constitute a single type, "PA-bions," which contrast *in their functioning* with t-bacilli, despite the widely varying sources of matter from which they derived. This opposition is discussed further in Chapter 5.

In a March 22, 1937 letter to Roger du Teil, Reich discussed dialectical materialist thought in dealing with the meaning of "spores," reproductive structures seen among fungi and some bacteria but often invoked, without any direct evidence of their presence, to explain apparent cases of life appearing from lifeless matter. In such cases, it is said that the spores somehow get into the infusions in question and germinate into living microbes. Reich responded,

> However, what I believe to be more probable, the "spores" are vesicles which form when matter swells. One must question the origin of the spores. It is metaphysical thinking to view them as eternally unchanged. It would be materialistic to determine their progression. Those were simply thoughts that intruded in my mind during the experiments. I adhere to dialectical materialism as the methodical basic principle of the experiments. It addresses the basic question of mechanism and/or vitalism. Essentially, both separate Life from non-Life into its own absolute realm without thinking about a transition, about the evolution of Life. It seems

to me that the tension-charge-formula solves a contradiction. Life and Non-Life have mechanics and electrical energy in common. However, the specific combination of both, the jump from the mechanical charge (vesicles) to electrical charge and further on to discharge and elimination does not occur in non-living matter. Life is, thus, a *special combination* of non-living mechanical functions. It includes a physical-chemical function in similar manner as non-living matter; it would *simultaneously* be a separate field in special capacity simply due to a combination of inorganic actions.[4]

Though he was careful to separate his experimental results from theoretical interpretation in his 1938 book on the bion experiments, *Die Bione*, Reich felt it important to include an entire chapter near the end of the book, "The Dialectical-Materialist Method of Thinking and Investigation," claiming it was crucial to conceptualizing and carrying out the bion experiments. In a talk on May 1, 1937, Reich asked rhetorically, "What is dialectical materialism?" And in its most basic sense, he said, "Dialectic materialism is a very simple, I would almost like to say, *the* simple, humane way of viewing things and events. This does not indicate superficiality. It merely means that man is part of nature, the object of processes and also the subject of social activity. He forms a unit with everything that he observes, works with, or creates." Outlining the history of the method, Reich said, "It was Hegel's brilliant idea to rip the philosophical reasoning of his time out of its static rigidity by delineating the law of change, the flow in all that exists. However, with Hegel," Reich said, even a natural phenomenon or process "was nothing more than the reflection of human 'ideals.' Marx succeeded in, as he put it, getting the dialectic method 'up and running.' According to him, human concepts are reflections of scientifically understandable processes and events. And a concept is then correct when it *immediately* reflects the objective process."[5] So far, Reich was covering ground that many biologists might agree with. But from the outset, and even in his 1929 paper, it is clear that for Reich the pursuit of a quantitative energy principle is also a key methodological focus. He seems to see this as an implicit conclusion from dialectical materialist science as he understood it. Only gradually did Reich come to recognize that other "dialectical materialist biologists" did not share this key element of his own approach.[6] We will therefore outline the logic of this chapter of *Die Bione* in some detail.

Reich saw numerous antithetical pairs in life functions: charge/discharge, swelling/detumescence, anabolism/catabolism, tension/relaxation, assimilation/dissimilation, growth/dying, positive charge/negative charge. His orgasm formula, which he hypothesized was the life formula per se, noted that in living things the function of mechanical swelling was followed by buildup of electrical charge. But the single greatest gap Reich said he found hard to explain was "to grasp all the conditions which lead from mechanical filling to electrical charge." He called this the problem of the "mechano-electrical leap."[7] If one wanted to understand the transition from nonliving substances to living matter, then, understanding how bions acquired an electrical charge was crucial.

Concepts from Colloid Chemistry

Recall from Chapter 1 that Liverpool biochemist Ben Moore made a connection between the idea of a specific life energy and the chemistry of colloids. Reich, too, interpreted his bion experiments through the lens of colloid chemistry; electrical charge was a fundamental characteristic of colloids, he pointed out; thus, an understanding of how colloids build up their charge might help solve the problem of the "leap." Colloids are distinguished from crystalloids such as inorganic salts, which dissolve to form true solutions. Colloids, by contrast, such as albumen, starch, or glycogen, have much larger particles and form gels or suspensions when dispersed in a solvent. Colloid chemistry therefore describes the behavior of those particles in a solvent or at a solvent-solvent boundary, based on physical properties of the particles, especially particle size and electric charge. From 1900 until 1940 or so, many chemists, biochemists, and biologists believed that most biological phenomena, for example, the antigen-antibody reaction, perhaps even the difference between life and nonlife, would be explained in terms of the behavior of colloids. Proteins are key to living systems. Hemoglobin carries oxygen in the bloodstream, antibodies defend against invaders, and enzymes are the crucial catalysts of all biochemical reactions. Since most proteins form colloid solutions, and since protein structure was thought by most chemists to be variable, with no constant molecular weight such as simpler organic molecules, proteins seemed ideal candidates to explain the variability of living reactions via behavior controlled by electrical charges on their surface. Between 1900 and 1930, specialized journals, professorships, and whole institutes of colloid chemistry were created, most of all in the German-speaking countries. Wilhelm Ostwald founded the field; his son Wolfgang, in Leipzig, and Wolfgang Pauli, at the Institute for Medical Colloid Chemistry in Vienna, were prominent researchers in the colloid chemistry tradition.[8]

In the mid-1920s, Hermann Staudinger in Germany and Swedish chemist The Svedberg began to argue, using such tools as the new ultracentrifuge, that huge proteins like hemoglobin actually did have a constant molecular weight and therefore behaved like the traditional molecules of structural chemistry—they referred to these as "macromolecules"—but this point of view gained ground very slowly, becoming a clear majority view only by 1940 or after. X-ray crystallographic studies of proteins, especially by W. T. Astbury at Leeds and J. D. Bernal in London, also solidified the evidence for macromolecules. Later Whiggish histories labeled the period as the "Dark Age of Biocolloidology" and emphasized only the conflicts between the two schools of thought. But during the time Reich conducted his bion experiments, it must be emphasized that colloid chemistry explanations were still a completely respected tradition of research and that many, not least Pauli, used colloid chemistry ideas in a complementary way with structural chemistry and "wished to encourage all attempts to link the two."[9]

Reich notes that some colloids, called lyophilic colloids, for example, proteins, can be dried out yet then produce again a colloidal solution when rewetted. This is crucial for his understanding of the "germ theory"; that is, it is precisely this ability of proteins that explains why a desiccated unicellular organism can be restored to life by rewetting: "the restoration of the "living, functioning colloid" is what constitutes the emergence from the "germ state."[10]

Reich found Pauli's work particularly useful and suggestive in thinking about the "mechano-electrical leap." The opposed positive and negative charges on colloids and ions, as well as the opposed pairs of biochemical substances (for example, lecithin vs. cholesterin) that were, as he put it, "functionally identical" with expansion and contraction, seemed to Reich to naturally suggest dialectical relationships. For example, recall that in analyzing the behavior of colloidal solutions in the bion experiments, Reich found only those bion preparations that remained colloidally turbid for more than a few minutes gave strongly electrically charged bions and successful growth of cultures.[11] Was there also, Reich wondered, a dialectical relationship between the electrical charge of colloidal solutions and the ability of the original substance to swell up in the fluid to produce the solution? He turned to Pauli's work to seek information that might help answer this question.

As mentioned, Reich connected the electrical charge of the bions with their ability to remain in suspension and with their individual motility, stating,

"The quivering, dancing motion and change of position of the bions is connected with the force effects of electrical fields."[12] Reich hypothesized that the electrical energy had been bound in the structure of, for example, earth crystals or coal chunks, but that swelling in fluid or heating to incandescence "breaks down the mechanical cohesion of the particles in the matter, releasing the energy" when individual vesicles break free. Thus, he reasoned, the vesicles acquire an electrical charge and their movement in place "can be explained as the result of electrical force fields of the individual vesicles acting on each other."[13] (Note that this movement is not Brownian, nor is it the same as the internal pulsation Reich saw in bions at high magnification.) Reich said his ideas on incandescent heating and its role in life's origin derive partly from Pflüger's cyanogen theory and partly from Svante Arrhenius's *Worlds in the Making*. Both describe the early Earth in an incandescent state prior to life's origin.[14]

Reich was interested in Pauli's demonstration that electrically uncharged protein particles swell up less in water than charged proteins do, because the greater the ability to swell up, the more stable the particle would be in solution. Reich, however, found advantage in seeing the relationship Pauli had reported in a different way. If, he said, one turns Pauli's expression around, then one could explain one of the most important observations from the bion experiments: "The more readily a protein substance can swell, the more readily it can build up an electrical charge, and the more stable it is in colloidal solution." Reich argued that "there are certain advantages to be gained by turning Pauli's explanation around. For example, it indicates how one can solve the difficult problem of why the swelling of a particle causes the buildup of an electric charge," that is, solve the problem of the "mechano-electrical leap."[15]

Reich also noted the importance of a membrane (e.g., formed by lecithin) or a gel in forming clumps of vesicles and in the motility of those clumps (as in his "pseudo-amoebae") if the vesicles were electrically charged. The mutual repulsion and movement of the bions *within the membrane* was what produced the crawling of the clump in a coordinated manner and the expansion and contraction of the clump as a system. The same was true of an amoeba formed from a clump of grass bions, or a protozoal cancer cell formed from a clump of bions in degenerating, oxygen-starved tissue, Reich thought. The expansion from localized forces of repulsion among vesicles would cause a bulging of the membrane at that point, but, Reich noted, that would produce an opposing counterforce of surface tension from the stretched membrane.[16] Reich had been thinking actively about the opposition of internal pressure versus surface tension, often with the visual image of a bladder filled and stretched, ever since his 1932 theoretical explanation of the desire for "bursting" in masochism. This image figures prominently in his thinking about the "functional identity" of psychological and somatic processes.[17]

Reich gave thought to other forces involved in the physical aspects of life functions as well. As the proteins, for example, the gelatin in bion preparations, swelled up, they took up a large volume of fluid and increased substantially in volume. If this swelling occurs in a closed space that prevents any increase in volume, then immense pressure can be generated, referred to as "swelling pressure." Numerous substances, including metallic ones such as gold, could form colloidal solutions, noted Reich, but the colloids that form life differ because they exhibit this swelling pressure. "The opposition between swelling pressure and surface tension is of eminent importance for understanding the division of cells into spherical objects," Reich observed.[18]

Reich was aware that a mechanist would immediately object that the locomotion, expansion, and contraction his bions exhibited were "merely surface tension phenomena of a purely physic-chemical nature." So he more closely examined statements in this area by Kaiser Wilhelm Gesellschaft biologist Max Hartmann, whose work Reich offered as an exemplar of the mechanistic viewpoint in modern biology. Since 1933, Reich had closely studied Hartmann's experimental protozoology work on "relative sexuality." Hartmann had shown that some "still-uninvestigated 'substances'" determined whether the gametes of numerous species of protist acted as male or as female when in the presence of another group of gametes, that is, that their sexuality was relative, not fixed by genetics. Reich had found in bion experiments that it was bioelectric charge rather than chemical substances that was responsible for the mutual attraction and clumping of bions. Thus he saw it as more likely that Hartmann's "unknown something" was related to bioelectric charge rather than to specific chemical substances. Reich thought a suggestive parallel could be seen in Gurwitsch's demonstration that the cell nucleus began to emit radiation as it underwent mitosis. Here, too, a critical force in a reproductive process was linked to energy rather than to a "chemical messenger."[19]

Reich had long been critical, he said, of the moralism and teleology in supposedly "modern" life science. Not surprisingly, his view was that dialectical materialism "denies the existence of a purpose principle in nature, i.e., the existence of a supernatural goal beyond the energy process that . . . determines development."[20]

For Reich, the "not just the sum of the parts" logic of dialectical materialism meant that the difference between living and nonliving matter was "not constituted by *the addition of something new* in living matter that makes it alive. Instead, the difference lies in a *special combination of functions* which are found singly in nonliving matter as well. . . . [Namely,] *the specific combination of the rhythmic alternation of tension, charge, discharge, relaxation, renewed tension, charge, etc. is the fundamental distinguishing characteristic of life.*"[21]

Like Oparin, Reich concluded that the origin of life was conceivable only as a series of developmental stages. But unlike Oparin and Haldane, Reich did not feel compelled to assume either that the process must always require millions of years or that it could no longer be occurring under the conditions of present-day Earth. Development, Reich concluded, was driven dialectically "by the presence of opposites within matter which cause an antagonistic contradiction." He went on: "The opposites force a change in the situation, and something *new* is formed. This *new something* . . . develops new contradictions, which in turn force further solutions, and so on. . . . In the mechanistic view, [Reich said, by contrast], the opposites are absolute and irreconcilable. In dialectical materialism, opposites are viewed as identical [i.e., different manifestations of a common, deeper substratum], and as a consequence one can develop out of the other. Hate is not merely an opposite of love, it can develop out of love; much conscious love is unconscious hate, and vice versa."[22]

He pointed out that because the psychic apparatus has a biological origin, two kinds of laws must be invoked. The first set of laws would cover what functions the psychic shares in common with the biophysical, such as tension and discharge, response to stimuli, and others. The second set of laws are those that come into play only at the more complex level of development that separates the psychic apparatus from simpler biological systems and that contrast it with the purely biological laws, for example, repression of instinctual drives, introjection, and projection.[23] Further, claimed Reich, "mechanistic science . . . represents the standpoint that development [can only be] a gradual process. . . . There are no *sudden* changes. The materialistic dialectic, on the other hand, recognizes that gradual development can become sudden development, that evolution prepares the way for sudden change in development. . . . Mechanistic as well as idealistic philosophy denies that it is possible for quantity to develop out of quality, and vice versa. In contrast, dialectical materialism asserts that not only can quantity convert into quality . . . but that this change-over is one of the fundamental principles of any natural process."[24]

Reich realized that a mechanist or one committed to teleology would object: But what about when a germ starts to develop, without having come in contact with some external force? In that case the principle driving development is contained inside the germ itself. So *"what gives rise to this inner contradiction which drives development. . . ? Even this inner contradiction must itself have formed at some time!"* Reich said this question was entirely justified. He said the key issue is, how did the inner contradiction come about, and what is its function? In all the scientific arsenal of dialectical materialism to date, said Reich, "this question has neither been posed nor answered."[25] And thus, he argues, it has been difficult until now (1938) to find a place in dialectical materialist thought for biological development.[26]

To resolve this problem, Reich argued, one must distinguish between three different kinds of antithetical relationships in nature. First is "dialectical system antithesis," in which two systems encounter one another (e.g., two planets or a planet and its satellite) and as a result both change direction. Second is the "dissociative antithesis." Reich illustrated this with the example of a magnet brought near a piece of iron. The magnet induces a north pole at one end of the iron and a south pole at the other end, that is, two antithetical manifestations internally, which are induced by an external force containing its own antithetical forces. Reich thought this also described electrical reactions and acid-base-salt chemical reactions, among others. The salt produced from a reaction between an antithetical acid-base pair was a *third* resultant that neutralized the tension that previously existed in the acid-base opposition. Unlike lifeless systems, Reich said, after such a neutralization reaction, *"life can generate new tension by itself."*[27] The third relationship found in nature Reich called the "genetic antithesis," which he said is found only in living nature, unlike the first two types that are seen in both living and nonliving systems. "A living cell dissociates before it divides, at first internally, through the process of mitosis. The copulation of two such internally dissociated cells signifies reunification of the antithesis," said Reich. He emphasized, however, that cell division, unlike the simple dissociation processes in nonliving nature, "proceeds in a geometric progression."[28]

Unlike Oparin and Haldane's basically biochemical approach, Reich claimed "a fundamentally dialectical materialist approach requires that the organism be examined as it is, that is to say, that life be studied in the living state," first and foremost. Moreover, said Reich, "this approach is diametrically opposed to the mechanical one in which, for the sake of reliability, living objects are killed in order to study life in the dead organism, a procedure that is bound to result in a mechanical view of life." He reasoned,

> If scientific research is to be truly productive, it should be continuously motivated and guided by the need to view the whole without losing sight of the detail. Mechanical concepts of life must of necessity be methodologically defective; they rely on the synthetic movements of the living substances becoming more complex and perhaps giving rise to life. The important thing about life, however, is not the complex substance but the complex function. Concepts such as Biogen, molecule, energid, etc. [see Chapter 1] are only practical aids to understanding. . . . They try to substitute the action of "substance" for the understanding of function. They tell us something only about the mechanical-chemical process, but they become metaphysical when called upon to explain function.[29]

This is certainly a more aggressive, fundamental critique of physical-chemical reductionism in biology than Oparin, Haldane, or Bernal was making in 1938 (or perhaps, ever). For Reich, however, it was this aspect of mechanism that led physiologists to describe the nerve pathway by which an impulse traveled and then to think that *explained* the impulse itself. And to viewing the organism "in sociological terms," with the brain as "the 'central agency,' the 'controller'. . . , rather like the ruler of a country."[30] But for Reich, since the brain is a phylogenetically recent structure and since numerous organisms have no brains but still live quite successfully, the brain is unlikely to be the *origin* of living functions. Indeed, the "tension/charge" formula basic to living functioning was present in all living tissue; it was the sine qua non of life, Reich claimed.

During the bion experiments, on May 1, 1937, Reich gave a lecture titled "Der dialektische Materialismus in der Lebensforschung" at the formal opening of the completed laboratory of the Institut für Sexualökonomische Lebensforschung [Institute for Sex-economic Biological Research].[31] Reich pointed out a parallel between the splitting up of mechanistic science and the "dissecting" of the living in biology, such that scientists lost sight of the fundamental unity of living functioning: "In the course of the development of the natural sciences, Life was split up into different fields, which are specifically separated from one another, and are so dealt with. Observation of Life [*des Lebendigen*] as *one unit* in spite of all of its complicated fragmentation suffered because of this. Fragmenting Life into branches is equivalent to splitting up sciences into mechanical specialized disciplines, one separated from the other."[32]

Reich noted that while dry-sterilized charcoal and soil particles were sterilized, when cooked in a solution that stimulates swelling they nonetheless produced growth when inoculated on agar, organisms completely different from those that grew from unsterilized coal or soil. Thus, life was destroyed by heat, but on the other hand new life was created by the same heat. The increased amount of life created, and its dramatically increased electrical charge and motility, both resulted if the charcoal or soil was sterilized by heating to incandescence in a Bunsen burner flame rather than dry sterilized at 180°C. Said Reich, "This fact might surprise the supporters of mechanism or vitalism; however, following the completed experiments up to now, it is clearly evident that two different life forms are produced in relationship to cooking: one that is destroyed through boiling, and another that forms due to boiling. Correctly expressed: We must differentiate between life that was already present through swelling or reproduction and *destroyed by heat* and the same or different life forms created above an otherwise fatal temperature limit. After it has successfully organized, the same life that was created through heat can later be destroyed by this same heat. This is also materialistic dialectic."[33]

Near the end of the talk, Reich considered the question, was a dialectical materialist science (or ought it to be) apolitical? He pointed out that the prevailing genetic doctrines, leading to eugenics, were inherently political. But, more importantly, he argued, science is an outgrowth of overall life processes and particularly of human life; it loses all meaning if it becomes cut off from man's life. Science, he asserted, must work *for life*, and "there is no such thing as a science that is apolitical, operating outside the realm of social processes." Thus, he argued, "it doesn't matter at all whether science is political or not; rather what matters is only *which kind* of social attitude lies in the scientific research."[34] Reich's lecture came just five days after the brutal fascist attack by the German Condor Legion that devastated the town of Guernica in the Spanish Civil War. Reich put the choice for scientists not so differently from how Bernal or Haldane put it at this time:[35] "One can incorporate the 'struggle for survival' as a basic principle in biology and from this explain the 'natural necessity' of the massacre of a city's entire population like in Guernica by pilots doing it for sport. As opposed to this, one can make the creative process of life the basic principle of one's own outlook by doing everything to create life, to help the living progress forward, and to preserve life."[36]

Examining Preparations in the Living State

Reich stated that "it was one of the principles of our work that we study only living, motile matter, because it is the *changeability*, the *functioning* and not the static element, not the structure, which are of primary interest."[37] In a more telegraphic jotting in his laboratory notebook on February 22, 1937, Reich described his emerging sense as follows: "We need not split up the natural process, to overdo it. Rather only to reconstruct it, to decipher it, to gain control over it—*to move it forward!!!!*" (See Figure 4.1.)[38] While it was not a majority view during the triumphant rise of electron microscopy, Reich was not alone during the period 1930–1950 in putting priority on observations in the living state. French biologist Pierre LeComte DuNoüy, for example, was skeptical of the artifacts introduced in killed, stained specimens, stating in 1931 that cytology based on these methods "is incapable . . . of explaining the most ordinary pathological phenomenon, such as the cicatrization of a wound or the growth of a tumor. . . . It is not by observing and dissecting the corpses in the morgue that we can hope to understand the economic life of the city of Paris."[39] And interestingly, others who made similar claims, such as South African bacteriologist Adrianus Pijper and American embryologists Warren and Margaret Lewis, also put a great deal of time and energy into time-lapse microcinematography.[40] In 1938 electron microscopy was still sufficiently in its infancy that such a claim would seem less out of the mainstream than by 1950. Even compared to pathologists' standard practices of killing, fixing, and staining tissues before observing them, none less than Rudolph Virchow himself had warned against pathology relying excessively upon dead specimens rather than focusing on the developmental process.[41] In Reich's study of developmental processes underlying cancer, he found observing blood, vaginal secretions, and other specimens in the living state ever more important. This is further discussed in Chapter 5.

With the "molecular revolution" ushered in by Bernal, Max Delbrück, Francis Crick, and many others who had crossed from physics into biology, attitudes in the life sciences shifted still more dramatically away from the ability to take seriously such claims about prioritizing observations on the living state. As Donald Fleming has so clearly shown, the physicists who

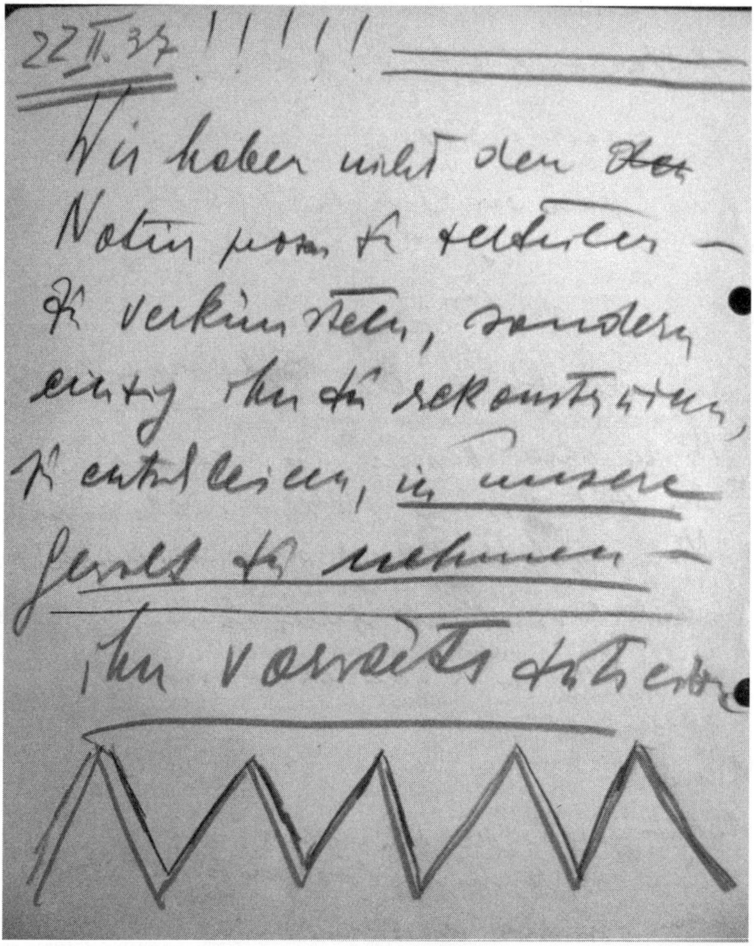

Figure 4.1: Page from Reich laboratory notebook, 22 February 1937. Courtesy of Wilhelm Reich Infant Trust.

colonized biology, beginning with Schrödinger's 1944 book *What Is Life?*, brought with them a more extreme reductionist view than had been seen before. And Watson and Crick's 1953 DNA structure swept the physicists' view rapidly into majority status.[42] Once their new brand of extreme reductionism acquired such status, James Watson and others worked to make all biology molecular and to convince others that "whole organism" biology was hopelessly antiquated and useless.[43]

Response of Haldane and Bernal

There is very little record of direct response to Reich's book *Die Bione* from the point of view of Reich's version of "dialectical-materialist biology." Reich's friend A. S. Neill on his own initiative approached J. D. Bernal and J. B. S. Haldane in 1938, to suggest they examine Reich's bion cultures and perhaps help safeguard them in the event of oncoming war. In mid-1938 Neill passed a copy of Reich's bion book to Bernal, who later did important work on the origin-of-life problem, and whose sons were pupils at Neill's school Summerhill.[44] (Neill also later tried, unsuccessfully, to interest H. G. Wells in Reich's bions.)[45] Haldane had given "guest scientist" lectures at Summerhill. Bernal had as well, and he greatly respected Neill's work as an educator and looked further into the matter.[46]

A well-regarded British crystallographer and physicist, Bernal was completing his book *The Social Function of Science* in the summer of 1938, just a few months after Reich's *Die Bione* was published. At about the same moment, geneticist Haldane was completing his book *The Marxist Philosophy and the Sciences*.[47] In these books we can see some of the state of their thinking about dialectical materialism and the sciences. Both men certainly shared Reich's passionate belief in a crucial role for the scientist in combating fascism. Bernal also lamented, as did Reich, the reemergence of mysticism in the sciences (vitalist ideas, etc.), which culminated in Nazi "racial science." Even if neo-vitalists like Driesch were personally anti-Nazi, as were many socialists or communists (such as H. J. Muller)—enthusiastic about what they considered a rational version of eugenics—they may not have realized the extent to which their ideas were drowned out by the fascists' co-opted versions of those ideas.[48]

Responding to Neill, Haldane was dismissive of the bion experiments from the outset, declining even to read Reich's published reports.[49] As Neill reported to Reich, "I tried to interest Prof. Haldane of London University Biometry Department. [Haldane had written a seminal 1929 paper on "The Origin of Life."] 'Too unorthodox,' he said. I never knew before how little these great scientists are."[50] Neill urged Bernal to read the bion book and tell him his reaction. Bernal's reply, of August 28, 1938, is brief but contains some tantalizing remarks:

> Eileen [Bernal] has shown me Reich's book [*Die Bione*] which I have read, not very thoroughly but enough to get the general hang

of it. I think he has got hold of some quite interesting things—the relation between vital processes and swelling—but it seems to me far too generalized and sweeping. It is relatively easy and other people have done it in even greater detail than he has to reproduce in inorganic systems the appearances of life—swelling, growth, and even multiplication, but this only shows that *these* processes are *general* ones. The *characteristic* features of life are *chemical*, and though I believe that these are just as reproducible, no one has come anywhere near it yet.

The dialectic approach is not to begin with lots of matter as we know them now and try to make them come alive—this is pure wish magic—but to reconstruct as nearly as possible the conditions on a primitive earth before life which needs rather elaborate cooperative research.

There are general tendencies in all living and nonliving systems but they only point to explanations[;] they explain nothing themselves.[51]

Another letter, undated but slightly later than the first, is similar in content.[52] These letters had been passed along by Neill to Reich and were found in the Reich archive. Bernal signed them only as "S." This was a habit to him since he was widely known by the nickname of "Sage" to friends. Bernal was critical of Reich's work, it is clear, but until now we have not known from Neill's account to Reich that there was at least some positive tone, nor of the details of how Bernal's sense of dialectical materialism differed from Reich's. Bernal thought Reich was on to some interesting ideas, such as the importance of the swelling of matter; nonetheless, he was not convinced Reich's sterile technique was not faulty (though, again, he never witnessed the experiments). Bernal was convinced that the "synthetic approach" to origin-of-life studies pursued by Reich, Alfonso Herrera, Oparin (with coacervates), and others was *not* "the dialectic approach." He favored the analysis of primitive Earth conditions, such as in the Oparin-Haldane hypothesis, and the later Miller-Urey experiments to which it led. Since Oparin himself had no problems pursuing the synthetic approach as well as the "analytic" one, Bernal's view appears to be his own (and perhaps Haldane's) personal interpretation of dialectical materialism vis-à-vis origin-of-life research.[53]

Perhaps the greatest distinction between the dialectical materialist approach to the origin of life versus simple mechanistic materialism can be seen in the disagreement between researchers in the 1950s and 1960s over "metabolism first" versus a "gene first" approach to life's origin. Most dialectical materialists, including Oparin and Bernal in the 1950s and 1960s, viewed "the origin of life as the result of a series of probable steps of increasing complexity, inevitably leading up to the living state"; by contrast, the mechanistic position, for many, "explains the origin of the first living thing in terms of the chance combination" necessary to produce the first self-replicating molecule or "gene."[54] Reich shared the same idea with Oparin, that "genes," in the "master molecule" sense of Jim Watson in the 1950s, were a somewhat mystical entity that carried older teleological/religious overtones into modern biology.[55] And Reich understood his bions to be stages in a transitional process, where the gradual establishment of the four-beat combination of functions from his orgasm theory represented the fully living state. On the other hand, Oparin and Bernal both thought (as did Engels before them) that such a rapid transitional process (a few hours or days) from nonliving to living was not believable and carried the discredited taint of "spontaneous generation"— only an evolutionary process requiring millions of years, and only under conditions more like those of the primitive Earth could actually produce something as complex as even the simplest cell.[56] Despite the fact that the "colloidal" approach was soon to be thought outdated, given his more sophisticated biochemical "basic antithesis of life" argument, perhaps if Reich had referred to his bions under the later conceptual framework of "protocells" rather than under the older framework (outdated by the 1950s in the eyes of "molecular" biology) of "synthetic cells," a greater degree of acceptance, albeit qualified, of his results might have been possible. Such a qualified "revival" of Alfonso Herrera's ideas is currently underway for at least the past two decades, despite Oparin's choice of Herrera as a "whipping boy."[57]

Reich claimed a dialectical-materialist approach required that one prioritize study of *movement* over structure or biochemistry. While Bernal thought Reich was on to some important insights, this, surely, is an area where he staunchly disagreed. Bernal was, after all, a crystallographer, whose primary focus, by definition, was on structure as the key that would explain function. Excited by Astbury's work at Leeds on the structure of wool and other proteins, Bernal was committed to the belief that macromolecules—not colloids—would explain the functions of life—the core belief of what later came to be called "molecular biology."[58] Reich's view that electrical charge is a key element of life is compatible with an older, colloid chemistry view of life, which the macromolecular view was beginning to challenge in the late 1930s.[59]

Bernal and his student Dorothy Crowfoot showed that pepsin crystals kept wet diffract X-rays to high resolution, proving they must have identical structures and thus—like Svedberg's work on hemoglobin using the ultracentrifuge—contributing to disproving the prevailing colloidal theories about proteins having variable structures.[60] Thus to Bernal, Reich's key "life formula" of "tension → charge → discharge → relaxation" was not the primary characteristic of life, nor was it an inevitable implication of a dialectical materialist approach to biology.[61]

Some Features of Reich's Unique Approach to Life Sciences

We should note again, in summary, that two features sharply distinguish Reich's research program from the beginning: his pursuit of an energy principle behind emotions and other life phenomena and, most importantly, that his central line of investigation was to understand the function of the orgasm. Reich was investigating questions of interest to many biologists at the time, such as the origin of life; furthermore, he made use of the dialectical materialist system of Marxist thought in his biological research program to find a middle way between mechanism and vitalism, as numerous biologists did in the 1930s (such as the British scientific socialists, as well as Alexander Oparin, Julius Schaxel, and numerous others). Nonetheless, it is important to understand how different these two features (focus on an energy principle and on the function of the orgasm) made Reich's approach to these questions from all other biologists of the time—so much so that, as I pointed out in the Introduction, by 1939 or so Reich realized these differences were more important than what he shared in common with dialectical materialism, and so he decided to call his approach by a different name—energetic functionalism (or orgonomic functionalism), to reflect its key features and to make clear that (as he put it a few years later) "energetic functionalism of today has as much to do with dialectic materialism as a modern electronic radar device with the electric gas tube of 1905."[62] This change in thinking paralleled Reich's move in his social thought during this same time period, from supporting traditional socialist or communist models of society to developing his own independent model of "work democracy."

The differences between Reich and Bernal in particular are instructive. After they took offense at Reich's critiques of their failures (1933–1936), many Communist and Socialist Party organizations expelled Reich and/or refused to distribute his writings.[63] Bernal never rejected the Soviet Communist Party, not even after the Lysenko triumph of 1948. (In this, and especially as an apologist for Lysenko post-1950, he was alone, even among the British scientific socialists.)[64] Haldane openly criticized Lysenko after 1948 but had been ambivalent about his work until then.[65] Thus, Reich's understanding of how truly unique his research approach was emerged and crystallized in the context of his disillusionment with Socialist and Communist Party politics, those parties' forceful rejections of his work and responses such as Bernal's and through the fierce campaign of opposition that developed in Norway to his bion research. In Chapter 6 we look more closely at those who became Reich's opponents in Norway.

As I stated in Chapter 1, the Lysenko affair by 1948 had thoroughly discredited "dialectical-materialist biology" in the West, so much so even today that one reviewer of a Bernal biography found it extremely difficult (through today's Whiggish lenses) to understand "how a man with such a marvelous analytical mind could come to terms with dialectical materialism."[66] In 1950 a Soviet biologist who supported Lysenko, Olga Lepeschinskaia, claimed to have shown the origin of complex cells from a cell-free fluid—much like Schwann's cytoblastema theory of the 1830s and 1840s—and she explicitly called this a "dialectical-materialist approach" to origin-of-life science.[67] Oparin himself at least tacitly supported Lysenko after 1948 when it became politically expedient to do so. And in 1951 he at least once spoke favorably of Lepeschinskaia's work, though it is very unlikely he believed it valid, given how bluntly it clashed with his own assumption that life could not possibly come into being by a sudden, brief process.[68] As a result, with the single exception of Bernal, all Western scientists after 1951 considered "dialectical materialist origin-of-life research" to be just as tainted as Lysenkoist genetics: merely Stalinist politics masquerading as science, with little or no legitimate scientific content. In this atmosphere, Reich's bion experiments could not be taken seriously by Western scholars, probably not even by Western historians of science during the Cold War years. This is an ultimate irony for Reich and his work, given that his 1935 *Masse und Staat* and his 1936 *Die Sexualität im Kulturkampf* were two of the earliest, most scathing critiques of Stalinism from a Western intellectual who had been a former enthusiast of the "Soviet experiment." Reich was publishing about "the God that failed" far earlier than many who had been involved in the German Communist Party (KPD) with Reich in 1930, such as Arthur Koestler, and far earlier than J. B. S. Haldane.[69]

Being an "Outsider" versus Being an "Independent Scientist"

Gary Werskey, in his group biography of the British scientific socialists, makes some insightful comments about their differences as well as what they share in common. Of the British scientific socialists, Lancelot Hogben seems in many ways to have most in common with Reich: a feminist, antieugenist, bluntly anticlerical/atheist, not quiet about his very radical politics; he took politics into the lab in his choice of research topic, especially sex. He did not confine his intellectual ideas about politics and sex to merely "smarty-pants" statements *outside* his scientific career or only in his personal life, as did Haldane and Bernal.[70] Also like Reich, says Werskey, by the late 1930s Hogben

> remained the perennial non-conformist, the troublesome gadfly of all "establishments," the nit-picking nemesis of any "true believer." Prior to the formation of a communist-led left-wing intelligentsia, his had been . . . a radical stance. From a point of near-heroic isolation, Hogben had kept faith with his own socialism. . . . Along the way, however, he had also cultivated some of the psychological and philosophical defenses we might have expected of someone who occupied such an exposed position. One of these was a tetchy disposition, which prevented Hogben from becoming too closely involved with any individual or institution. This pattern allowed him to keep his own counsel, though the price he paid for this independence—the periodic loss of close friends and his detachment from the labour movement—was obviously a heavy one.[71]

It is worth noticing that Werskey, though sympathetic to his subject, puts this "outsiderness" on Hogben, rather than on all the forces that isolated him—much as many authors do with Reich.

In one other respect, Reich shares an important piece with Hyman Levy alone among the British scientific socialists, who realized (while all the others denied) "that the social needs might enter into the very construction of scientific theories. One writer who did not share this notion was H. G. Wells. . . . 'What worried Wells,' Levy would later observe, 'was that I embedded science in society. . . . To Wells science was something which impacted on society. . . . He didn't like the idea that science wasn't the essence of clear thinking.' Nor did Bernal, Haldane, Needham, and even Hogben, all of whom were as apt to isolate science from its social context as a leader writer in *Na-*

ture trying to cap an argument for more research funds from the state."[72] Perhaps Reich shared this view with Levy because of his insight as a psychiatrist into human motivations; perhaps because of another major thing he shared with Levy: hands-on experience with the misery, the daily problems of working class people—this made Reich aware of the social constructedness of ideas for, for example, capitalist purposes, of the role of the authoritarian family as a factory for citizens subservient to state authority.[73] Reich saw modern bacteriology in this way and warned du Teil about it. But that did not seem to prevent Reich from believing in the possibility of truly objective science.

Reich described his sense of a Marxist, dialectical materialist science in a lecture he gave on April 18, 1936. The task of scientists was to be "radical," he argued, in Marx's sense of "going to the root of things." In that sense, he said, any science that takes itself seriously is necessarily radical. Thus, "strictly speaking, there cannot be such a thing as a reactionary science. All the things that call themselves science and are at the same time in the service of reactionary ideologies, can easily be shown to be not science, but mysticism and superstition. Science and the scientist should not be expected at all to adhere to this or that political credo; but they should profess to being radical, that is, they should go to the root of things." Reich lamented that true scientific work "today is being hampered to an incredible degree. Our times are characterized by mysticism and blind obedience to authority in the masses of people. Science is in great danger; hundreds of scientists have had to leave Germany just because they were scientists. Hardly any of them had any political affiliation; they were just scientists." When scientists are driven into exile, yet the general public takes no rational action to counter the general irrationality, Reich said, still others will have to seek refuge elsewhere. Being cautious and hoping to hide from fascist irrationalism, he warned, would do no good. "Those who did not dare to draw the conclusions from their findings had to flee also. If that is so, it is better to be consistent. The scientist has only one correct social task: *to continue his search for the truth in spite of everything*, not to give heed to any restrictions that the bearers of life-negating ideologies may try to impose on him. . . . We have to be prepared and ready for a time when a rational order, when all those who will work and will themselves determine their existence, shall need us."[74] It is useful to recall that by this point, that is, months before Stalin's "show trials" began, Reich foresaw the same fate for scientists under the Soviet regime—even cautious ones—that had already befallen German scientists under the Nazi regime.[75]

For anyone considering Reich, the problem of "outsiderness" is even more extreme than for Levy or Hogben, because over time the pursuit of his research aim to study libido or psychic energy led him to cross many disciplinary boundaries. Inevitably, this meant repeatedly leaving behind former colleagues who were not comfortable crossing the boundary, for example, from psychiatry to physiology or from psychiatry to sociology or anthropology.[76] Reich's ability to move across disciplinary boundaries and to leave behind colleagues who were not comfortable following is usually seen by critics and biographers as a weakness or a problem: Reich too egotistical to realize he is not competent in physiology (or, later, in bacteriology, or in physics). Biographer Myron Sharaf says Reich habitually pushes people away and calls his outsiderness "latent," implying that it is a neurotic problem Reich is unaware of—because, in Sharaf's portrayal, Reich is a zealot on a "mission."[77] Sharaf implies that Reich rationalizes this neurotic behavior by seeing himself as capable of greater intellectual mobility than those around him. Sharaf gives less weight to the possibility that Reich is isolated because many scientists treated him irrationally. Benjamin Harris and Adrian Brock have a somewhat more nuanced view, that Reich's "rivalry" with Fenichel or others is what pushes them away.[78] After enough such rivalries, a zealous egotist would surely always become an "outsider."

Reich's is hardly the only ego of such a size to ever play a part on the scientific stage. Bernal encouraged friends and colleagues to call him "Sage," and despite his intellectual blind spots (e.g., Lysenko and Stalin) is still considered to have made brilliant contributions. Which view of Reich is more persuasive?

Perhaps there is insight to be gained by using a different analytical framework: the "independent scientist." Chemist Peter Mitchell strove for this status, knowing his unorthodox views on chemiosmosis would have a better chance of successful development if incubated in a nursery somewhat buffered from the usual scientific grant process. James Lovelock still more forcefully argued that any young scientist must strive to become financially independent of the academic tenure and grant mill if she was to have any hope of pursuing truly original ideas and seeing "big picture" ideas the existing paradigm could not see (or accept). In both cases, this required the scientist to have an independent source of funding.[79] Reich had to support his research entirely on his own income as a therapist, along with donations from students and a few wealthy admirers of his work.

Reich's ideas ranged across even more disciplinary boundaries than Lovelock's, and did so twenty-five to thirty-five years before Lovelock. Given this, and because so many organizations and individuals found Reich's ideas threatening—the threat proportional to Reich's large stature within psychoanalysis and within socialist and communist circles in Europe—Reich was forcibly excluded by those organizations. Some individuals, such as Otto Fenichel and Reich's estranged first wife, Annie, went so far as to spread rumors far and wide, particularly in America, to which both immigrated, that Reich was psychotic. Annie Reich harbored deep personal animosity toward her ex-husband for the rest of his life and tried to turn his two daughters against him. Fenichel, a former friend of Reich's, had nonetheless been a rival and a key participant in the backroom machinations at Lucerne in 1934 that led to Reich's expulsion from the International Psychoanalytic Association (IPA). Thus it might not be too much of a stretch to imagine he might have had a bad conscience toward his former friend. Fenichel not only claimed Reich was mentally ill but fabricated the fantasy that Reich spent time in a mental hospital. While completely untrue, this rumor still circulates widely today. Prominent Freudians Ernest Jones and Anna Freud (Sigmund's daughter) could also have had a bad conscience about their treatment of Reich. They were the chief architects of Reich's expulsion from official psychoanalysis.[80] In his 1953 biography of Freud, Jones disguised the truth, stating that Reich resigned voluntarily from the IPA at Lucerne, that is, that it was Reich's initiative to part ways. Jones actively rewrote the history of psychoanalysis, and rendering Reich's contributions less important appears to have been one of his goals.[81] Because Jones's biography is widely considered an "official history" of psychoanalysis, its influence, combined with the rumors of Reich's mental illness, has for almost sixty years misled readers about Reich's important role. After Reich's expulsion from Marxist parties, with a few exceptions he became persona non grata in Communist circles. The net effect of all this is that it is overwhelmingly likely that those first hearing about Reich today will be inclined to put the "outsiderness" *on him*. That is, they will insist he isolated himself, or was so unreasonable that it is only his own fault if people distanced themselves from him, despite the fact that he did not voluntarily initiate leaving any of these organizations.

Reich's ability to move on and continue working after such expulsions and to leave behind marriages that no longer functioned, colleagues who could not follow his latest theoretical developments, organizations that blocked his work or showed themselves too concerned with their own power base to "get it" about the Nazi threat or about Stalin—all of this speaks of an extraordinary ability to function independently in pursuit of where the logic of his work led. The Beethovens and Picassos of the world find society obstructs them, tends to destroy them. They have many enemies. Yet of Beethoven we do not immediately assume this means *he* must have been mentally ill. Picasso had a volatile sex life with many women partners (as did Bernal), yet this makes nobody hesitate to recognize the genius in his work. Indeed, we recognize that it is their very ability to function independently of many social conventions and personal ties that makes work of such geniuses possible. Perhaps it is worth considering whether this description helps us better make sense of Reich than an account that historical evidence shows is based on falsehoods, misrepresentations, and outright fabrications.

A Role for Intuition in Research

As I mentioned in the Introduction, Reich noted frequently that he was strongly led in research by intuition. He takes pains to explain that intuition alone cannot lead to scientifically valid conclusions. Nonetheless, he argues it is essential for truly pioneering research, and he states it is an inherent part of using the dialectical materialist method as he understood it. Summarizing the necessary balance to be struck, Reich said, "The main rule is: do not automatically believe in anything; convince yourself of something by observing it with your own eyes and, having perceived a fact, do not lose sight of it again until it has been fully explained."[82] He continued:

> Provided one has good critical self-control, one can risk making "unscientific" leaps. . . . For example, it was essential at the beginning of my work, even if it was not intended, to mix the substances together *without sterilizing them*. If I had started off with a high degree of sterilization, the distinction between the relatively immotile nature of non-sterile matter and the motile nature of extremely sterile matter immediately following production of the preparation would not have revealed itself. In the first experiments, the swelling and, in particular, the culturing of extremely sterile crystals of coal seemed even to me to be a little "mad." Yet it was precisely this leap into a highly unscientific procedure that made it possible to explain the spore theory. Such leaps are only valid if control experiments are carried out after-

wards. However, scientific research is often stifled by an excessive amount of such controls.[83]

As an example of such excess he considered stifling, Reich recounts how his first coworkers on the bioelectric experiments, trained physiologists Wilhelm Hoffmann and Hans Löwenbach, were fixated on tiny spikes on the skin surface electrical excitation curve that corresponded to the heartbeat of the subject. These spikes were about one-twentieth the size of the ups and downs of the curve caused by the subject's emotions, yet the physiologists practically ignored the much larger phenomenon before their eyes. "This phenomenon troubled me greatly," said Reich; "I tried to explain it . . . by assuming that even the most conscientious scientist is so tempted to give in to playfulness and foolhardiness that he must protect himself from weakness by relying too much on equipment and by performing excessive controls. . . . To my mind, this is an exaggerated attitude."[84] Because his chosen research question was the function of the orgasm, working in a sex-repressive culture Reich was forced to learn early to look as much at the attitude of the researchers as at their actions. The theoretical attitude of one of the physiologists, Löwenbach, was missing some of the most basic points of the experiment, Reich pointed out. Löwenbach raised a common concern about controls; but in this case Reich believed that the objection was indicative of malice and contempt for the experiment whose central subject was sexual pleasure.[85]

Reich made clear that he fully understood the role of having exactly the required equipment. For example, he said, when one sees an object migrate under the microscope in an electric field, one cannot just assume the movement is cataphoresis. One must have equipment that permits reversal of the current; if that step causes the movement to reverse, then one can distinguish genuine cataphoresis from movement merely caused by streaming of the fluid. Yet, on balance, he concluded, "But the apparatus should not be more important than the phenomenon which one wishes to understand. In the course of the present work [the bion experiments], I had to learn to keep an open mind on some very questionable—indeed, very improbable—phenomena, and at first without carrying out any control tests, so as to let them create their own impression on me. It never fails to surprise me what results one can arrive at by consciously adopting this highly unscientific approach, always providing that the playful leaps are later verified by strict controls."[86] At one point early in the experiments, Reich wrote in his diary, "The detail-technique that has developed has erased all tracks leading to the vital source of life. It must be a *very simple* matter—the *origin of life*—or an overrated idea? A feeling of certainty that my thinking is correct is directing my work. I experiment as if I were in a trance. Do amoebae (motile plasma) really originate from the swelling, formation and dissociation of inorganic matter? Stay calm, wait, be cautious. Mistakes don't matter—just don't get upset. Scientists have chopped up life into pieces. They isolate the details and then want to reconstruct a *whole* from the isolated details, like a machine from its parts. And yet the Orient did see one aspect correctly—namely, that living matter is not a machine capable of being constructed."[87]

Later, in the fall of 1936, after having a strong hunch that cancer cells form from disintegrating tissue in the animal body just as amoebae seemed to be organizing from disintegrated moss tissue, Reich first saw a film of live cancer cells made by Alexis Carrell, at a cancer conference. He was stunned that the cancer cells in motion looked exactly as he had imagined them in his intuitive leap. In his diary, he confided, "No doubt about it. I'm right. 'Migratory cancer cells' are amoebic formations. They are produced from disintegrating tissue and thus demonstrate the law of tension and charge in its purest form—as does the orgastic convulsion. Now money is a must—cancer the main issue—in every respect, even political. It was a staggering experience. My intuition is good. I depend on it. Gave a report to [Ferdinand] Blumenthal in an outburst of zeal. Was absolutely driven to buy a microscope. The sight of the cancer cells was exactly as I had previously imagined it, had almost physically felt it would be."[88] His intuition was a process Reich felt *physically* in some cases. From his psychiatric work as well as his study of the autonomic nervous system and from the bioelectric experiments, this made sense to Reich, even if he was still surprised at times by how correct his intuitive guesses could be. As he put it,

> It is already possible to measure the state of the ego as a reflection of the vegetative excitations active at any one time. A person feels the contraction of the sympathetic nervous system directly when he stands close to the edge of a sheer precipice. We experience pleasure as expansion, as stretching, widening. . . . Admittedly, measurements and replicate experiments still have the last word in science. But when I see an amoeba stretching and the protoplasm flowing in it, I react to this observation with my entire organism. The identity of my vegetative physical sensation

with the objectively visible plasma flow of the amoeba is directly evident to me. I feel it as something that cannot be denied. It would be wrong to derive scientific theory from this alone, but it is essential for productive research that confidence and strength for strict experimental work be derived from such involuntary, vegetative acts of perception.[89]

Reich often made helpful use of tangible, physical models or imagery in his thinking, particularly to explain feelings. Recall the "tense pig's bladder" image in understanding masochism, and, from even earlier, his image of "reaching out" emotionally and the physical erection of the penis as identical with the sticking out of a pseudopod from an amoeba. The "pig's bladder" image sounds a lot less odd when one recalls that Reich grew up on a farm and regularly saw animals slaughtered, a background steadily less likely for biologists ever since.[90] In this context, Reich's upbringing in contact with a lot of living creatures in their natural setting, their sexual activity, the organs of animals seen during slaughtering, and other such observations could be seen as quite an asset for a researcher in life science.

One of Reich's most important contributions to psychotherapy was reading the patients' facial and bodily expressions as direct expressions of their inner feelings, of which the patients themselves were largely unconscious. The bioelectric experiments forcefully validated his readings of this "expressive language of the living," as he later described it.[91] So it is of a piece with this earlier work that Reich's intuition was so strongly connected to his involuntary vegetative reactions that when the cancer film so strikingly verified his physical image of what cancer cells would look like, he felt compelled to buy his own state-of-the-art research microscope immediately. Yet Reich was arguing here for a rather nuanced view, rooted in what he knew as a clinician and from the bioelectric experiments, and raising the issue of the importance of the state of the observer's autonomic and sensory responses in her ability to perceive certain things in nature. He later, in 1949, developed this line of thinking in much greater detail in his book *Ether, God and Devil*.

In many ways, Reich's use of his "intuitive" feelings as a research tool resembles Evelyn Fox Keller's description of Barbara McClintock's method of working. Interestingly, Reich's clear-cut imagery ("pig's bladder, erection "reaching out," etc.) is even more concrete and tangible than Keller's description of McClintock's experiences of sitting, Zen-like, under a tree thinking and suddenly picturing the movement of genetic elements clearly. Reich, for one thing, found in the bioelectric experiments that the degree of pleasure felt by a subject depended not only on the stimulus (e.g., gentle vs. rough stroking), but even more so on the state of *readiness* of the organ (e.g., the tongue) or the whole organism to be excited. Thus, the same tongue might be stimulated once by sugar solution but far less stimulated after multiple exposures to the same sugar or after a sudden unexpected salt solution rendered it "cautious" to subsequent stimuli for a time.[92] Reich felt this kind of difference *from one person to another* (e.g., the functioning of their sensory organs) might also lend itself to being examined and quantified by experiment. When he saw in the bioelectric experiments that patients whom he knew from therapy to be "emotionally lively" showed quantitatively greater changes in skin surface voltage in response to emotions than "more vegetatively rigid" patients, his first impulse was to try to study the phenomenon. He asked the more rigid patients to breathe more deeply and found this often increased the voltage changes on the skin of their abdomen as well as their subjective perception of the strength of the emotions they felt.[93]

Modern researchers tend to deal with individual differences among experimental subjects by working with large groups and relying on the average or mean result, seeing such differences as a problem, methodologically. Reich, motivated by the question of the energy underlying emotions, by contrast saw those differences as a research opportunity. He began to suspect that emotionally rigid people reacted to the bions, that is, being faced with lifeless material taking on motility, with involuntary fear and revulsion. This was connected, he suspected, with the fear, revulsion, and guilt feelings that many people had developed about involuntary sexual movements.[94] At least one American academic (Leo Raditsa) thought this was Reich's most salient characteristic: Reich saw the obvious that everybody else unconsciously tried very hard not to see. He saw the obvious in the scientific study of emotions: that a patient would unconsciously shake her head "no" while saying "yes," or that genuine sexual gratification cured neurosis. But Reich said he also saw the obvious in people's reactions to that work—even if they were scientists—when their own characterological structure was built upon guilt, fear, and avoidance of the importance of sex.[95] Making the same point about Reich's unusual ability, one of his physician students later observed that "as a corollary, we may also assume that the capacity to comprehend Reich's theoretical conclusions and the often intense emotional reactions to them must depend to some extent on the biological structure of the inquiring individual. How to contend with this 'subjective' factor in any attempt to arrive at an objective evaluation of his discoveries is a problem that will not soon be solved."[96]

Reich's Attitude toward Existing Scientific Disciplines: Demand for Immanent Criticism

Reich wrote du Teil on October 4, 1937, describing how the bion experiments unexpectedly led him into research on cancer, saying intuition had a prominent role. Reich sensed in du Teil, especially after the latter worked at his lab in Oslo in July and August 1937, a serious attitude, far more open-minded than usual among academics in Reich's experience. Thus, even though somewhat conflicted, Reich felt he could broach such questions directly in his correspondence with the French scientist. These exchanges tell us much about Reich's sense of his own experiments and of his strategic thinking about publishing his work: "As great as my desire is for discussion with you, my willingness to hold back is even greater, if that is the right thing, as my intuition above tells me. I am sure you have noticed that in our circle, instead of superficiality, insignificance and uncalled for politeness, a direct openness steps in. The following one only asks his very best friends."[97] Reich went on to ask du Teil to tell him bluntly whether he wanted to be kept up to date on the details of the new emerging cancer research thread, or whether that would distract him from his own important work on replicating the culture experiments. If du Teil felt that way, Reich said he would just apprise du Teil of the major new developments from time to time. "This would have its advantages but also its disadvantages, because the whole work approach is so strange and unique, that easy contact is lost if the methods are not continually and mutually discussed," Reich said. To elucidate, Reich used an analogy: "I have landed in a jungle and I have built a wide road, but from time to time I am tempted instead of continuing to build the road straight ahead, on the basis of a vague intuition, to penetrate sideways into the jungle. I can hardly give account to myself for why I am doing this. The way this work is developing here on the cancer problem is just such a new jump sideways. But I am assured that it absolutely proceeds in a correct fashion. The logic which unfolds itself in front of us . . . is sometimes almost spooky. I know that you understand me."[98]

Du Teil replied to Reich on October 8 in the same frank tone, which showed he was not completely comfortable moving so fluidly into yet another new realm of science. Because their letters give much insight into their ideas about doing science, I will quote at length from this exchange. As du Teil expressed it, "when we enter a domain as vast, complex and deep as that of cancer, which has already given rise to hundreds and millions of remarkable studies to which people have devoted their entire lives, and where the greatest bacteriologists break down before problems of detail accessible only to a few initiated people, it is absolutely impossible to discuss these questions with them unless we have a culture as deep and complex as theirs."[99] Du Teil felt quite anxious, already facing a very difficult challenge to convince bacteriologists about large, paradigm-changing implications of the bions and their culturability. The only thing helping mitigate his anxiety was that at least his bion work was in a realm of research (bacteriology) where the questions raised were directly relevant to the techniques in which those scientists felt competent. Du Teil felt urgently that, strategically, Reich ought to challenge only one major specialty in the sciences at a time. Moving into cancer research, he thought, was so different a specialty, with such an enormous body of specialized techniques and knowledge, that he feared to raise those questions before first winning the already very difficult battle with bacteriologists.

Reich may well have felt similar anxieties. But he found the logic of the work so compelling that it outweighed his anxieties. He moved ahead. Perhaps his training as a physician sensitized Reich more to the degree of human suffering the cancer problem caused. Probably, too, his personality played an important role. The experiences he survived—not least World War I— had given him a high degree of self-confidence, or at least trust in his own instincts. Each new breakthrough in his work, from almost the beginning of his formulation of the orgasm theory and the concept of orgastic potency, had produced significant conflict with his scientific colleagues. Yet Reich felt motivated at each turn by, as he saw it, the logically compelling advance in treatment and in understanding the basis of human illness—sufficiently that he pushed forward with the next logical step. Then too, while many scientists criticized this as impulsiveness and arrogance, at each step, other important scientific authorities agreed with Reich and encouraged him to press on, for example, Malinowski's support of Reich's foray into anthropology or Friedrich Kraus encouraging Reich's "basic antithesis of vegetative life" thesis and his attempt to test it via the bioelectric experiments.[100] Furthermore, Reich's psychiatric and psychoanalytic training made him gradually come to believe that unconscious discomfort with sexual matters played a large role behind his opponents' supposedly scientific criticisms. He came to see human emotional rigidity as a significant part of scientists' discomfort with what we would now call "paradigm changing" discoveries, as well as scientists' tendency to view crossing disciplinary boundaries inherently as arrogance. All these forces seem to have made Reich more willing than others, even du Teil, to tolerate the anxiety produced when the logic of the experiments moved him into new areas such as cancer research. Whether this represents arrogance on Reich's part, bold independence, or both in some measure, the reader will decide for himself.

Du Teil advised Reich that his own feeling was: do not move into a different new area of specialization unless you first master all the tools and theory of that area. Otherwise, in the very first conversations with cancer specialists, they will see gaps in your knowledge and conclude that there is no reason to take your new ideas seriously—that much less so if those ideas dramatically clash with major existing paradigms in their field. It is essential to cross such boundaries "only with *all the weapons of the technician and the specialist*," warned du Teil. And even then, he told Reich, do not try to convince all the specialists at once; rather, "you should aim for a technician who is a modest executant, having never made [important] works of his own and who, consequently could attach himself to yours without reservation." And start with bacteriology, advised du Teil, where the struggle to convince his French colleagues was still far from won: "I advise you then, in the interest of the development of these works, to attach a young bacteriologist to your laboratory, who, just coming fresh out of university, would be perfectly au courant with all the latest works on cancer and tuberculosis, in order that he may translate what you have found in technical terms accessible to all the technicians."[101]

Du Teil said he used the director of the bacteriology laboratory in Nice as a source of precise technical information and orientation in specialist theories, adding, "He provides me with arguments that sometimes I lack. For the rest, I worked and worked very seriously every day on a manual of bacteriology, where I found fundamental information without which I could have done nothing." Having such a resource person allowed him in a few weeks to make "an approximate identification" of the bions (i.e., how a mainstream bacteriologist would classify them) and "to reply with precision to *very detailed* and *very technical* questions which were asked of me at Pontigny and at Paris. To the point that Dr. Bonnet was very astonished to learn by chance, when I was leaving him, that I wasn't a doctor. Bacteriology has become a science of extreme complexity and the question of the culture media alone, for example, demands months of study."[102] When he was questioned about technical matters by bacteriologists, du Teil said, "If I had not been able to reply to them, I felt very clearly that they would no longer have taken my experiments seriously. It is necessary to speak to people in their language and to be with them on equal terms at all levels. I'm all the better able to tell you this since I know that you know my friendship toward you and will take this as a mark of my friendship."[103]

Reich thought du Teil naïve about scientific politics—about which Reich had learned in the school of hard knocks—and again replied frankly. He wrote on October 15 but in the end thought better of it and did not send the letter, filing it for the future. He opened by complimenting du Teil on his ability to build bridges to scientists from other fields; Reich was perhaps thinking he had become too guarded, too often anticipating an irrational response and wondering to himself whether du Teil's style might offer some lessons for him. However, he urged du Teil to consider his caution, since the outcome of these contacts was not yet known: "If these colleagues truly keep their promises and carry out the experiments until they succeed, then we will certainly be home free as far as the first matter is concerned—namely, the production of bions. These remarks do not apply, or at least not yet, to the second matter—that is, the cancer question." Reich was distinguishing cancer in order to address du Teil's strategic warnings about how to deal with specialist topics:

> I would like to state briefly that I am totally immersed in this subject and did not get into it by chance. In the same way that the bion experiments developed logically from my clinical orgasm formula, so the studies on cancer tissue and the experiments on animals are developing equally logically and consistently. I sometimes sit here completely dumbfounded and cannot believe that this logic is possible. I would like to stress that although I work with a great deal of intuition, this intuition is backed up by very solid clinical and experimental facts. I am helped in all this, last but not least, by the cognitive method I use [i.e., dialectical materialism]. . . . Because of the close contact that we have now, fortunately, established with the world of bacteriology, all of us have entered into an extremely dangerous phase.[104]

By the last remark, Reich meant that, as he had found to his chagrin with Fischer, Thjøtta, and others, once contact was made with such specialists and they felt invited in to publicly express an opinion about the experiments, because his own results could be so potentially upsetting to established theories, they would be strongly inclined to dismiss those results. Thus, any small flaw or failure could trigger a disastrous, sweeping dismissal from such specialists.

> If the work that we now do is successful, then we will certainly triumph. If anything goes wrong at any point, however, the setback will be far greater than the disadvantage of having no contact with this particular scientific field would have been.... The bacteriologists with whom you are dealing in order to obtain confirmation of the bion experiments are highly specialized experts and I know that I could not engage in a discussion with any of these people in their particular fields. Although I am familiar from my medical studies with the fundamental aspects of bacteriology and its methods, the field of bacteriology has become much broader and deeper, and I would therefore not undertake to engage in any discussion of specific bacteriological problems. There can be no doubt that this is a correct assessment of the situation.[105]

But Reich pressed for a crucial caveat. He claimed he had created *a new field of research* in sex-economy and its extension into laboratory biology; thus, he insisted, "it is equally certain and clearly established that the best modern experts in bacteriology cannot argue with me on the basic problems of my clinical experience and my specific experimental work." For example, bacteriologists for the most part insisted that fixed, stained specimens had priority in characterizing microbes. In sex-economic bion research, this was directly in conflict with the theoretical basis of the field of research. So if their criticism was rooted in such different methodological assumptions, they were making the same mistake du Teil was accusing Reich of: not knowing enough about the specialist field they were criticizing. Reich put this pointedly:

> In this connection, I would like to express a feeling I have always had whenever my enemies, or people who review my work from afar, described me as a systematist, a philosopher, or a synthesist. I do not feel I am any of these three things. I am a scientist working in a field where I must use methods of research and thought developed entirely by myself that are not known in any of the existing specialized fields. I am referring

to dialectical-materialist sex research. Recently I had the opportunity to say to a cancer specialist [Leiv Kreyberg] who was very arrogant to me in the presence of several other physicians: "Dear colleague, I fully recognize your exclusive competence in the field of cancer research in its present form, but I would ask you to take note of the fact that I claim absolute authority in all questions of sexual functioning and its relation to vegetative life. I do not know about your cancer problems, but you are not knowledgeable about my field of expertise either. I hope that in the interest of the matter at hand we can arrive at an understanding."[106]

Reich said that he had held this basic attitude for many years, as a result of conflicts he had weathered within the science of psychoanalysis. "When I was struggling sharply in the early'30s, defending Freud's scientific psychology of the unconscious against opponents such as trained psychiatrists, experimental psychologists, neurologists, etc., outspoken people with whom I could not debate in their special fields, I learned a crucially important consideration that helped in my later work."[107] Experimental psychologists, Reich recalled, tried to refute or to ridicule the Freudian method of investigation, but not to understand it in its own terms or what it might bring that was valid, which Reich called "immanent criticism." Reich recalled one psychiatrist saying, "It's nonsense, when Freud said that the energy behind kleptomania was sexually derived.... I have seen many kleptomaniacs, and they don't see any connection with sexuality." To du Teil, Reich said, "Imagine, my dear friend, that one had yielded to that position at that time. He's right, with his method of investigating, he can claim a clear conscience that kleptomania had nothing to do with sexuality because kleptomaniacs don't know anything about it." But to anyone experienced in using the analytical method of investigation to solicit unconscious material, "it was quite clear that compulsive kleptomania is a substitute for sex, for love or masturbation. At that time we took the position: 'We do not want to interfere in their sphere of work, but if they want to discuss our work with us, we urge them to learn first our research method.'"[108] From Reich's point of view, he was still today taking that same position in scientific sex-economy (including bion research) vis-à-vis bacteriologists: "We must demand from the experts criticism grounded in our work [*immanente Kritik*]: to judge our work they must, if they are at all honest, temporarily abandon their own methods of thought and adjust themselves to our position and criticize us from the standpoint of *our* work. This does not mean that we would assume a superior air and declare that we know everything better and that the others are fools."[109]

Reich emphasized that he agreed with du Teil in important ways. "For example, you are absolutely right when you say that I must very soon employ a young bacteriologist who will try to situate our results in the overall bacteriological scheme of things. However, if this should sometimes prove impossible, then we would first of all have to check to see who is right—existing bacteriology or ourselves. (I am reminded again of the diagnosis of staphylococci that was made in the case of structures that beyond doubt are not staphylococci.)" He assured du Teil he would do "everything possible to accelerate, and bring about this linkup. . . . But I cannot and must not adjust myself to the ways of thinking and the methods used by bacteriologists lest I lose the thread of my own study methods and thought processes."[110] Reich reminded du Teil of conversations they had on these matters when du Teil visited Oslo in July and August 1937, just two months previously. His unusual frankness is testament to the confidence Reich placed in him, and his hope for honestly working through their differences on this—as he saw it—critical matter of professional collaboration.

> I have the feeling, and we have already talked about it here, that you are too optimistic in your assessment of the willingness of the scientific world to accept these matters today. Please permit me to retain my skeptical attitude until you succeed in convincing me that I am wrong in holding to it. I assure you that I will do nothing to hamper you in your attempts to convince the bacteriologists and biologists. On the contrary, I will make all the material and all the controls available to you so that you will be fully familiar with our work. I shall also be happy to learn of any objections, even awkward ones, you report to me. But you will not be able to make me surrender my skepticism about the willingness of the majority of scientists. I know all too well how the highly specialized worker is anchored in his own problems and ways of thought and I will not abandon myself to any dangerous illusions on that score. I accept the fact that I shall have to go on working for ten years and more, quietly but firmly convinced of what I am doing but without any means of arriving at a breakthrough. Please believe me that I do not want to play the

> role of martyr. I merely wish, using all the means available to me, to prevent you, a valuable and cherished coworker, from getting into difficulties by entertaining illusions about the possibilities of a quick breakthrough. Let us wait calmly and see what happens. We do not have to somehow raise ourselves up to participate on the same scientific plane as the current bacteriological or other disciplines. In our own specific fields, we are equally good scientific authorities.[111]

In many ways, Reich's argument here poses the question, do we expect the same attitudes toward disciplinary boundaries from what Thomas Kuhn called "normal science" to apply toward pathbreaking research that, by its very nature, challenges existing disciplinary boundaries? But just the same Reich constantly reminds du Teil that he *is* trying already, as he sees it, to follow the useful parts of du Teil's advice: "I have employed the following people in my laboratory: a specialist in chemistry and physics [Odd Wenesland], a trained bacteriological laboratory assistant [Vivi Berle], and a trained film maker [Kari Berggrav]. And in Dr. Havrevold I have found an extremely well-trained expert on internal organs and, in particular, on internal secretion. As soon as financial circumstances permit, I will be very happy to follow your advice and employ a bacteriologist and a cancer specialist. As you correctly observe, in this way I will discover in advance what the objections might be and will be able to take part in discussions armed with the necessary counterarguments. At the moment, unfortunately, I am struggling with colossal financial difficulties."[112] Reich decided not to send this letter to du Teil. In the end, then, Reich and du Teil were far apart, but not as totally far apart on strategy as it may at first appear. Money was a major issue; Reich poured a lot of his own income as a therapist into the experiments and received occasional support from a few wealthy donors. But the differences between the two men would emerge in more dramatic relief when their opponents turned up the heat by igniting a major public controversy over the bions. This is described in Chapter 6.

Reich's psychoanalytic insight amplified his view of the role of the irrational among scientists, but Reich was hardly alone in this view. Many researchers studying mitogenetic radiation at this time, for example, also felt the opposition to it was based as much in irrational prejudice as upon experimental evidence, as is discussed further in Chapter 7.

Du Teil Complains to Havrevold

Roger du Teil, however, had by late 1937 privately begun to attribute a significant share of Reich's difficulties to what he saw as Reich's "personality issues," differences about strategy in which du Teil felt Reich clung to his own philosophical position with irrational stubbornness. He thought he observed some of this during his August 1937 visit to Oslo. Reich disagreed in their letters, keeping the focus on the irrational in science. But by January 7, 1938, du Teil wrote to Reich's student and experimental assistant Odd Havrevold, to air his concerns to one he considered a confidante close to the situation. Havrevold had written du Teil to apprise him of attacks on Reich in the press and in Oslo psychiatric circles in November and December 1937. Du Teil sympathized, saying, "The disquieting news that you shared with me makes me sad, but it doesn't surprise me. In general, the price of discovery is persecution." He elaborated: "The attacks that you cite are not new and, likewise, neither are a part of the arguments you reported. All psychoanalysts, without exception, have had to deal with accusations from the moral middle-class that they abuse their female patients. The necessary occurrence of transference and the neurotic character of the ill continuously provide new material for these accusations, and I do not believe that one can seriously dispute with you about them."[113] However, du Teil suggested he thought Havrevold a level-headed person who could be a good influence on Reich, to help avoid pitfalls du Teil thought Reich's strategy and his own personality might bring upon him.

Du Teil argued this issue had several aspects. First, "I have told my friend, Reich, again and again both verbally in Oslo as well as in letters to him since my return home about my concerns regarding his overall espoused attitude and the methods he uses to distribute his reports. I already pointed out the dangers of mixing general policy with science, and this in such a systematic way, to him in Oslo. And although I agree with him on most issues, I have always protested against mixing the two areas." Du Teil said Reich had later written to agree that he would, at least for the time being, try giving up such a broad range of publication—the wide-ranging mix of politics, sociology, pedagogy, and science in his journal *(ZPPS)*—and that Reich "intended to henceforth retreat to pure science." But, du Teil observed, this alone would not likely solve the problem entirely since Reich had already written so much about politics and sociology and was still the editor of a journal *(ZPPS)* that "epitomizes to the entire world the systematic destruction of accepted, general morality and society. It does not matter which view one takes about this matter, one must recognize that that is the fact and that he started the offense. It is daring, and I always admire his courage, but it brings with it the risk of an unequal material battle. It would only be surprising if society would not defend itself. The way it does it is probably not exactly correct, but war is war."[114]

Du Teil argued that his own experience in France—sticking entirely to the bions—was lauded by many different labs and had gained fame for Reich in France without any blowback because of political and sexual implications. Thus, it was clearly a better approach, he reasoned. Second, however, he argued, what mattered was that he had not put the experiments themselves forward in the primary place of importance. "This emphasis I have placed on a general theory," he continued, "for which the experiments, in spite of their importance, only mean a verification of a specific point, as if biogenesis were merely of secondary importance to us. The world *accepted* the announcement in this form. I did not pose as an enemy of Pasteur, rather as the continuation of Pasteur, who had been too narrowly interpreted up to now, and even then I got everyone on my side." Furthermore, "I have avoided denying facts that are still generally accepted, such as Brownian movement. Such a denial would have alienated the entire world." Du Teil attempted to include Brownian movement, among other movements, as "one of the defining elements of life." Citing recent press coverage, he bragged, "In short, I strove to specifically use all of the old truths as support in order to build a new theory into the old ones, or rather vice versa. That was successful to such degree, that a newspaper article compared my synthesis with that of Louis de Broglie."[115]

By contrast, du Teil said that Reich worked exactly the other way around. "He continuously repudiates everything that was known before and . . . he does it even where it is unnecessary. To deny Brownian Movement means, for example, to intentionally and unnecessarily shock all physicists; this gives them the impression that one has only superficially examined the question. . . . In short, Reich's impersonal and sometimes aggressive behavior is, in my opinion, the reason he has provoked animosity. Added to this is the combative behavior he has embraced in the social part of his endeavors."[116]

Third, du Teil said he had repeatedly urged Reich to consult professional bacteriologists, especially since the technical minutiae of one experiment were complicated, and this complexity increased as one moved to the next experiment based on the first, then to a third suggested by the second, and so on. Du Teil did not accept Reich's insistence that the bacteriologists use his concepts and terminology—since his sex-economy was a different science than bacteriology, and the bion experiments were part and parcel of that other science. "Reich and the bacteriologists do not speak the same language," said du Teil; Reich "speaks in general terms and sees from above a whole. The others [bacteriologists] refer to underlying particulars. By systematically neglecting this, he gives the impression that he knows nothing about it."[117] Du Teil had expressed this to Reich before and was frustrated that Reich continued basically as he had. Reich had apparently not convinced du Teil that academic authorities such as bacteriologists were, more often than not, using terminological minutiae as a smoke screen to disguise their basic irrational antipathy to seeing their turf (microorganisms) become connected inseparably to Reich's orgasm theory. Nor did du Teil sound as though he accepted Reich's claim that, should bacteriologists wish to criticize his work in a basically different field, they must first familiarize themselves with the necessary background, his orgasm theory; otherwise (as Reich saw it) their criticisms were not "immanent"—by which he meant not pertinent to, not inhering in the correct context.

Finally, du Teil urged, "Reich displays a much too impatient nature in this issue. Under the auspices of having discovered a world with a single glance, he demands that everyone else recognizes and understands this as quickly as he, and this demand is in itself contradictory. There is no letter in which I was not forced to stifle his haste, and I still had to remind him in the last letter that certain vaccines or serums . . . had to be tested more than ten years . . . before they were announced or even released." As an aside, du Teil remarked, "I have the impression that it is the necessity of stabilizing his financial situation that drives him to seek an immediate, resounding success. I consider it extremely pernicious when material issues affect the course of the experiments by causing haste."[118] Du Teil lamented his own difficulties, having had to borrow extensively to set up his own laboratory for the bion replication work. But since he felt supreme confidence in eventual triumph, because of the basic fair-mindedness of his scientific colleagues, he was not as worried about this debt.

Du Teil was only one of many over the years who pointed out Reich's impatience with those who claimed to be supporters but who could not keep up with Reich in understanding the importance of his experiments. This, of course, could be a characteristic of an arrogant person who overrates his own work; however, it could also be the mark of a paradigm-changing intellect and the loneliness such a person must inevitably feel.

Du Teil, in closing his letter to Havrevold, became reflective. Reich's personality was, after all, he thought, what made Reich capable of such major insights and paradigm-changing discoveries. So du Teil concluded, somewhat pessimistically, "Even if Reich would recognize our diagnosis as valid, could he be able to change his behavior? I do not believe so. His temperament is as it is, and one must accept him as he is. Such persons are destined to be simultaneously idolized by some while being burned [at the stake] by others. Indifference or torpidity regarding such people is impossible. And I am very much afraid that there is no way to heal this, for the very same conditions would appear again in any other country."[119]

It is unclear whether Reich knew about this letter to Havrevold at the time; the letter did eventually end up in Reich's archives. But Reich continued discussion of many of these issues with du Teil in April 1938, after a renewed press controversy began about bion experiments. That controversy is discussed in Chapter 6.

By now, perhaps many readers sympathize with du Teil and think what Reich is asking so unconventional, so beyond their concept of science, that they cannot help but hear his demand as arrogant. But it is worth viewing the question Reich raises through Thomas Kuhn's conceptual framework. In addition, perhaps considered through the lens of the "independent scientist" category, Reich would seem no more arrogant than James Lovelock when he first proposed his Gaia hypothesis. He himself may not yet realize that his conceptual framework and his methods of work might already be so foreign a language to a conventional bacteriologist that none could have worked successfully with his team because no understanding was possible across that language barrier.[120] Nonetheless, Reich had the wisdom to anticipate, for example, the possibility of du Teil's losing his job because of his work with Reich, which came as a complete shock to du Teil when it actually occurred in June 1938. This argues that in at least one important way Reich's view of scientific politics is not simply paranoid, but on the contrary, quite prescient; du Teil's view—however reasonable it sounded—turned out to be naïvely optimistic. Might this not argue in favor of Reich's approach to science, in which one pays careful attention to the motivations of one's opponents, rather than assuming the evidence from the laboratory will carry the day on its own (at least when one's science clearly challenges basic features of an established paradigm)? Du Teil had himself earlier observed that Reich's (as du Teil saw it) aggressive personality was inseparable from who Reich was and essential to how much Reich had already accomplished scientifically against substantial resistance. Not all those declared "outsiders" by official science are "cranks" (a term used toward Lovelock in the early days of Gaia theory) or "pseudo-scientists." Not all those who wander (across disciplinary boundaries) are lost, as Peter Mitchell's "wanderings in the garden of the mind" showed.

Let us turn next to how Reich's bion work led him into cancer research and a radical break with the dominant theories in that field as well.

5

REICH'S THEORY OF CANCER

The bion experiments, Reich tells us, led him directly to a new theory of the origins of cancer cells. In addition, he injected many of his bion preparations into experimental animals such as mice and found that some of them instigated cancer tumors while other bions seemed to help cancer mice resist tumors they had previously. For the most part, this chapter discusses Reich's developing ideas about cancer only until the time when he first immigrated to the United States in late 1939. This is a necessary prehistory for his later, much more well-known experimental work developing the orgone energy accumulator, whose use was at the center of the Food and Drug Administration's legal case against Reich. The accumulator and its experimental use on cancer mice and human cancer patients are beyond the scope of this book; the reader is referred for more information to Reich's book *The Cancer Biopathy*.

"Life Rays"

In the 1930s, Reich was aware, as were most educated people, that radium was in widespread use to treat cancer tumors but also that, paradoxically, working with radium and other radioactive isotopes with insufficient protection had caused cancer in some researchers.[1] Gurwitsch's "mitogenetic rays"—said to cause increased mitotic division in other cells nearby—had been reported in many actively growing tissues, particularly onion root tip and yeast cells; they had also been found at elevated levels from cancer tumors. Robert Kohler has pointed out that biologists in the 1930s were "imbued with expectations of a 'new biology'" because of these developments and such discoveries as heavy water, which it was hoped might have medical uses.[2] Spencer Weart has noted that ideas about "life rays," often associated with "magnetic" sexual attraction, were also widespread in popular culture and had been reinforced by scientific reports such as Butler Burke's claim of "radiobes," Frederick Soddy's talk of radioactivity and its "half lives" in quasi-biological terms, and more recently by Gurwitsch's mitogenetic rays.[3] As noted in Chapter 1, Reich speculated, half-unconsciously, about a life energy ever since hearing Kammerer discuss it in his University of Vienna days in 1918 or 1919.

It is quite interesting, then, to note a public lecture, "Character and Society," that Reich gave on April 18, 1936 to the Norwegian Student Organization in Oslo—just as he was embarking on the bion experiments. Near the end of that talk, Reich said, "The war which is at our door will bring about changes of a magnitude as yet inconceivable. For the time being the scientist can do no better than prepare for the time when the great change shall come. So many are busy trying to find ever better and more effective death-rays; let us hope that coming events will move the scientists to search, at last, for the *life-rays*."[4] As he completed the write-up of the bioelectric experiments just a few months previously in January 1936, Reich had written in his diary, "Sex radiation—if each body radiates, then the sexually excited organ must radiate as well. It only has to be discovered, but how?"[5] Clearly, the motif of life rays was emerging from Reich's unconscious—where it had been working beneath the surface, as he later acknowledged about Kammerer's influence on him.

In his book *Nuclear Fear*, Weart suggested "life rays" and "sex rays" were a deep-rooted, fundamentally irrational motif that led numerous researchers astray so that they imagined "N-Rays" and other forms of radiation from 1900 to 1940 or so, where later experiment showed none existed. For those predisposed to judge Reich's "orgone energy" as imaginary, this kind of storyline has been so tempting that it is adopted rather uncritically. It has similarly been used to dispose of "mitogenetic radiation," despite the fact that its existence was confirmed and quantitatively measured in many labs all over Europe.

But Weart's hypothesis of irrational "wishful thinking" is somewhat Whiggish or retrospective an explanation and need not be the reason Reich was interested in such ideas or took "life rays" as a serious hypothesis. Radium research and its relationship to cancer were very prominent in the "Red Vienna" in which Reich came of age intellectually.[6] Radium treatment of cancer had become big business as early as 1916.[7] And in 1936, though some labs had begun to report negative findings about mitogenetic radiation, most still considered it an established scientific fact.[8] In addition, widely respected Berlin protozoologist Max Hartmann's theory of "relative sexuality" similarly was still taken seriously by many biologists, and Reich reasoned, after his bioelectric experiments, that differing bioelectrical charge on the cells was the underlying cause of their ability to act in some cases as male, in others as female. So respected, peer-reviewed scientific research seemed to point from many different directions to some kind of "life rays" and/or "sex energy." How were these ideas connected to Reich's evolving understanding of cancer?

To begin, let us return to Reich's early, educated guessing (which I discussed in Chapter 4, in the section on intuition in research) in May or June 1936 about a parallel between the origin of cancer cells and of amoebae from bionous disintegration of plant tissue. In his notebook, Reich jotted, "*Dialectic of death and life*: 1. When the metazoan dies, the protozoan appears."[9] And shortly thereafter, on June 15, 1936, he added, "4. *Cancer brings together the vesicles*, organizes them, but disrupts their motion. Cf. the general, combining, paralyzing function of cancer. . . . Products of decaying material in the water . . . are without doubt the vesicular grains, energy units. Probably the same thing happens in the course of digestion: the food is broken down into energy vesicles."[10] Then in the spring of 1937, Reich had been struck to find one of his packet amoeba bions in the fluid he was examining from a mouse tumor.[11] It seems clear he thought there was some kind of connection between bion formation and cancer.

In this regard, Reich's bions can be viewed as a twentieth-century revival of one of the ideas often held by nineteenth-century advocates of spontaneous generation: J. F. Lobstein, William Addison, John Hughes Bennett, and others, for example, believed small but microscopically visible bacteria-sized particles they called "histological molecules" were released by disintegrating or unhealthy tissues in the body and that these "molecules" were involved in the creation of cancer tumors and other pathological growths.[12] Hughes Bennett later became convinced that these same "molecules" could cause spontaneous generation of protists in infusions.[13] British microscopist Lionel Beale disputed the possibility of spontaneous generation, yet nonetheless agreed that these microscopic particles (he called them "bioplasts") could be the source of disease. Beale used magnifications of up to 3000× during the 1860s and 1870s to study these "bioplasts."[14] It is unclear whether Reich was aware of this earlier literature.

Warburg's Theory of Cancer

Reich was greatly stimulated in his thinking by the famous Viennese biochemist Otto Warburg's theory of cancer as a result of anaerobic metabolism.[15] In 1923 Otto Warburg began experiments on inefficient respiration (glycolysis) of cancer cells related to oxygen deprivation in the mitochondria. Warburg believed that the loss of ability to respire aerobically was crucial in the development of the cancer cell, that the shift to the much less efficient glycolysis was a key characteristic that distinguished the cancer cell. (Glycolysis yields only two molecules of ATP (adenosine triphosphate) from each molecule of glucose consumed, whereas the usual process of glycolysis plus aerobic respiration yields thirty-eight molecules of ATP per glucose molecule respired). As Reich's experiments on the natural organization of protozoa progressed, he was struck by the fact that it was only dead grass or moss, collected in autumn, that would undergo bionous disintegration and reorganization of the bion heaps into protozoa such as amoebae and colpoda.[16] As his June 15, 1936 lab notebook entry shows, he was groping toward the thought that there might be a parallel between the disintegrating moss and disintegrating human or animal tissues in cancer, since dead or dying grass or moss would have a much lower amount of energy than fresh grass.

By September 22, 1936, wondering about the relationship between bions and bacteria, Reich began to examine staphylococci and streptococci from tissue samples of a student of his who suffered from chronic osteomyelitis. This was Jørgen Neergaard, a Danish sociologist and political scientist—brother of the biologist Paul Neergaard, discussed in Chapter 3—and a member of Reich's Danish study group, led (after Reich's move to Norway) by Ellen Siersted. Neergaard had a bone-marrow deficiency in both legs since childhood; he agreed to Reich's study as his illness worsened through 1936, and he died on February 2, 1937.[17] In the tissue samples Reich saw bionous disintegration occurring, similar to what he had seen in the dead grass and moss. This confirmed his speculation from June about whether unhealthy animal tissues might be subject to the same disintegration, while healthy tissues had a high enough charge and cohesion to resist breakdown (much like the new, fresh spring grass and moss that never yielded amoebae).[18] By November, as noted in Chapter 4, Reich's intuitive sense of this connection had grown so strong that when he attended a conference on cancer and there saw a film by Alexis Carrell of live cancer cells, he was struck that they looked exactly like the amoeboid cells he expected. He also injected the bacteria grown from Neergaard into experimental mice, some of which he was injecting with various types of bions from his experiments, to see whether they had potentially harmful or helpful effects medically.[19]

Reich was also much struck by the fact that a high proportion of cancers in women appeared in the breasts and genitals, and in men prostate, testicular, and colorectal cancers were common, suggesting a connection with what he called "muscular armoring" of those parts of the body caused by fear and guilt about sexual feelings.[20] This line of thought made Reich wonder, he said, whether one key factor that could set animal tissues into the process of bionous breakdown was insufficient oxygen supply. Chronically contracted muscles would restrict blood flow, and thus oxygen availability to those tissues, he reasoned. And "in countless cases, this inhibition must be present for decades before it takes the form of cancer."[21] If the cells in those tissues lived at reduced oxygen levels for decades, Reich believed, Warburg's theory was relevant. "Warburg discovered that the various cancer stimuli [coal tar, soot, etc.] have one common feature, namely the production of a local oxygen deficiency, which in turn causes a *respiratory disturbance* in the affected cells," Reich reasoned, noting that the respiratory disturbance is a property common to all malignant tumors including the Rous sarcoma.[22] However, Reich did not accept Warburg's idea that "the cancer cell is just a normal cell assuming a different kind of growth because of oxygen deficiency." His observations on bionous disintegration and reorganization into protozoa led him to believe, rather, that "biologically, the cancer cell is basically different from the normal cell; it is nothing but a protozoal formation."[23] Reich did not bring up the cancer work in his book *The Bion Experiments*; he thought it premature since the cancer experiments were still developing rapidly in September 1937 when the book went to press. Lest the reader think it was perceived as unheard of in the 1930s for pathogenic cells to be created within the body ("endogenous infection"), Reich reminded the reader in the bion book that "in the case of tuberculosis, the endogenous nature of the Koch bacillus has long been asserted."[24]

Reich sought out cancer patients or their physicians or pathologists, willing to donate samples of live cancer tissue to examine. In a few samples he was able to obtain, such as sputum from a lung cancer patient, Reich did find cancer cells that looked very much like protozoa, often caudate-shaped protozoa in appearance.[25] When he watched bions clump together in precancerous tissue, at first the bions formed spindle-shaped cancerous cells.[26] At a more mature stage of development, they assumed the caudate, or club-shaped appearance, with a more or less sharply pointed "tail-end."[27] The entire process of development and maturing of experimental cancer in a mouse took eight to fourteen months.[28]

By August 1937, from a sterile sample taken from the interior of a sarcoma tumor, Reich obtained a kind of bion not seen before. It was much smaller than previous bions, only about 0.25 to 0.6 μm in length, lancet shaped, and very fast moving.[29] When gram stained, these bions were red (gram negative), unlike PA bions that usually stained purple (gram positive).[30] Reich called it the S-bacillus (for its origin from sarcoma). Later, after injection experiments in mice proved these bions were highly pathogenic—even capable of initiating tumors in healthy mice—he changed the named to T-bacillus (for *Tod*, the German word for death). When du Teil volunteered to replicate the experiments producing these new bions, Reich cautioned: "For this, please use only sterile cancer tissue, meaning avoid tissue that lies on the surface and, instead, surgically remove metastasis from cadavers or use those from freshly operated [patients] taken from interior tissue under sterile conditions. A bit of tissue with liquid is seeded onto an agar plate. Two types of structures always grow from this: a bluish-gray growth that in no way differs from carbon bions; second, a greenish-blue shimmering [growth] consisting of short, rapidly moving rods."[31] Reich advised that the two growths should be separated using sterile technique. Then, if du Teil wanted to repeat the mouse injection experiments, he would have the necessary pure cultures. Reich emphasized that for the study of cancer, blood tests would be needed, as well as the examination of live cancer tissue at high magnification, saying, "I tend to shred sterile cancer tissue on a specimen slide and to examine it directly following the operation, within 12 hours at the latest, at 3000× magnification. One can study cancer cells very well by doing this. They are nothing more than jerky moving clusters of vesicles which are only a bit contractile. Studying the border tissue is very important. I have made several photos where the organization of decaying vesicular cancer tissue can be directly observed."[32] Reich sent du Teil a tube of S-bacillus culture and asked him to see how a traditional bacteriologist would identify it. He speculated that a bacteriologist might classify it as *Bacillus pyoceaneus* (now called *Pseudomonas aeruginosa*), though he was not certain.[33] Reich *was* certain its cancer-causing properties were unknown to bacteriology, even if they thought the organism fit one of their taxonomic pigeonholes.

Later Reich found that t-bacilli resulted not only from sarcoma tissue. Cultures of identical microbes resulted from the degeneration of certain other bion types; also, they were produced from putrefaction of a wide range of protein-containing materials. On agar they grew quickly at 37°C and had "a sharp acid and ammonia-like smell." When injected in large doses, they could kill mice in twenty-four hours; in smaller doses they could induce formation of tumors or other infiltrating growths over eight to fifteen months.[34]

Reich's emphasis on the importance of observing cancer tissues (as well as grass infusions) in the living state cannot be overstated. He saw in the two settings, he said, the same process: slow stagnation and energy starvation led to bionous breakdown of the tissues, creating PA bions (and in cancer tissue, also T-bacilli). Then these PA bions clumped together to form amoebae or other protists in the grass infusions, or amoeboid cancer cells in the animal or human tissue. Reich studied cancer tumors in mice, including some tumors he had induced by injection of T-bacilli. He sent many fixed, stained sections of these tumors to professional pathologists to see if their identifications confirmed his own (and confirmed his T-bacilli could cause tumor formation in otherwise healthy mice).[35] The pathologists did confirm Reich's identifications of the tumor types.[36] But in addition to their textbook pathologist's classification, Reich had seen evidence of the tumors' origins in the living state. He pointed out that traditional pathologists' study of cancer tissues missed the process of their bionous formation completely, because it was the motility of the live cancer cells, their bionous structure, and resemblance to amoebae in grass and moss infusions that allowed Reich to recognize this basic feature of the cells. They also missed the bionous disintegration process because traditional cancer researchers rarely used magnifications above 1000×.[37] Despite some relatively recent tools, such as staining with nonlethal fluorescent dyes (that bind to cell surface receptors), that allow cancer researchers to study live cancer cells, the parallel Reich pointed out with grass and moss infusions is still completely unfamiliar to cancer research. Since modern video cameras mounted on microscopes allow much greater ease of replication and documentation of the formation of amoebae from grass and moss infusions, it should not be terribly difficult to establish whether there is any substance to Reich's claim of a parallel with the formation of cancer cells.[38]

By September 1937, Reich realized that his incandescent charcoal or soot bions (type 3e by his designation) injected into mice also produced tumors. He thought this was connected with the fact that exposure to soot had produced cancer in chimney sweeps and exposure to coal-tar caused tumors in experimental animals such as rabbits.[39] In his diary, he speculated that the bions were "pure carbon" and their presence produced an excess of carbon that the cells could not combust fully. Exposure to tar, smoking, and 3e bions all produced this same excess of carbon in the tissues, Reich reasoned, which would consume oxygen when it was respired, producing CO_2. So perhaps, he speculated, "when there is an excess of C or a deficiency of O the tissues suffocate."[40]

By October 14, 1937, Reich began to feel he was clarifying the role of the S-bacillus in the cancer process. He laid out his sense of a stepwise process in his diary, noting, "I am on the track of" how cancer gets started:

1. Local oxygen deficiency due to poor respiration makes tissue decay.
2. S bacteria form.
3. The tissue regenerates itself as a tumor.
4. The S form the metastases.
5. Sarcoma S is more dangerous than carcinoma S.[41]

In other words, Reich felt the S-bacilli were the source that initiated metastases of the original tumor, which itself formed from the protozoal cells produced by clumping of PA bions.

By late October or early November, he had seen red blood cells disintegrating in blood samples from patients with cancer. His theory then progressed to the following point:

> I further thought that the red blood cells were destroyed through some type of internal asphyxiation process. They decompose and so produce S-rods. The main problem continued to be which chemical process was responsible for the greenish-blue shimmer. Should it perhaps be connected to my earlier hunch that we were dealing with the production of hydrocyanic acid as the end product of decomposing matter, and that this had more to do with it than I first believed?
>
> These observations followed the direct and inevitable, immediately following necessary assumptions:
> 1. S-bacillus is really the cause of tumor-formation.
> 2. It is present in the blood before any trace of cancerous growth and tissues.
> 3. It is the by-product of the internal asphyxiation of red blood-cells as they decompose.
> 4. *A cancer tumor is not really a symptom of the illness, rather: The cancer tissue reacts to the attack by* S-rods with vesicular decay, formation of moving, bounded cells, such

as the formation of cancer cells, meaning with reactive proliferation.
5. The S-rod is, thus, the actual disease factor of cancer.
6. Operating on the tumor or radiating it with nuclear substances does not really reach the source of the illness, rather only one of the effects of the illness and the organism's defenses.
7. Metastasis, that is the spread of the illness from one part of the organism to another, is actually done by S-rods.[42]

Reich's developing understanding of the role of the red blood cells in the cancer process is discussed further in the following section.

Help from the Oslo Radium Hospital?

From March through August 1937, Reich had obtained some samples of sterile cancer tissue from the Oslo Ullevål Sykenhus [Ullevål Hospital], from both surgical and postmortem procedures. He asked for living tissue (not frozen or preserved), but sterile samples only, to avoid seeing microbes that were the result of air infection. The hospital sent Reich samples from esophageal, colon, lung, and skin tissue, but "the primary focus of [his] examination was on metastatic liver, testis, and osteal tumors."[43] Reich reported, "I used magnifications between 1600 and 3500 almost exclusively for observation purposes. Previously I had only observed stained cancer cell preparations under a 200–300× magnification. I had never before observed living cancer cells. When I saw the first living cancer cell, it was as though I had been observing them all of my life. They differed only slightly in appearance and in motion from the soil, charcoal, and mixed bions that I had studied over and over again."[44] He cautioned: it was imperative that "tissue be microscopically examined not later than one hour following the operation, that it is kept fresh in the meantime, and is not exposed to possible airborne infection." If cancerous tissue were not examined until hours later or after being frozen, he warned, "then the typical club- and spindle-shaped formations appear only sporadically. The tissue decomposes into unrecognizable clusters without specific characteristics. Examining living tissue at magnifications of at least 2000×, better yet 3000×, is crucial if the problems raised during bion research are to be made understandable."[45]

One of the sterile samples Reich received from Ullevål Hospital was testicular cancer tissue, immediately after it was surgically removed from a patient. In this he saw thick masses of vesicle clusters that moved in a jerky fashion and were very reminiscent of the pseudo-amoebae he had created with gelatin and earth bions. He also observed nucleated cancer cells of different shapes; the nuclei and vesicles within these cells "radiated sharply in blue light." He mixed sterilized packet amoebae from his Preparation 6cI bion mixture with the cancer cells; twenty-four hours later the bions had clustered around individual cancer cells, which were partially immobile and deformed. This immobilization and breaking open of the cancer cells happened to a far lesser extent or not at all in sterile controls to which PA bions had not been added. After long and repeated observations, he concluded that the packet amoebae (PA bions) were penetrating and killing the cancer cells.[46]

Reich was careful to use only sterile cancer tissue samples for these observations, because he knew "that cancer researchers assumed an extreme 'susceptibility for infection' in cancer tissue," that is, that cancer tissue was very often found to have bacteria growing in it after several days, despite having been collected in sterile fashion. By March 1937 Reich concluded, based on his observations, that this interpretation no longer seemed correct: the bacteria were arising endogenously within the cancer tissue, not because of infection by germs from without.[47] "It appeared to me to be more probable that what was previously described as 'infected cancer tissue' were actually formations of vesicularly decomposed malignant tissue. . . . The actual pathogen in cancer is, thus, not so much the generation of formed cancer cells as it is the *bionous self-decomposition of tissue*." However, Reich said he at first found this idea "so absurd and contradictory" to all he had been taught in medicine about cancer that, although he made a mental note of it, he then did not remember the idea for months.[48] His memory was probably restimulated in July or August 1937, when he discovered S-bacilli. These will be discussed shortly.

In September 1937, Reich contacted the Radium Hospital, the primary center for cancer treatment in Oslo, to ask if he could obtain samples of blood from cancer patients to help him in his research. His request was referred to Dr. Leiv Kreyberg, pathologist and researcher at the Radium Hospital (see Figure 5.1).[49] Kreyberg had worked on the genetics of cancer susceptibility in multiple generations of tar-painted mice.[50] On Tuesday, September 21, Kreyberg came to Reich's laboratory (in his home) to see Reich demonstrate some of his work. Reich's assistant, the neurology and endocrinology specialist Dr. Odd Havrevold, was also present.[51] By coincidence, that same day the newspaper *Tidens Tegn* carried an article by journalist Odd Hølaas

Figure 5.1: Leiv Kreyberg, ca. 1970. Courtesy of National Library of Norway.

describing a press release by three Oslo University professors, Otto Lous Mohr, Einar Langfeldt, and Klaus Hansen, all of whom described Reich's claims of "spontaneous generation" as impossible (though, as Reich noted in his diary, none of them had seen his experiments).[52] This reaction was prompted by the publication a few days before of Reich's "Dialectical Materialism in Biological Research," on Saturday September 18. This was the first published description of Reich's bion work, but it was the text of a lecture Reich had given to a student group, not any kind of formal scientific report on the details of the methods or results of the experiments.

Reich demonstrated for Kreyberg the preparation of incandescent carbon bions (from blood charcoal, designated 3e by Reich), at which he said Kreyberg was amazed, though, Reich thought, also objective and serious. Reich showed him the film of Preparation 6 bions, and at Kreyberg's request allowed the pathologist to take with him a charcoal bion culture on egg medium that had been created earlier. According to Reich, Kreyberg promised he would "identify the 3e bions in his department at the university, w[ould] supply us with blood taken from cancer patients," and would help identify the types of tumors Reich's experimental mice obtained. The entire meeting lasted about one hour.[53]

Soon after their initial meeting, Reich sent Kreyberg samples of tumors that had been induced by the S-bacillus and by 3e bions in his own experimental mice. The tumor samples were embedded in paraffin "in a scrupulously correct manner" for sectioning, but Reich lacked a microtome—necessary for extremely thin sections—so he was requesting that Kreyberg diagnose the types of these tumors.[54] Reich at this time also began to study the blood of healthy people, to contrast with his studies of blood from cancer patients. Ten of his colleagues and friends volunteered to donate blood samples, and when he autoclaved small amounts of this blood (he called this Preparation 10), he found that the bouillon-KCl fluid in which it had been autoclaved now contained blue PA bions. Reich studied these bions over time and found that many of them could be cultured.[55] (Two strains of the blood bions on which a lot of further work was done were called by Reich HAPA and BOPA bions.) In addition, he got a major surprise: his S-bacilli could sometimes be found in the blood of healthy patients. He recorded in his diary: "Everything is changing. I find S in the blood of *healthy* young people! Either the entire S business is wrong or—or—Unavoidable conclusion: It is possible everybody has S in them. This would be the 'cancer disposition'—that S can form spontaneously in blood, *but from what??*"[56] Reich later concluded that the presence of S-bacilli (T-bacilli) in the blood was a diagnostic sign of some insult to the tissues that the body tries to fight off. In cases of weakness (e.g., immune suppression or chronic long-term muscular armoring), that ability to defend oneself against the toxic effects of the S-bacilli would be compromised, possibly leading to disease, such as cancer. The differences between blood bions from different people whose state of health he knew well started Reich thinking about the possibility of using this as some kind of diagnostic measure of health. This will be discussed further later in this chapter, as it led to development of a series of blood tests referred to as the Reich Blood Test.

Reich had been keeping du Teil apprised of developments in his cancer work in a very general way but had not gone into great detail about the experimental work, for fear of sidetracking the Frenchman from his work on replication at labs in Nice and in Paris. As mentioned in Chapter 4, after six weeks of hearing how quickly the cancer work was developing, du Teil wrote back, on October 8, 1937, sounding a bit scared of the prospect of the work moving so rapidly into such a huge new research area: "When we enter a domain as vast, complex and deep as that of cancer . . . all questions of bodies [i.e., antigens] and antibodies, of genes and antigens, of toxins and antitoxins—which can act even after sterilization, and sometimes *only* after sterilization—render these questions appallingly complex and require us before announcing a new find or discovery in the order of things, to have obtained the agreement of a bacteriologist since we are lacking in personal and specialized knowledge—a bacteriologist who could speak the same language as the others and discuss the objections with the others on the same level, taking into account all the latest discoveries and the present state of the question."[57] Du Teil felt that just convincing the bacteriologists was a major problem, but at least the techniques of the bion experiments were techniques bacteriologists had familiarity with. And on that front, du Teil felt he was making real progress with Decourt, Bonnet, and Debré, as described in Chapter 3. This was the source of his ardently urging Reich to hire a professional bacteriologist, as discussed in Chapter 4.

Reich tried to reassure du Teil, who had less biology background than he did, saying, "I find it to be a great advantage to be a medical doctor and thus familiar with the basics of biology, even when I am not a specialist in the area and am lacking completion in each [specialized] department. But that is easy to learn as I go along, from the necessity to refute objections. I have learned very much about special techniques these last four years. You do not have to fret."[58]

Kreyberg Becomes an Opponent

Meanwhile, Kreyberg had written to Reich on September 23 in a rather dramatic about-face, saying that he had "carried out a stain of the [3e bion] culture from egg medium in a smear preparation. The result is completely consistent with gram positive staphylococci, which I will not hide from you, look just like the result of a common [air] infection of the medium."[59] Reich nonetheless went ahead with sending the mice tumor slides to Kreyberg for identification. Twelve days later, on October 5, Reich agreed to come with Havrevold to a meeting Kreyberg had convened to discuss his diagnosis of the slides: "I reached an agreement with Dr. Havrevold yesterday. We will arrive on Friday, 8 October, at six in the afternoon at your location. I will bring the protocol of my experiments with me. Perhaps I can also bring fresh, living cancer tissue. However, I would like to mention outright that the work taking place in France as well as my own work here put limits on my resources. I don't believe that this will affect our critical, factual inquiries."[60] Reich still seems to have believed Kreyberg was at least open-minded, perhaps thinking his request for blood samples from cancer patients was to be discussed. Instead, when Reich arrived, Kreyberg declared the tumors in Reich's mice were not cancerous; he "expounded his view in an extremely professorial manner, proclaiming that it was nonsense to speak of a carcinoma when all we were dealing with was inflammatory granulation tissue. He asked me whether I could diagnose a cancer cell under the microscope, to which I replied yes, because I really am capable of that. I then asked him whether he had ever seen a living cancer cell at magnifications of 2000× or 3000×. It turned out that he had not."[61] The confrontation escalated, both men feeling on the spot in front of witnesses to avoid being shown incompetent, rather than remaining in a level-toned exchange about their differing opinions about Reich's slides and about the cancer process.[62]

Du Teil had written Reich on October 8, saying that one reason he so strongly urged Reich to hire a professional bacteriologist was that "in this way, you'll avoid replies like that of the cancer specialist [Kreyberg] you quote to me in your letter before last. Or if such a reply has occurred, you'll have at hand someone who can provide you right away with all the technical arguments to reply to it."[63] Reich wrote a very long letter in reply to du Teil but in the end decided not to send it. Reich gave his version of the interaction with Kreyberg as an example of why one should not give authority to judge their work too readily to representatives of "official science":

> This cancer researcher [Kreyberg] is afraid that if I am allowed to examine his motives for even a moment, his status as a specialist will be questioned; but he has absolutely no reason to think that, because I very kindly and politely offered him the possibility of cooperating with us. Now he is trying to "finish me off." But the physicians who were present and who have to struggle with the practical work of treating patients were on my side. One elderly doctor said to him, after I had spoken: "Yes, why shouldn't there be any cancer cells in granulation tissue?" And that is also correct from the standpoint of current cancer research. It is absurd to strictly distinguish granulating and proliferating granulation tissue from cancerous tumor tissue when, in addition, it is

known that cancer can develop very easily from granulation tissue. But this example shows quite clearly that we are exposed to great danger—namely, the danger of colliding with all the forces of conservatism in science.[64]

Reich was probably unaware that in the fall of 1937 Kreyberg was being proposed by Professor (also at that time the dean) Otto Lous Mohr as a candidate for a university professorship.[65] So the stakes for Kreyberg, of appearing mistaken in public about a question in his core area of expertise, were extremely high at this moment in time.

The issue of granulation tissue is one upon which there was genuinely scientific disagreement. Inflammatory granulation tissue is often found adjacent to cancer tissue, as the doctors at the meeting observed. Reich wrote Kreyberg on October 12, summarizing their differences of opinion. In this letter he pointed out that there were several reasons he argued there is not a sharp distinction or boundary between the granulation tissue and the cancerous cells. First, Reich said he had seen, at 3000×, moving cells near the margins of the tissue in question that were definitely cancer cells (hence his challenge to Kreyberg about whether the latter had ever seen live cancer cells at 2000 to 3000×, not only those in stained slides). Second, Reich said he had observed cells in live preparations that appeared in all forms along a spectrum from clearly granulation tissue to clearly malignant cancer—there was no obvious, sharp morphological boundary.[66] As Reich put it in a document in his files,

> In principle, there is no difference between infiltrating and destructive cancerous tumor growths on the one side and unlimited proliferation of granulation tissue tumors on the other. The difference is an artificial one dictated by the important practical need a surgeon has for a diagnosis. Biologically no such boundary exists. The reason why cancer tissue has no boundaries and infiltrates surrounding regions as opposed to granulation tissue could be related to the following: Destructive tumor proliferation had the prerequisite that neighboring tissue was damaged enabling further destruction through vesicular decomposition possible. There was, so to say, a stimulus there for further permeation and proliferation. If surrounding tissue is undamaged, then it is resistant to infiltration by spreading cancers, and the tumor remains encapsulated.[67]

Reich found some agreement in the cancer research literature with his point of view, though opinion was divided. In the midst of the discussion, Reich wrote du Teil, asking if he could find out what French cancer specialists currently thought on the issue of whether there was a sharp, discrete divide between the two tissue types.[68]

Kreyberg had given Reich in the meeting a slide of tissue from the hibernation gland of a mouse. This tissue looks so similar to cancer that Kreyberg used it in his pathology teaching as a way to test a novice. Reich had not seen such tissue before and mistakenly diagnosed it as cancerous, to Kreyberg's delight.[69] In his October 12 letter, Reich told Kreyberg he had now examined this slide at 3000× and could now state clearly that it was dormant, unlike the samples from his mice in which motile cells were present.

Reminding Kreyberg of the area in microscopy where the Norwegian had come up wanting, Reich added, "You claim that one cannot diagnose cancer cells purely through microscopic observation. I believe firmly that I have learned to distinguish in fresh cancer tissue true cancer cells from other cells that more closely resemble bacteria." This was not merely to score points, however. Reich tried to clarify that it was directly connected to what he thought was an important contribution to research made by his bion experiments: "In my explanation of one of my fundamental research methods, I told you about my observation about coal and earth crystals, in which I saw and filmed motionless systems becoming motile. Because in the cancer disease a motionless area can transform itself into a motile one, which is generally known, it seems to me that this observation I bring from a realm of research outside cancer, is of importance."[70] Reich reminded Kreyberg of other issues he still hoped to discuss with him, given Kreyberg's expertise: metastatic development and Reich's idea that cancer was fundamentally a disturbance of the autonomic functioning of the organism (he later called such a disturbance a "biopathy"). Finally, Reich said he still wanted to hold Kreyberg to his promise to help him obtain samples of blood from cancer patients for his research, saying, "We will stick to our agreement, to wait until I have clarified more in my experiments."[71] In that matter, Reich was to be disappointed. Kreyberg advised the director of the Radium Hospital not to allow Reich access to blood from any cancer patients, and Director S. A. Heyerdahl, MD, wrote to Havrevold, "I have no reason to believe it justified for Dr. Reich to use blood of cancer patients, so I am denying his request for it."[72] Reich found it necessary to ask whether du Teil could help him obtain such samples from a French hospital.[73]

Why did Kreyberg's relationship with Reich sour so suddenly after September 21, 1937? Could his belief that Reich was misled by air infection fully explain his sudden wish to distance himself from Reich? Or only the clash of two large egos, escalating unexpectedly under the contingent conditions of a public meeting? Was Kreyberg unable to take seriously Reich's emphasis on observing tissues in the living state, because of his training and commitment to primacy of fixed, stained specimens? (If this were so, he would have been like the vast majority of pathologists, according to the critique of pathologist and medical historian Leland J. Rather.)[74] All these factors may have been in play; however, there is at least some reason to believe this change of front by Kreyberg may have also been related to Kreyberg's relationship with Mohr, his hopes for a professorship, and the burst of bad publicity over Reich started by Mohr in the *Tidens Tegn* article on September 21. Other issues may be involved as well. This question is further discussed in Chapter 6. In any case, it is clear that once Kreyberg decided to distance himself from Reich, he sought to do so very energetically and very publicly by convening a meeting and confronting Reich there about his abilities as a laboratory researcher on cancer. And, as will become clear, once he took an antagonistic stance, Kreyberg only became more hardened in that stance as time passed. By March 1938, he was exaggerating his role in examining a single culture Reich sent with him for identification, claiming this meant he (Kreyberg) had controlled Reich's bion experiments and found the bions to be nothing but staphylococci from common air-infection.

Reich felt compelled to document in a letter to Nic Hoel and Odd Havrevold his recollection of what had actually occurred during Kreyberg's visit on September 21, 1937: "The most important results of the session that followed are recorded in Dr. Kreyberg's letters from October 12, 1937. Shortly before, I had requested cancerous blood. For this purpose, a meeting with Dr. Kreyberg in the presence of Dr. Havrevold took place in my apartment. Here, too, there was no 'control' completed by Dr. Kreyberg. I showed him the film of Bion Preparation 6 and, in his presence, created a blood-charcoal-bion preparation. On this occasion, he took a culture created earlier with him and diagnosed it as staphylococci. Now many bion cultures can truly be mistaken for staphylococci without that actually being the case. Dr. Kreyberg never asked me about the detailed conditions of the bion experiments."[75]

Nic Hoel's later account of this episode is worth recounting for contrast, though she was recollecting events in 1952, fifteen years later and, more importantly, after she had had something of a falling out with Reich over his

Figure 5.2: Nic Waal. Courtesy of Wilhelm Reich Infant Trust.

more recent orgone theories. She was remarried at that time and using her new married name, Nic Waal (see Figure 5.2). When interviewed by a U.S. Embassy official investigating the scientific validity of Reich's past work in Norway, Waal gave the interviewer the impression that she felt

> that most of Dr. Reich's troubles stemmed from his personality rather than from the biological theories he espoused.... She said that the claim that Reich did not know how to use a microscope was based on a visit he paid to Professor Kreyberg in order to show Kreyberg some slides that he had made illustrating his bion theory. She said that when Reich visited Kreyberg, Kreyberg instead of examining Reich's slides, gave Reich several slides containing cancerous tissue and asked Reich to examine them and describe them for him. She said that Reich instead of admitting that he knew nothing about Kreyberg's slides and keeping the visit on the subject for which he had come, arrogantly and foolishly attempted to describe Professor Kreyberg's slides. Reich was of

course, she said, completely wrong in his descriptions of Kreyberg's slides, *at least in the usual medical sense,* and from this incident, which was largely the result of Reich's own impulsiveness and intellectual arrogance, the story that he could not use a microscope . . . arose.[76]

She acknowledged that some of the negative reaction to Reich's biological work was actually because of hostility or revulsion about his sexual theories, but she thought that less important than the personality issues.

She and Reich had a relationship based on frank, open criticism.[77] The embassy official felt she had a fairly objective view because, while having been a supporter of Reich, she was still willing to be so frank about—as she saw it—his personality shortcomings. It bears repeating, by contrast, that Hoel's recollection could also be substantially colored by her later difficulties with Reich. She seems to be mixed up, for example, in not recalling that Kreyberg called a public meeting to confront Reich. By contrast, in June 1938—still in the middle of the bion experiments—she wrote to Ellen Siersted in Denmark, "I have not followed the very culturing of the bions because I have too little time. But I myself have watched the experiments made by Reich and I have compared these with what appears under the microscope. You know that I have been serving for twelve months as a physician at the hospital (in Oslo) and as such have responsibility for both the laboratory and the microscopy—some swindle or charlatanism cannot be considered."[78] However, the 1952 interview does seem to indicate the possibility that at the time of Kreyberg's confrontation with Reich in the autumn of 1937 she may already have felt disappointed in the way Reich reacted to Kreyberg's aggressive challenges and had begun to see this as arrogance on Reich's part.

Chronic Shrinking, the Concept of a Biopathy

Reich's study of cancer tissue led him to think that not merely armored respiration ("chronic inspiratory attitude") created the conditions necessary for a tumor to form. Reich began to suspect that a chronic contraction of the autonomic nervous system developed, which led to actual shrinking of the tissues and of the autonomic ganglia. The well-known wild, out-of-control mitosis seen in stained cancer cells had to be accounted for; yet any connection with the bionous disintegration of tissue was not obvious. Reich was struggling to understand whether this wild cell division was related in some way to biologist Richard Hertwig's concept of the "nucleus-plasma relation." It had long been known that most cells are a certain particular size when they divide, but there was no clear explanation of why reaching that size should trigger mitosis. According to Hertwig, the ratio in size of the nucleus to the cytoplasm is "normal" just after cell division. However, the cytoplasm at first grows more rapidly than the nucleus. Hertwig postulated this disproportion produces a "tension" in the cell that triggers a burst of more rapid growth by the nucleus and, immediately thereafter, stimulates the cell division.[79]

Reich was more interested in the relationship of the *energy* levels of the nucleus versus the cytoplasm than their relative sizes.[80] He was aware of the discovery that in tumors "the *mitogenetic radiation* of the suffocating cell nuclei intensifies greatly. This was confirmed by Klenitzky, in the case of carcinoma of the uterus."[81] Gurwitsch, Reich pointed out, had shown the presence of elevated levels of mitogenetic radiation in tumor pulp. Elevated mitogenetic radiation levels were thought to be an indicator that something had changed within the cancerous tissues. Mitogenetic radiation triggered increased mitosis in surrounding tissues. Reich interpreted this to mean that *"the affected cell nuclei attempt to compensate for the failure of the total organism"* by radiating more intensely.[82] He pictured cells undergoing a reaction somewhat like the whole organism when it underwent convulsions as death approached. The dying off, suffocating of the tissues would, he thought, lead some individual cells to disintegrate into bions. But those that could resist disintegration would violently convulse via mitogenetic rays, leading to excessive cell division. Thus, Reich understood the increased rate of mitosis in cancer tissue to be a defensive reaction against decades-long stagnation, shrinking, and dying off of the tissues from lack of oxygen and, thus, lack of energy. In the end, this led to a chronic, long-term shrinking of the organism.

In summary, Reich thought the cancer process developed as follows:

1. When Reich inoculated t-bacilli onto a culture medium, they grew but they also caused the protein of the culture medium to disintegrate into PA bions. In a living organism, t-bacilli are produced as a result of tissue damage (rather than by being injected experimentally from without). But next:
2. The t-bacilli bring about the bionous disintegration of surrounding tissues, production of PA bions being a defense against the t-bacilli.
3. If the PA bions are weak (in a weakened organism), more of them must be formed to cope with the t-bacilli. These larger numbers of

bions clump together to form amoeboid cancer cells. Thus, the cancer cells are actually formed as part of a bodily defense against an initial local infection with t-bacilli.
4. Eventually, in a final stage of cachexia of the cancer disease, the cancer cells break down again into t-bacilli. In this stage, t-bacilli are produced in such massive quantities that their toxicity overwhelms the organism, causing general putrefaction and t-intoxication of the blood. This is the final stage of the cancer disease, in which the patient develops bedsores and a terrible odor of putrefaction, and eventually dies.

Over the next several years, Reich worked out a concept of some diseases, including cancer, as "biopathies" or "biopathic diseases." By the term "biopathy," Reich meant chronic rigidity of the musculature *and* the emotions, contraction, and eventually resignation from any hope of pleasure, all of which lead to impairment of the basic function of pulsation of the total organism. That includes impaired respiration, impaired peristalsis, impaired sexual excitation and discharge, and impaired pulsation of the cells themselves. This, in turn, leads to further damming up of biological energy: stasis. This, according to Reich's concept, can lead to a number of biopathic syndromes: hypertension, bronchial asthma, anxiety, cancer, and others. "Why one or another syndrome occurs is not really known, but it is the disturbance of the basic pulsatory function that they all have in common," explained Chester Raphael, MD, one of Reich's students, in clarifying the concept of biopathic diseases. "In some cases there is an intensification of the energetic flow—as in hypertension. In other cases there is a weakening of that flow, leading to gradual deterioration of the organism, a local product of which is tumor formation."[83] Under the conditions of restricted pulsation over many years, a slow process of degeneration sets in, the signs of which may not be detected for a long time but which eventually leads to disintegration of the tissues into bions and T-bacilli. In cancer, shrinking of the total organism over time, often with very rapid weight loss in the final stages, is a key feature. Hence, Reich referred to this process as the "carcinomatous shrinking biopathy."

Since the fundamental disturbance of pulsation is the disease, according to Reich, the medical community's excessive attention to the tumor and failure to recognize it as merely a local product of the disease process is a grave mistake. It is a mistake too, said Reich, to think that a tumor mysteriously appears in an otherwise healthy organism. Concentrating all energies upon elimination of the tumor, without understanding the process of biopathic shrinking from which it arose, means that, as often as not, a new tumor will be produced—not because of metastasis, but out of the same tissue disintegration process that led to the first tumor. A cancer surgeon usually remains focused entirely on the tumor and if he removes it can feel accomplishment. Reich's reaction, as concisely summed up by Raphael, was, "'So what?'—Said not with a feeling of derision, but with a feeling of the enormity of the problem." Raphael stated, "It has to be pointed out repeatedly that cancer is a disease of the total organism."[84] In this regard, Reich's approach to disease, and to the "disposition to disease," fits squarely into the reconsideration of patient "temperament" or "constitution" seen in the biomedical holism of 1920–1950 so well characterized by Lawrence, Weisz, and Harrington.[85]

Even in the case of cancers where some irritant or carcinogenic chemical (or t-bacilli) plays a role in starting the process of cancer formation, Reich felt that the underlying biopathic disturbance of pulsation was what amounts in some patients to the "disposition to cancer." Why doesn't every smoker develop cancer? Reich would argue that an important part of the answer is that those whose total organism pulsation is disturbed or diminished have tissues that are less able to resist the irritant and its pathological effects. Cancer research often assumes that the "immune function" that helps some resist carcinogens is limited to B-cells, T-cells, antibodies, and other immune molecules. For Reich, the strength of the immune function—like the strength of respiratory function, peristaltic function, and orgastic function—is rooted in the degree to which free pulsation of the total organism proceeds relatively uninhibited. "The weakening of the tissues due to oxygen lack, is what renders them vulnerable to such [carcinogenic] stimuli, which would otherwise have no significant adverse effect."[86]

In the cancer biopathy, on the cellular level, the same process takes place as in the total organism. Basically, the chronic contraction of the total organism prevents normal respiration, and disrupts the energy metabolism of the cell. "The cell contracts, just as the total organism does, and begins to shrink. The excess of carbon dioxide that accumulates because of the disturbed metabolism causes a condition similar to the suffocation of the total organism. The nuclei, being the stronger element of the cell, become extremely overactive as the protoplasm suffocates, with the result that an excessive number of mitoses occur." Finally, the nuclei disintegrate into bions, as do all tissues eventually. This affects the neighboring cells as well—not least because some of those bions are T-bacilli. This results in heaps of PA bion vesicles, "out of which protozoa—the cancer cells—develop."[87] This is essentially a process of life reverting—under very depleted energy conditions—to the lowest biological level, where only protozoa and bions are able to function, but not complex metazoans.

Some critics of Reich's cancer theory, most notably the writer Susan Sontag, have claimed that Reich's concept of cancer as a disease originating "from sexual starvation" or "starvation for pleasure in life" amounts to "blaming the victim" for her cancer. Sontag wrote this in her well-known book *Illness as Metaphor*, written while in the midst of her own battle with cancer in the late 1970s.[88] As her son, David Rieff's posthumous biography of Sontag reveals, she was profoundly depressed and found launching this attack, among others, beneficial to her own state of mind.[89] Sontag's reaction is an understandable one, but not necessarily relevant to whether Reich's theory is actually valid. Only fair testing by the scientific community could resolve the latter question. Does describing a process in which a patient's withdrawal from life leads to disease amount to blaming the victim, or could it merely be a difficult kind of biological reality for a very ill patient to accept?

Reich's Thoughts about Research for Profit

By December 18, 1937, Reich wrote du Teil to report that he had begun "the first cancer-therapy tests," that is, PA bion injections "with the terminal female patient about whom I have already written you." It is worth noting here—particularly given the widespread false claim that Reich promised he could cure cancer—that from the outset of such experimentation, Reich had "hopeless case" patients who volunteered to participate in experimental treatments sign a consent form that acknowledged "Dr. Reich charged me no money and promised no cure."[90] (This was true of the bion injection experiments and of later experimental treatment with the orgone energy accumulator.)

In this letter, Reich recounts a recent conversation with his friend Sigurd Hoel, in which Hoel drew his attention to "the enormous danger that looms from the various trusts set up to manufacture and sell radium for the treatment of cancer." It may have been the first time Reich became aware of—as he put it—"the practices and the brutal determination of the representatives of these trusts and of the people who profit from them."[91] Radium production for medical use had already become a factory-scale business in France by 1916 and had grown since worldwide.[92] Reich had lengthy discussions with friends and supporters on this topic and came to realize that if his "method of cancer treatment were to succeed, it would completely pull the rug out from under the cancer radium industry." Because thousands of people around the world worked in this industry, Reich anticipated they would turn against his work: "Not because they are bad people but simply because they would fight quite unconsciously for their existence and against the need to undergo difficult retraining." Another "very smart and foresighted friend" suggested to Reich that the most dangerous attacks would come from those who were making large amounts of money from the radium industry.[93] In another context, Reich made a similar comment: "Could Edison, the inventor of the electric light bulb, expect recognition from the manufacturer of gas lamps?"[94]

Reich had just learned of a journal article critical of the Union Minière du Haut-Katanga for Radium Production, a Belgian trust in the Congo, where mines for radioactive ores were numerous. The article claimed that trusts such as this one monopolistically kept the prices of radium high so that their profits were on the order of 1,000 percent. Reich said he had no way of confirming the details of this claim, but opined, "to the extent that I know 'business,' they are probably correct." He was quite earnest in trying to get du Teil to think seriously about the potential future problems for their bion work this implied. Reich drew an analogy with the huge tuberculosis sanatorium at Davos, Switzerland (where he himself had convalesced from the illness in January–April 1927) and countless similar institutions all over the world. This huge business, Reich argued, "exists only because tuberculosis cannot be cured." Anticipating du Teil's suspicion that Reich was just toeing some "party line," Reich said, "It is not an ideological capitalist or an ex-communist speaking in me but a physician who, under all circumstances, clings to his function and does not wish to be destroyed in the process."[95] He stated again that many radium therapists would be unconscious (or largely so) of their own motivations but would feel the need to oppose the bion and cancer work all the same.

Reich emphasized that others who supported his work had gotten him to realize this danger: "Because I am usually very naïve and these facts had to be pointed out to me, it would be wrong to claim that my behavior is paranoid." This characterization of himself clashes with Reich's insight as a psychiatrist into individuals' behavior. But it might still be true: it would not be the first time a brilliant psychiatrist was unable to transfer everything he knew from the treatment room over into his own personal or family life. More importantly, he urged du Teil that the crucial implication of the problem he was pointing out was that du Teil should exercise the utmost caution in "whatever reports you make to anybody about my ongoing cancer work."[96] One can read Reich's anxiety that du Teil will not be convinced, still hovering in his closing line: "In the deepest conviction that you will not misunderstand this letter and that you will assist me with your advice, I remain . . ."[97] Reich continued over the years in his attitude that any possible health benefits to humanity from his research on cancer and, later, orgone energy should be protected from individuals with profiteering impulses.[98]

Cancer Model Experiments

In 1938 and 1939, Reich began trying various model experiments with different bion types, in an attempt to better understand the cancer process. In particular, he found his Preparation 13 HAPA blood bions useful. When this preparation was autoclaved, then blood charcoal or charred soot was added, along with 1cc of blood serum or sterile egg white, round cell-like structures were created, interpenetrating the blood and soot bions, which Reich dubbed BLUKO I cells (from *blut* = blood + *kohle* = charcoal). Reich reported to the French Academy of Sciences in October 1938 that these BLUKO cells were a model for cancer production, showing the process of excess carbon interacting with blood bions, to produce a new type of cell.[99] In Preparation 14, Reich created another model of his concept of the process of cancer cell formation. In this experiment, BLUKO III cells also developed from Preparation 13 that was grown on egg medium IV. The BLUKO cells resulted from charred soot inoculated into the slit-open egg medium. They originally grew along with rods; however, they could later be isolated. This culture grew solely on egg medium IV, 7.4 pH, at room temperature and decomposed quickly into rods if kept at body temperature. The growths were creamy and grayish-white. The size of the cells varied tremendously, and they appear as cell associations. If deteriorated or decomposed, they revived quickly with addition of fresh KCl solution, likewise when exposed to electrical voltage pulses. In trials with injection of BLUKO bions into mice, Reich expected to see tumor growth. This occurred with BLUKO I, but the results were not yet in for BLUKO III when Reich published his description of these bions.[100]

Most of this work, through about April 1939, was published—albeit in somewhat sketchy form, because of Reich's imminent departure for America—in his July 1939 booklet *Bion Experiments on the Cancer Problem*.[101] Reich wrote an earlier draft of this report, from October 1938 and covering developments until that time (it lacked the discovery of SAPA—sand packet—bions), which he enclosed with his October 18 letter to the French Academy of Sciences.[102]

Even earlier, in February 1938, Reich had serendipitously begun another cancer model experiment, using another bion type: SEKO cells (which, it will be remembered from Chapter 3, were produced from Preparation 12, BOPA bions, and which had also caused tumor development when injected into mice). Reich had produced the SEKO cells by first mixing sterile blood serum with his usual autoclaved bouillon+KCl solution, and then added incandescent charcoal. They had central structures resembling cell nuclei with chromatin threads. Reich thought the SEKO cells were "nothing other than strongly organized colloid materials that have absorbed the finest charcoal particles. The charcoal particles clearly form the structure that is similar to the chromatin-framework structure of cells. Charcoal and organic substances must have mutually *infiltrated* each other."[103] In some subsequent experiments, however, he had been unable to reproduce the SEKO cells.

In a separate experiment at this time, Reich reported that "to vary the soot-incandescence experiments I added ashes, soot, and blood-charcoal, all heated to incandescence, to bouillon+KCl and immediately shook the mixture. I heavily inoculated a fresh egg-culture medium using this preparation after only 3 minutes, so that its surface was covered with a layer of liquid." Dense culture growth regularly appeared twenty-four hours later. Reich "was astounded to find cell formations with blackish margins the form of which strongly reminded me of SEKO structures in the growths so produced, which were merely a variation of soot-bions. They were in the ashes as well as in the blood-charcoal preparation."[104]

This led Reich to think he now understood why his attempts to reproduce the Preparation 12 SEKO cells had sometimes failed. In those instances, he "had mixed bouillon+KCl with incandescent blood-charcoal and then later added the serum. In the Seko experiment, however, [he] had first mixed the serum with the bouillon+KCl and then added the incandescent charcoal." Reich concluded from this, "It must apparently be the case that the incandescent charcoal immediately penetrated the egg-white. I could now see this infiltration in the egg-white foundation." The albumen had been reduced to "various vesicular shapes that are firmly attached to one another."[105] Reich saw the excess carbon actively penetrating protein and causing it to undergo vesicular disintegration as a model for cancer formation where coal-tar, soot, or similar substances were the instigating carcinogens. The bions in the egg white organized into protist cells, Reich told du Teil, just as the bions in disintegrating tissue in the human or animal body clumped together to form amoeboid cancer cells.[106] He referred to this new "cancer model experiment" as experiment 14.

Reich now thought he could outline the steps of development of a soot- or coal-tar-induced tumor as follows:

> 1. Very fine soot- or other carbon substance particles, infiltrate the blood stream.
> 2. There they change, probably into carbon-bions, under the influence of blood components, especially serum.
> 3. The so-formed carbon-bions implant themselves in tissues especially suitable to serve as breeding grounds, such as in our model experiment.
> 4. There now follows a mutual infiltration of tissue substance and carbon-bions.
> 5. The carbon-bions make organic substances mobile, living and organize them to protozoan structures, to cancer cells.[107]

Reich felt he now had to revise his earlier theory that the incandescent soot bions simply withdrew carbon from the egg medium upon which they were growing. Rather, the "incandescent soot bions more or less bring the egg-medium to life and to proliferation. There is no one-sided involvement, rather a mutual infiltration from bions and culture-medium."[108]

Reich repeated this experiment, first lacerating the edge of a sterilized egg medium with a sterilized spatula, then drenching it in either Preparation 12 or Preparation 13 bions. In each case, after twenty-four hours of incubation the medium was covered with a proliferating bion culture, with numerous charcoal particles penetrating into the slits in the egg medium. After several days, a thick growth formed that included many rod-shaped bacteria, as well as "tumors on the damaged egg-substances and egg fragments." By this, Reich meant structures that "contain fine, organized, developed, more or less symmetrically shaped and spasmodically moving, nucleated cell structures displaying the characteristics of cancer cells." But there were, in addition, "rounded cells with eccentric, vertical vesicular shapes, and a brightly shining nucleus."[109]

More on T-bacilli

In an experiment in February 1939, Reich found that T-bacilli were not killed or inhibited by potassium cyanide (KCN) like all other bions (and like living cells); more than that, they were actually strengthened by KCN.[110] Recall that Reich had been thinking for some time about Pflüger's hypothesis that there was an important connection between the cyanogen (CN) radicle and the origin of life. Putrefying protein matter, from whatever source he sampled, always yielded T-bacilli. The cultures were always acidic.[111] Reich had found that when KCN was combined with egg white, it produced the same effect as T-bacilli. He reasoned, "If T-bacilli are always produced during the decomposition of organic matter, then the theory that the last products of decomposition, amides and nitrogen and carbon, are present in organisms in the form of T-bacilli cannot be refuted." Thus, Reich further explained, "the potassium cyanide T-trials begin with the assumption that potassium cyanide immediately destroys organic life" and "if potassium cyanide and T-bacilli are functionally the same, then they should not contradict one another, rather one must strengthen the other."[112] Reich tested t-bacilli bouillon cultures from three different sources, each of which had produced malignant tumors when injected into mice. Each was mixed with a small amount of KCN crystals. The characteristic blue-green color of the cultures intensified, and at 3000× magnification in darkfield, Reich said he observed "fine, flitting rods, which looked exactly like T-bacilli" actually separating from the disintegrating KCN crystals. When t-bacilli were added to the KCN directly, they were not killed. Indeed, after twenty-four hours and again after several days, the solution still produced successful, live t-bacilli cultures when inoculated onto agar.[113] Reich later noted that the relationship between T-bacilli and cyanide, while still obscure, was definitely important. As a poison, cyanide acts by shutting down respiration, as Warburg discovered. And Reich noted that numerous mice injected with t-bacilli died "with typical manifestations of suffocation such as hypervenous blood and respiratory paralysis." He saw what he considered a clear connection also with the suffocation of the cancer cell's metabolism, adding, "Here is a vast field of exploration for the biochemist."[114]

Reich laid out a plan for further experiments to be conducted, to test different variables—using calcium cyanide, for example, rather than potassium cyanide, to see if the contraction or dehydration effect of the calcium would alter the outcome in any way. Another possibility was to introduce CO_2 into fresh animal tissue to see whether it would facilitate development of T-bacilli.

Through the late winter and spring of 1939, Reich said he was struck by the ability of newly discovered SAPA bions to paralyze and kill T-bacilli even more forcefully than previous types of PA bions.[115] In previous experiments, he reported that PA bions, including BOPA bions, could "penetrate into cancer

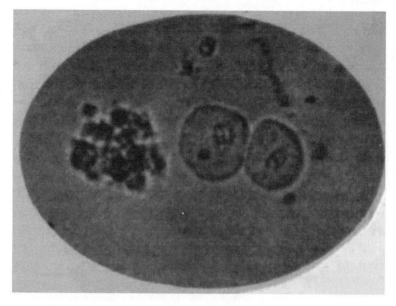

Figure 5.3: SAPA bions with cancer cells. The SAPA bion on the left is 2–6 μm across. Courtesy of Wilhelm Reich Infant Trust.

tissue, destroy cancer cells in a microscopically observable way," and break up masses of tuberculosis bacilli. Injections of packet-amoeba (PA) bions had been tried, in combination with injections of S-bacilli, to see whether the PA bions could kill S-bacilli in an infected animal.[116] Reich found the S-infected mice did not die if they received PA bions as well.[117] The SAPA bions paralyzed cancer cells even more rapidly when placed next to them on a microscope slide (see Figure 5.3).[118] This led Reich in the autumn of 1939 to begin a series of experiments in which he injected SAPA bions into mice with tumors, to see whether there was any beneficial effect.[119]

Reich said he began to think his T-bacilli might be related to the still-unidentified agent of the Rous chicken sarcoma. Ernst Fränkel had found this unknown agent was bound up with the red blood cells and the globulin protein. Reich noted that the T-bacillus could indeed originate from degenerating erythrocytes. The "something" other researchers were looking for in Rous sarcoma was to be found "in the blood prior to the existence of cancer cells and also develops through the disintegration of cancer cells," he noted. The famous cancer researcher Ferdinand Blumenthal had noted that "in all cases there is proof that something emanates from the cancer cells that transforms previously normal cells into cancer cells." Reich was struck that his T-bacilli had just these characteristics.[120] Still, with the presence of T-bacilli much as with the presence of any other irritating factor or chemical carcinogen, Reich believed that the general state of the tissues (including the red blood cells)—especially their energy level—was the key factor in whether the organism could resist the noxious influence or whether it instead succumbed, with tissues degenerating in response to the carcinogenic influence.

The data on the Rous sarcoma agent helped direct Reich's attention to a possibly important role for the red blood cells in the cancer process (or in resisting it). In some of his experiments with SAPA bion injections in cancer mice this idea was confirmed. (See Chapter 7 for how SAPA bions were created and the properties of the radiation they emit.) In the SAPA bion experiments, blood or blood serum that had been charged by contact with SAPA bions and their radiation was injected, rather than the bions themselves. Reich concluded from these experiments that the SAPA bions helped lengthen the survival of cancer mice by a marked amount, more than any previous PA bion type. However, the effect of the SAPA radiation on tumors did not appear to be via direct contact of the bion with the tumor, but rather by way of the radiation charging the red blood cells and those conveying that charge to weakened tissues throughout the body. Of all the variations of this experiment, injection of red blood cells that had been previously charged by contact with SAPA bions had the greatest therapeutic benefit for cancer mice.[121] Importantly, when the tumors underwent bionous disintegration after contact with SAPA bions, the cause of death for most of the experimental animals, determined by Reich upon autopsy, was that their kidneys, spleen, and liver had been overwhelmed while trying to clear out all the detritus from the dissolved tumors.[122]

Reich eventually directed much more attention to the study of the red blood cells, hoping their condition could be used as an indicator of health or incipient disease, even in people who showed no overt symptoms of illness. Since his concept of biopathic disease indicates that a slow decrease in energy over many years or decades precedes tissue disintegration and overt local symptoms such as tumor formation, Reich believed a measure of that decrease in energy to the tissues could help with early diagnosis.

As a result of this research, he concluded that a battery of different tests was helpful in assessing the strength of the red blood cells and, by proxy, of the ability of the entire organism to resist biopathic influences. This group of tests came collectively to be known as the Reich Blood Test. It includes an assessment of the rate at which red blood cells disintegrate into bions in warm physiological saline solution once removed from the body and observed under the microscope. The shorter the disintegration time, generally speaking, the greater the concern for possible future health problems. If T-bacilli appeared as "spikes" projecting out of the disintegrating cells during their disintegration, Reich considered this a still more serious indicator of incipient biopathic disease. The behavior of the erythrocytes when subjected to autoclavation was also part of the Reich Blood Test. Another part was culturing the blood in nutrient media to see whether t-bacilli or decay bacteria grew. A detailed description of the test procedures was published in 1952.[123]

AIDS and Endogenous Infections

What relevance might Reich's concepts have for more recent understanding of disease? In a 1982 lecture, Reich's student Chester Raphael, MD, suggested that *Pneumocystis carinii* pneumonia and other opportunistic infections experiencing sudden increases in frequency might represent further examples of endogenous infection. *Pneumocystis* pneumonia, cytomegalovirus, cryptosporidium, and other "parasitic infections" are associated with immune-compromised and often terminally ill patients (with cancer, AIDS, and other diseases). Prior to the recognition of AIDS in 1981 and the identification of HIV in 1983, *Pneumocystis* pneumonia was—like Kaposi's sarcoma—known in the medical literature but quite rare. There were sudden outbreaks in Europe during the chaos and economic deprivation that followed World War II; it occurred predominantly in undernourished, marasmic infants. The outbreaks gradually subsided with the economic recovery that took place after the war. Protein-calorie malnutrition is strongly correlated with the illness. To mainstream medicine—which considers the disease to be caused by infection by the *Pneumocystis* organism, transmitted from without—this suggests predisposing factors like malnutrition act as immune-suppressing conditions. In 1950 *Pneumocystis* pneumonia had never been seen in North America. Its appearance there was first connected with the use of immunosuppressant therapy (such as organ transplant recipients, cancer patients, and children with innate immune deficiency disorders).[124] In AIDS or cancer with immunosuppressant therapy, *Pneumocystis* pneumonia thus stems from the immune-compromised condition rather than from the primary disease itself.

Raphael pointed out that the facts about *Pneumocystis*, both clinical and epidemiological, also can be explained via Reich's theory of endogenous infection. The original immune suppressing conditions would damage the basic function of total organism pulsation, and that most fundamental damage would set in motion compromised immunity *and* bionous tissue disintegration. The bionous disintegration would lead to the formation of the fungal cells of *Pneumocystis*, rather than its being a "parasite" acquired by infection from without. *Pneumocystis carinii* has never been found in nature outside of infected organisms (organisms for study come only from sick animals); its exclusive known habitat is in the lungs of man and lower animals, globally affecting most mammalian species. In addition, according to Raphael, close to 100 percent of initially healthy experimental animals will manifest the disease after two to four months of corticosteroid administration.[125] Both Reich's theory and that of mainstream medicine have explanations for these facts. Occam's razor might suggest to at least some observers that Reich's theory explains this fact more simply than assuming either that "it must exist somewhere in nature and we just have not found it yet" or the organism gradually evolved until it could survive only in animal lung tissue; it can still be infectious and spread rapidly under the right conditions of malnutrition, despite never producing symptoms or being detectable in healthy patients. (*P. carinii* "cysts" were, however, reported in asymptomatic guinea pigs infected by *Trypanosoma cruzii*.) Pentamidine drugs can kill off the already formed *Pneumocystis* fungi; however, if Reich's theory were correct, more of them would continue to be produced from the underlying condition and the bionous disintegration of tissues.

In recent years, an argument with some similarities has been made regarding a possible connection between *Treponema pallidum*—the syphilis organism—and AIDS. The AIDS-syphilis connection has been argued by Alan Cantwell, Harris Coulter, and even as prestigious a biologist as the late Lynn Margulis.[126] The extent to which diseases related to immune suppression involve microbes that can be infectious once they appear—by whatever means they are first generated—suggests possible new areas of relevance for Reich's theory of endogenous infection via protozoans and t-bacilli. Depending as it does on the possibility of microbes coming into being "without parents," like so much else about the bion experiments it faces a steep uphill battle against existing cancer biology.

6

OPPOSITION TO THE BION EXPERIMENTS

First Attacks in the Press, September–December 1937

As I described in Chapter 5, after the publication of Reich's article "Dialectical Materialism in Biological Research" on Saturday, September 18, 1937, a critical newspaper article by Odd Hølaas appeared in *Tidens Tegn* on Tuesday, September 21, in which Oslo University professors Otto Lous Mohr, Einar Langfeldt, and Klaus Hansen were quoted as declaring the bion experiments nonsense. In a follow-up article the next day, Mohr was quoted saying, "Let anyone believe who wishes to do so. I, for my part, should not have to state that I consider this total rubbish. In my opinion, it would be sad for something like this to be taken seriously in this day and age" (see Figure 6.1).[1]

Tidens Tegn printed a letter from Reich asking that the papers contact him first before reporting on his scientific research, though at least partly the problem was caused by the fact that the first published report on the bions was a somewhat informal talk given to a students' group, which lacked most of the crucial scientific details Reich was including in the book in preparation on the experiments. Reich's letter stated,

> In your paper's issue 219 from Tuesday, the 21st of this month, a lengthy article appeared that informed readers about the basic features of my scientific endeavors.
>
> With my letter today, I would like to justify my request that, in the future, you contact me concerning my work prior to publication. I wish to avoid the impression, which newspaper articles can easily do, that I am launching articles in the press for propaganda purposes.

Figure 6.1: Otto Lous Mohr, ca. 1940. Courtesy of National Library of Norway.

I naturally cannot and will not prevent *published* results of my scientific efforts from being widely distributed to the general public; however, in the interest of not disrupting work and to avoid misperceptions, I would like to emphasize just a few points:

1. It is implacable and damaging to the issue when one causes a public uproar, before a complete review of the scientific material and without simultaneously providing the foundation needed to form an assessment. I haven't even published the bion experiments. The content of your article is from a publication that appeared on Saturday, the 18th of this month, in the *Journal for Political Psychology and Sex Economy*, but there is nothing there stating that Dr. Havrevold or I consider the answer to the cancer problem confirmed. I beg of you to please consider how

sensational and exciting such a report must feel to a cancer sufferer.

2. It is in the interest of the subject matter that professional scientists have the time and the quiet necessary to grapple with the new data. The quiet atmosphere of objective work is destroyed through premature publication. The experiments here will be replicated at the University of Nice and soon at the Biological Institute of the French Academy of Science for control purposes; however, this task requires time.

I would appreciate it very much if you would print this letter unedited in your newspaper.[2]

No sooner had Reich's letter gone into the mail on Wednesday, September 22, however, when in *Tidens Tegn* that same day (i.e., a day before Reich's letter was published) appeared an editorial cartoon titled "Vistemann på Kampen" [Wise man from Kampo; Kampo was a neighborhood in Oslo] (see Figure 6.2). The cartoon portrayed Reich as a cult leader, worshipped by devoted followers as he demonstrated complicated scientific experiments. The caption read, "High priest Wilhelm Reich reveals the mysteries of life to his followers" In the cartoon, Arnulf Øverland and Sigurd Hoel are in the front row of worshippers, Rolf Stenersen right behind Hoel, Odd Havrevold next to him, and another, probably Odd Wenesland, is behind Stenersen. The sarcastic implication was that the experiments were so much mumbo jumbo and handwaving, done behind drawn windowshades, to keep the followers spellbound. Reich opined in a letter to du Teil that the highly specific details in the original article suggested to him that somebody in his own lab (who also saw Reich for therapy) had contributed information to those trying to sabotage his work:

> A highly personal issue between one of my patients and the author of this article is hidden behind this journalistic trick. For neurotic reasons and with a great fondness for work my patient developed the idea that it was dangerous to be together with me because I am an intense individual and captivate everyone's attention. As I just found out today, he has been bragging to his friends—amongst them the mentioned journalist—that even though Sigurd Hoel, Øverland, and Havrevold were influenced by me, he was not. And what happened? He stepped right into

Figure 6.2: September 1937 *Tidens Tegn* cartoon (Arnulf Øverland, Sigurd Hoel in front row of worshippers, Rolf Stenersen right behind Hoel, Odd Havrevold next to him, another, probably Odd Wenesland, behind Stenersen). Courtesy of Wilhelm Reich Infant Trust.

> the trap [the journalist set for him], because on the next day in the same newspaper there was a caricature showing me cooking at the stove and the four above-mentioned people, including him, were worshipping me. That's how these things come home to roost.[3]

Indeed, this initial episode was the end of press coverage of Reich's bions until March 1938.

A few more critical letters from psychiatrists and psychoanalysts, including Johann Scharffenberg, Ragnar Vogt, and a former supporter of Reich, Ingjald Nissen, appeared in various papers through the autumn of 1937, mostly critical of how far Reich's therapy had departed from traditional psychoanalysis; this exchange gradually died down by the beginning of 1938.

Rockefeller Foundation Support

In a case in which the experimental outcome is underdetermined in this way, it is appropriate to explore other possible motivations for objections to Reich's work, such as the perennial theological implications of experiments that suggest a naturalistic origin of life. In this case, much more local circumstances are also suggestive. Until 1937 Reich's financial support for his laboratory came from his own earnings, contributions from students, and three wealthy private donors, Rolf Stenersen (an Oslo writer and stockbroker), Lars Christensen (Oslo ship-owner and whaling magnate), and Constance Tracey (of London).[4] A total of about 30,000 Norwegian kroner (N.K.) had been invested up to the end of February 1937, with monthly operating costs of 1,300 to 1,750 N.K. (300–400 U.S. dollars, at 1937 value).[5] As he began animal injection experiments with bions in early 1937, pursuing a hypothesis that this could be helpful with cancer, Reich estimated that the expenses of running his lab would increase substantially if he were to effectively pursue those injection experiments, for example, keeping a much larger number of experimental animals and constructing an adequate facility to house them.[6] By October 19, 1936, Reich was contemplating applying to the Rockefeller Foundation (RF), which during the Depression years was the only granting agency supporting life sciences research on a large scale in Europe and the United States.[7]

In January or February of 1937, the famed anthropologist Bronislaw Malinowski, a colleague of Reich's who greatly respected his work and who was now living in London, contacted the RF office in Paris to urge foundation officials to consider funding Reich's bion research. In late February, foundation officer Tracy B. Kittredge interviewed Reich on a visit to Oslo and encouraged him to apply for a grant. On the same trip, Kittredge visited other RF grantees in Oslo and asked them their opinion of Reich's bion work, including geneticist Otto Lous Mohr and physiologist Einar Langfeldt at Oslo University. Though Mohr had not visited Reich's lab or seen any of the experiments in progress, it is likely he had heard of Reich's experiment's (and Thjøtta's negative opinion of them) through Thjøtta or through Mohr's friend, Medical Faculty Dean Kristian Schreiner, who had expressed skepticism about the bions to Havrevold in February or March 1937. (Schreiner had also in the past been a supporter of Leiv Kreyberg.)[8] When the University Medical Faculty was asked by the Norwegian government in spring 1938 to pass on whether the importance of Reich's research justified an extension of his visitor's visa, by that time the faculty had heard enough through the rumor mill to convince them that "there was no reason to extend his stay."[9] Kittredge's inquiry to Mohr and Langfeldt produced the reply that in their opinion, as a laboratory scientist, Reich was "a charlatan."[10] As a result, the foundation turned down Reich's grant application fairly quickly after he submitted it in early March.[11]

Mohr's opposition to the bion experiments (or to Reich) is worth considering in greater depth. He was a prominent member of the international *Drosophila* genetics network.[12] His wife, Tove, and his mother-in-law were Norway's leading reformers for sex and birth control education, and Mohr himself was a left-leaning critic of the excesses of overly simplistic eugenics ideas.[13] Soviet geneticists wanted him to be one of the "progressive" critics of fascist race biology when they were planning the 1937 Moscow International Genetics Congress (which in the event was canceled because of the rising power of Lysenkoism among Soviet geneticists and the disarray into which much of Soviet science was thrown by this and by the Stalinist purges and show trials of 1936–1937). For whatever reasons Mohr had concluded Reich to be a charlatan, then, he would probably be quite exercised in attempting to discredit Reich. Otherwise Reich might provoke some public debacle and do harm to Mohr's dearest causes (sex education, scientific opposition to eugenics, etc.) because of his own famous association with those same causes.

Mohr had had a recent run-in—in which there were many parallels—with famed sex reform advocate Max Hodann, whose grand schemes Mohr thought were out of contact with the kind of intimate local politics required to actually accomplish concrete legislative changes.[14] That Hodann was also a supporter of Reich would not have escaped Mohr's notice in a small town like Oslo.[15]

Other Norwegians interested in liberal sex-reform measures might feel their project threatened by Reich's more radical—and, as they saw it, scandalous—initiatives. As an indication of this, one need only consult the official ruling of the medical division of the Norwegian government's Social Department in 1938, declining support for the extension of Reich's residency permit in Norway. This document shows the point of view of a liberal official of the Norwegian Labor government; it states that Reich's development of character analysis and his published biological research on sexuality are considered important and legitimate scientific contributions. Furthermore, Reich's work with educators of young children was justified since his theory proved that psychological damage at a young age could cause sexual pathology and attendant neurotic problems later in life. However, the writer (identified only by the initials "H. B.") takes strong exception to Reich's having spoken publicly before youth groups and encouraging them to develop an independent sex life and to use contraception. It almost sounds as though the writer could not believe Reich actually meant this and must have been misinterpreted because he spoke too carelessly: "Apart from the fact that the foundation of this teaching is sound, it is understandable that propagating this amongst laymen must occur with the utmost restraint—especially when the lectures are for youths. Judging by the reports in his own journal and what he publically expounded, it does not appear as though he had always shown sufficient caution. For example, a few of his lectures were repeated in such a way that they had to sound as though he was appealing to youths under the age of 16 to lead an active sex life while using preventive measures. There is reason to believe that the sex education mentioned here could be detrimental—not the least because it can damage already on-going, healthy sex education work in this important field."[16]

A protégé of Mohr's, Leiv Kreyberg, who studied the genetic predisposition to cancer in mice, joined in the attack on Reich, as described in Chapter 5, and in claiming that Reich's sterile technique must be faulty.[17] This probably gave Kreyberg helpful public visibility in the fall of 1937 when Mohr was trying to ensure that Kreyberg would be appointed to succeed the retiring professor of pathological anatomy at Oslo University. (By December 5, Mohr was proudly celebrating having helped Kreyberg secure the post.)[18] By early 1938 Kreyberg was lobbying the Paris professors who had been so impressed the previous summer by du Teil's demonstration of bions, even traveling to Paris for the purpose, trying to convince them that Reich was a charlatan and that they should distance themselves from any public support for his experiments.[19] As mentioned earlier, during this time period, RF officers visited Scandinavian labs on several occasions, gathering gossip on which researchers looked promising, which were thought to be mediocre, progress on projects that had been funded by the RF or were being considered for funding, and other such matters.

This networking process was the essence of how decisions were made about who would receive funding from the RF Natural Sciences Division. Another interesting example in this context is of a visit by RF officer W. E. Tisdale to the lab of Albert Fischer, a longtime RF grantee and thus someone whose opinions the RF relied on, in Copenhagen on November 16, 1936, just three weeks before Reich visited Fischer to demonstrate some of the bion work.

Fischer gave positive assessments of some research, but of Professor H. Schade of Kiel, Tisdale's diary records, "Fischer knows very little of his work, but what he does know is enough to justify a very low opinion of it. The biological work of S. which parallels Fischer's is all bad."[20] This attitude of Fischer's is remarkably similar to what Reich lamented about Mohr, that is, willingness to pass judgment, even very harsh judgment, about someone whose work he "knows very little of." The fact that the RF network put great weight on the recommendations by former grantees like Fischer allowed those grantees to use such pronouncements in this extraordinarily self-serving way. Note especially Fischer's comment on the work that parallels his own, where one would assume the greatest need for caution in regarding such a judgment, considering that it is a direct competitor that he is trying to discredit.

Another officer's diary passage relevant to the foundation's opinion of others who later participated in the Norwegian press campaign against Reich records a visit to Oslo by foundation officer H. M. Miller on January 12–13, 1937. He reports on numerous researchers thought to be worth watching as up-and-coming prospects, including Dr. Leiv Kreyberg, at that time pathologist at the Radium Hospital but not yet professor of pathology at Oslo University. The diary states that Kreyberg was about thirty-five years old, working on the genetics of susceptibility to cancer, and was "a very attractive chap who has recently come back from a flying visit to Madison at the invitation of the American group, to lecture on his work at the conferences held there . . . said he has some 30 generations of brother and sister matings of tar-painted mice. . . . K. has some hopes of finding a stipend" for a German exile researcher, then at Fischer's laboratory in Copenhagen. Another researcher Miller thought worth watching was bacteriology professor Theodor Thjøtta. Miller's report had the following to say about Thjøtta and his interest in obtaining Rockefeller funding: "Prof. Thjøtta—Rikshospitalet, Institute of Bacteriology. A very pleasant person, probably early 50s, who previously had a year of experience with [Oswald] Avery at the Rockefeller Institute [in New York City]. He succeeded in renovating and equipping his relatively small but very efficiently arranged institute with some 50,000 kroner, 35,000 of which came from private sources. The largest donor was a young shipping magnate whom T. had cured of asthma. . . . He would like to have some annual support for the research of his institute. Most of this along more or less classical bacteriological lines, including work on the flora of the respiratory tract, and culture of other bacteria. HMM [i.e., Miller] was discouraging in light of the present NS [Natural Sciences] program."[21]

The point of this last comment was that the Natural Sciences program had a very strong emphasis on interdisciplinary "borderlands" research projects, and on those that recruited or developed physical and chemical methods for solving problems in biology (ultracentrifuge, electrophoresis, electron microscopy, etc.), especially to the study of fundamental underlying "vital processes," in the vision of Warren Weaver, the director of Natural Sciences.[22] Because Thjøtta's work did not fit this model, Miller told him his chances for RF funding were poor.

On a visit to Copenhagen on the same trip, Miller "met Dr. Brekke from the Psychiatric Institute of Prof. Jessing [sic, i.e., Rolf Gjessing] in Oslo, a not-too-impressive young Norwegian on casual meeting. He is working on the effects of estrin and testosterone on metabolism of the uterus."[23] This serves as an example of what kind of research on sex the psychiatric field engaged in at the time. The subject is one that the RF could certainly find interesting, though the importance of the officer's impressions of the person seems to be great enough to override the inherent interest of the research topic.

Mohr, too, sought RF funding during the period of Reich's bion experiments. He had been a grantee for many years and hoped this would continue. He also pushed for money for Einar Langfeldt and his assistant Ragnar Nicolaysen to establish an Institute of Nutrition Research at Oslo University and to get Nicolaysen appointed as the new professor of biochemistry and nutrition. He discussed his requests in depth with RF representative W. E. Tisdale during the latter's visit to Oslo of May 18–19, 1938. Tisdale reported,

> Mohr talks of his own work. He has been made director of the Genetics Institute, to succeed Prof. K[ristine] Bonnevie. This institute is not an institute or department of the university, but is rather under the Collegium of the University, a committee of Deans and the Rector. M. is now 52, retires at 70. His research interests are on *Drosophila* and mice as experimental animals. Problems are concerned with chromosome deficiency, lethals, and problems of gene dominance, using chiefly salivary gland chromosomes of *Drosophila*. He thinks it politic and wise, as well as interesting, to spread his interests as widely as possible, and he therefore has associated himself with Prof. Fölling in the study of pyruvic acid as an indicator in certain psychiatric problems, with Dr. Saeboe, an ophthalmologist, on certain problems of the

cornea; with Prof. Tuff of the Veterinary H.[igh] S.[chool] on some curious developments in the silver fox industry. . . . In the Anatomy building Mohr has three large rooms for the Institute of Genetics. Two are on the first floor, and one, for animals, is in the basement. His budget is N.[orwegian] K.[roner] 2,600 per year, plus other local gifts, sufficient to run the Institute, but not sufficient for the provision of cages, and other accessories. He inquired as to the possibility of R.F. aid for equipment and accessories which he guesses as N.K. 10,000 to 15,000. He was told we would consider a proposal from him and the Rector; if this arrives, I am prepared to present the request.[24] [At this time, 1 U.S. dollar = 4.4 N.K.]

Thus, every laboratory scientist who was an opponent of Reich in public debates about the bion experiments—Mohr, Kreyberg, Thjøtta, Langfeldt, and Nicolaysen—was a direct competitor with Reich for the limited pot of RF grant money in the "small town" of Oslo. The scientists all knew each other and supported one another's work (and the Norwegians knew Fischer); not so Reich, whom they all saw as an outsider. None of them but Kreyberg ever visited Reich's lab (and he did so only once) or saw his experiments carried out; thus, it seems at least pertinent to wonder whether or to what extent their firm conviction that Reich's experimental work was "charlatanism" could be related to their concern over a competitor for funding. The decision of the Oslo University faculty not to recommend a renewal of Reich's residency permit was made, however, amid even more provocative circumstances. In March 1938, Reich's enemies began a renewed campaign in the newspapers to relentlessly attack his work.

Johann Scharffenberg

Another prominent figure who attacked Reich in the Norwegian press was Johann Scharffenberg, a psychiatrist famed for his frequent participation in public polemics, including his support for eugenics and the 1934 Norwegian Sterilization Law (see Figure 6.3).[25] Historian Nils Roll-Hansen has described Scharffenberg as follows: "A staunch individualist with a critical scientific spirit. His unorthodox views and untiring polemics had considerable impact on Norwegian public debate throughout his long life [1869–1965]." He adds, "Most famous, perhaps, are his candid attacks on Hitler and Nazism both before and during the German occupation of Norway, 1940–45. A series of

Figure 6.3: Johann Scharffenberg. Courtesy of National Library of Norway.

newspaper articles in 1933, warning that Hitler was 'a psychopath of the prophetic type—on the verge of insanity' resulted in considerable German diplomatic pressure on the Norwegian government to stop him."[26] Scharffenberg's motivations for participation will be discussed below. Here it is worth pointing out that since January 1932 Scharffenberg had been arguing forcefully for a sterilization law in Norway, which he hoped would lead to "a full-blown eugenic social program," including "a stronger element of coercion" for sterilization than in most public discussion in Norway at that time. As Roll-Hansen has shown, Scharffenberg "expected more radical measures to be demanded in the future, 'when the public has understood that human breeding must be made as efficient (rasjonell) as plant and animal breeding.' But one had to proceed gradually. According to Scharffenberg, both his own proposal and that of the commission for revision of the penal code were more radical than the few laws that had so far been passed in Europe, i.e., in Denmark and a few Swiss cantons."[27] Scharffenberg, interestingly, opposed eugenics based on race, which he repeatedly called unscientific. His argument was for *social* eugenics, that is, sterilization of what he called "unsocial" individuals, "people not fit for social life," like schizophrenics and people living in extreme poverty. He held that those who could not afford to raise children were not fit to have them; their children would eventually become the responsibility of the state.[28]

As in other countries, mild forms of eugenics had wide support, all across the political spectrum.[29] In Norway, eugenic proposals were somewhat more moderate than in Germany but were supported by members of the Norwegian Labor Party, the Liberals (Venstre), and the Conservatives (Höyre). "Most of them were what we can call Social Democrats in a broad sense ... [and] they were committed to the construction of a social welfare state."[30] It may seem paradoxical from our retrospective vantage point that one who thought Hitler a psychopath nonetheless supported a fairly strong form of sterilization—from a liberal, even "socialist" perspective. That is why it is worth reiterating that one of the most important things historians have shown about the eugenics movement was that it had broad appeal to and support from all across the political spectrum, in the service of a wide range of different agendas. In the United States, for example, eugenics seemed a good idea to capitalist robber barons, the Ku Klux Klan, organized labor, Protestant ministers and temperance reformers, Harvard genetics professors, Progressive Era social reformers, and many more. Scientists were not—at least not all—passive bystanders, wringing their hands over "social misuse of oversimplified scientific ideas." Many of them, such as Charles Davenport in the United States, were leaders of the movement.[31]

Scharffenberg's ideas were a bit more extreme than the Norwegian mainstream. He felt there was an important role for "genetic factors as causes of mental illness and retardation as well as various types of social misbehavior from vagrancy to theft and other kinds of crime. He had worked in mental hospitals and was for many years physician at the state penitentiary in Oslo." Scharffenberg insisted that "he stood clearly on the side of those geneticists who worked 'strictly scientifically.'"[32] (So, of course, did most German "race biologists.") He lamented the long history of impractical and/or ethically unacceptable eugenic proposals, going all the way back to Plato. Yet one of his main criticisms of the 1933 German Sterilization Law was that it allowed sterilization only for purely medical and genetic reasons and not on social grounds. Scharffenberg felt, for example, that those who were mentally retarded for reasons other than genetics should also be sterilized; the mentally retarded—he thought—ought never to be allowed to raise children because they were unfit. On the other hand, in a more liberal view, Scharffenberg considered that the German law included too many traits on its list of those caused by genetics. "Alcoholism, for instance, is usually caused by environment and not by heredity, he maintained. Still, he considered the [German] law to be an interesting experiment, the effects of which ought to be followed with attention, 'provided it is applied with reason.'"[33] Of course, history shows that it was not.

The Norwegian Press Campaign Resumes

After the publication of *Die Bione* in February 1938, Johann Scharffenberg began writing a series of critical letters against Reich in the newspapers that bordered on slander. He said that during the bioelectric experiments Reich had arranged for sex between mental patients at Dikemark psychiatric hospital as part of the experiments.[34] He also wrote to some of Reich's supporters, such as Odd Havrevold and Nic Hoel, suggesting to them how negative it would be for the future of their careers if they continued their association with Reich.[35] Leiv Kreyberg also contributed many items attacking Reich; numerous letters in support of Reich were also printed.[36] The episode quickly exploded, with thirteen different newspapers, all across the political spectrum, participating, and no fewer than 164 articles or letters published between March and December 1938 when the newspaper controversy finally died down. (Compare this with only nine articles, two of them by Scharffenberg, in the September–December 1937 press controversy over Reich's work.) The publicity led several professors at other European universities to contact Reich and offer support.[37]

The left-leaning papers had taken relatively little part in the frenzy of attacks on Reich before autumn 1937 (recall *Dagbladet*'s favorable article about du Teil's presentations in Paris in August 1937); however, in the major press campaign that exploded in the spring of 1938, the Norwegian Labor Party's organ, *Arbeiderbladet*, was a major venue for attacks on Reich. Many of Reich's critics, including Mohr, Kreyberg, and Scharffenberg, were supporters of the Labor Party and the Socialist Physicians Association. Norwegian historian of psychoanalysis Håvard Nilsen has pointed out that in the wake of Trotsky's expulsion from Norway in December 1936, the neighboring Soviet Union was strong-arming the Labor Party government to put pressure on other Trotskyist sympathizers. Reich was widely perceived as such a sympathizer because of his harsh criticism of the Soviet Communist Party in *Masse und Staat* and in *Die Sexualität im Kulturkampf*. Much of Trotsky's discussion of the authoritarian family structure in *The Revolution Betrayed* sounds quite similar to Reich's critique. Thus, suggests Nilsen, this could be why socialist leaders in Norway turned their back on giving Reich any formal support by late autumn 1937.[38] It certainly is the fact that they did so. And they thus offered no serious opposition to fascists in Norway, such as Jørgen Quisling—brother of Vidkun Quisling, the leader of the Norwegian Nazi party—who targeted Reich as a "Jewish pornographer" for his sexual theories.[39]

Reich wrote a letter to Havrevold and Nic Hoel—both of whom were being pressured by Scharffenberg—to set the record straight about Scharffenberg's claims about sex between mental patients. Reich stated, "With the approval of Dr. Gjessing [medical director of the Dikemark Asylum], my oscillograph was moved to the mental hospital in Dikemark, where my assistant at the time, Dr. Löwenbach, was to measure skin potentials on catatonics. Naturally, the potentials of the erogenous zones were also supposed to be measured, but in fact this was never done. In the course of a visit with Dr. Gjessing, we talked a lot about Freud, psychoanalysis, sexual stasis and its role in mental illness."[40] Gjessing was interested in Gabriel Langfeldt's research on the metabolism of nitrogen compounds in catatonics, in which Reich expressed interest and about which he later corresponded with Gjessing.[41] Although Reich was direct with Gjessing regarding his research on the function of the orgasm and the bioelectrical experiments and he did not hesitate "to assert that totally ignoring the sexuality of schizophrenic individuals is one of the main reasons we are unable to obtain a clear picture of the problem of schizophrenia," Reich insisted flatly, "I never made a request of the kind referred to above [i.e., Scharffenberg's accusation], and it would never occur to me to do so." According to Reich, Gjessing "was very interested in my sexual-physiological electrical experiments; he ordered the illustrations from my report [on the bioelectric experiments] for himself and never made any negative comments to me."[42] Furthermore, Gjessing never publicly confirmed Scharffenberg's claim that he (Gjessing) had told Scharffenberg Reich tried to arrange for sex between mental patients. Nonetheless, salacious claims of that kind take on a life of their own, and Scharffenberg's claim continued to circulate despite Reich's denial.

Wilhelm Hoffmann, a physiologist who had worked with Reich in the early stages of the bioelectric experiments but later sharply criticized Reich's experiments, wrote to *Dagbladet*, still generally critical of Reich as a scientist but corroborating Reich's position that Scharffenberg was really going too far in his assertions about sex between mental patients and that the idea was at most expressed as an idealized experiment that could prove a valuable point if such a thing were not obviously ethically impossible. Hoffmann stated, "Personally, I have never regarded Reich's 'suggestions' about sexual experiments at Dikemark as a serious matter. He expressed wishes, not proposals. I have put such little weight on this that . . . I have therefore with all my powers tried to minimize the significance attributed to these remarks, when I have been confronted with them both in 1935 and now in 1938. Reich has in my opinion not [sic; this word was clearly mistakenly inserted since it contradicts the entire sense of the passage] suffered unjustly, since the facts have been blown out of proportion."[43] It is noteworthy that, despite having fallen out with Reich on most scientific issues, Hoffmann wrote that Reich *had* suffered unjustly. *Dagbladet* printed the letter with a typo error ["not suffered"] that greatly confused Hoffmann's message. When the paper printed a "correction" the following day, it was hardly visible on the page.[44] Whether this strange event was purely accidental or maliciousness against Reich by the editor, we have no way of knowing.

In an April 16 article in *Arbeiderbladet*, Scharffenberg next cast doubt upon du Teil's scientific credentials. He found no "University of Nice," and wondered aloud if du Teil was a professor at all. But Scharffenberg had mistakenly looked under the wrong name. Du Teil's university, in Nice, was officially called the "Centre Universitaire Mediterranean," part of the University of Aix-Marseille. This prompted Reich to write du Teil, that Scharffenberg "casts doubt in an extremely insulting way on your competence to deal with scientific matters. I am including a translation of this passage."[45] This does not answer the question of whether du Teil was actually a professor, however. I will discuss du Teil's status and Scharffenberg's suspicions further later in this chapter.

Du Teil wrote to *Arbeiderbladet* on April 26, clearing up the status of the Centre Universitaire.[46] That same day, Scharffenberg wondered aloud in *Arbeiderbladet* whether Reich was actually a medical doctor as he claimed. Reich's lawyer wrote Scharffenberg in response, inviting him to come to his office to inspect Reich's University of Vienna diploma for himself.[47] The fact that the communication came from a lawyer was intended to put Scharffenberg on notice that further such malicious rumor-mongering would lead to a libel suit. However, once again, after the rumor was out in the press, it continued to circulate for a long time.

Danish journalist Gunnar Leistikow—a supporter of Reich—wrote about this episode four years later, "One of Scharffenberg's colleagues revealed the fact that Scharffenberg's admitted aim in writing these articles was that of getting Reich out of the country. Public opinion turned more and more against Scharffenberg; the 'Friends of the Right of Asylum,' an organization of which he was one of the founders, even considered his expulsion from the organization."[48] Leistikow further elaborated, "A woman teacher out in the Norwegian country[side] who had followed the press campaign against Reich, wrote an article against Scharffenberg in which she hit the nail on the head and named Scharffenberg's real motive, which, although known to those trained in sex-economy," had never been mentioned in the press. "She had heard rumors," she wrote (thus taking off Scharffenberg's method of defamation by "hearsay"), "that Scharffenberg was an ascetic and, furthermore, at an age when even the devil himself takes to the monastery. This hit home. Shortly thereafter, admitting publicly his incompetence, Scharffenberg gave up his campaign."[49] Whether this was the correct explanation, or whether contact from Reich's lawyer was the key factor, Scharffenberg's public statements and rumor-mongering about Reich did decline after May 1938. I found no evidence that Scharffenberg "publicly admitted his incompetence," however.

March 1938 *Open Letter Defending Reich*

Shortly after Scharffenberg renewed his attacks on Reich in the press, an open letter to the Faculty of Oslo University was printed in most of the newspapers. It was penned by Sigurd Hoel, but signed by a large number of well-known Norwegian writers, scientists, artists, and other public figures, asking the faculty not to be influenced in their decision by the senseless statements being made in the press. Prominent sociologist Christian August Lange (usually known as just "August Lange" (see Figure 6.4); his brother Halvard Manthey Lange, later Norway's Foreign Minister; famed antifascist poet Arnulf Øverland; and Karl Evang (later head of the Norwegian government's Health Directorate), among others, signed the open letter defending Reich's right to stay in Norway. Hoel included an appendix of those who signed, independent of any consideration of Reich's science, based purely on the principle that Norway should support asylum for those persecuted for their ideas. The

Figure 6.4: Christian August Lange (usually called simply August Lange). Courtesy of National Library of Norway.

newspapers published the list of signatures on April 27, 1938.[50] As Gunnar Leistikow later put it, "one of Norway's most prominent writers [Hoel] pointed out that all the excitement about the bion experiments was only a smokescreen for machinations to get Reich out of the country."[51] At the same time but independently, prominent anthropologist Bronislaw Malinowski published a similar letter of support for Reich, as did British educator A. S. Neill. Malinowski stated,

> I have known Dr. Wilhelm Reich for five years, during which period I have read his works and also on many occasions had the opportunity of conversation and discussion with him, in London and Oslo. Both through his published work and in the personal

contacts he has impressed me as an original and sound thinker, a genuine personality, and a man of open character and courageous views. I regard his sociological work a distinct and valuable contribution to science. It would, in my opinion, be the greatest loss if Dr. Reich were in any way prevented from enjoying the fullest facilities for the working out of his ideas and scientific discoveries.

I should like to add that my testimonial may have some additional strength, coming as it does from one who does not share Dr. Reich's advanced views nor yet his sympathies with Marxian philosophy—I like to describe myself as an old-fashioned, almost conservative liberal.[52]

As Neill's letter put it, "To me the campaign against Reich seems largely ignorant and uncivilized, more like fascism than democracy." Neill wrote, "The question is: Is Reich a useful person who is bringing new knowledge to the world? For my part I feel that he is."[53]

Hoel's letter also pointed out some of the irrational features in the campaign, for example, that Reich was being attacked by Nissen and Scharffenberg as a psychoanalyst though he had long since ceased to be one. Reich was accused of being a "Red," though the Communist Parties had long since excommunicated him (the Danish Party had done so, even though Reich never belonged!). Finally, Hoel claimed that "Reich had incurred the wrath of all kinds of reactionaries and neurotics because his science dealt with the subject of sexuality."[54] Some might object that, because Leistikow and Hoel were students of Reich's, they were "Reichians" and therefore biased. A pertinent question on this point is, does one speak of "Einsteinians," or of students of relativity theory?[55]

Dean Kristian Schreiner and Odd Havrevold

On April 11, 1938, Schreiner (who had become dean of the medical faculty in December 1937) wrote to Havrevold, responding to his complaint that Oslo academics were behaving unprofessionally, targeting Reich with gossip. Schreiner took the pose of innocence, denying (rather implausibly) any gossip at all: "You write that it is unfortunate that a certain group here in Oslo has become the earpiece and the carrier of rumors and gossip about Mr. Reich. I must say that I have not heard any gossip thus far. I asked several of my colleagues who might have knowledge of this, and it is not the first time that such insinuations have been made; however, no one could give me even the smallest explanation."[56] Referring to "Mr. Reich" rather than Dr. Reich, Schreiner was subtly but unmistakably collaborating in Scharffenberg's accusations that Reich was not even a legitimate MD. Acknowledging indirectly, however, that he and others were discussing Reich—and skeptically— Schreiner added, "My colleagues and I are only interested in eye-witness statements and known facts, such as publications. It was primarily the faith and enthusiasm that you displayed during our conversation about the phenomenon Reich almost a year ago that stoked my initial interest." Schreiner reminded Havrevold that he had been skeptical of the bions a year earlier and had at that time urged the younger doctor "to immediately present all details to a trained bacteriologist and physicist, for neither you nor Dr. Reich had the essential qualifications necessary to perform a critical evaluation."[57]

Schreiner was disturbed, however, that Reich did not turn to specialist bacteriologists and physicists: "Instead of immediately seeking out the best help, Mr. Reich turned the issue over to an unknown Natural Philosopher [du Teil] with no special qualifications in the areas of biology, physics, and chemistry, which are of special concern here. That, to the greatest extent, weakened his credibility with me." In addition, he said, Kreyberg's harsh critique of Reich's ability with a microscope—following the October 8, 1937 meeting over "granulation tissue" versus "cancer cells"—weakened his confidence in Reich still further.

Schreiner told Havrevold that the only reason he had originally taken any interest in the bions was "solely because of our old friendship and to protect you in some possible way from an issue, that, as I understood it, would demand all of your time and effort only to fail in the end." And then, rather more ominously, Schreiner ended the letter by saying, "Mr. Reich the person, his dos and don'ts according to the known facts—are absolutely irrelevant to me. . . . But I think that it is a damned shame that a man such as you throws away the most precious years of his life on such a fruitless endeavor."[58] Recall that this was the dean of Oslo University Medical School, speaking to a much younger doctor who still hoped to have a career as a prominent physician in Oslo.

Since December 1937, Reich believed that Otto Mohr was "moving decisively for refusal of my residency permit."[59] Kreyberg and Scharffenberg actively joined in this effort. In February 1938, Reich met with Dean Schreiner to discuss this situation. At this time, Reich was still laboring under the illusion that Schreiner was open-minded toward him and his experiments. He did not perhaps realize that Mohr, Kreyberg, and Schreiner were close colleagues and mutual supporters when he wrote in his diary, "The director of medicine in Oslo [Schreiner] is afraid of the three idiots who want to get me out. He's afraid of me too; during our talk today he practically put into my mouth the words necessary for me to survive."[60] Then again, Schreiner may indeed have feared the consequences of letting a world-famous psychiatrist know his true feelings. After reading Schreiner's April 11 letter to Havrevold, however, Reich realized that Schreiner was lined up with his opponents and would probably give him no more help in resisting their efforts to prevent renewal of his residency permit.[61]

Reich and du Teil: Turning Up the Heat

In the midst of this, Reich wrote to du Teil, using the behavior of Kreyberg and Thjøtta as an object lesson. By then Reich seemed to conclude that there was very little fairness among Norwegian scientists toward an outsider like himself: "You will probably remember that when you were here I warned you about specialists. At the time, I had the impression that you took this a little amiss and later interpreted it as an expression of my quarrelsome nature. I can assure you that about eighteen years ago, in the battle over Freud, I experienced to the limit all the dreadful sides of human nature that are now manifesting themselves, even though at the time it was not my own cause that was involved."[62] Reich warned du Teil, "I have always had the feeling that you are too optimistic, that you believed it would be an easy matter to convince the authorities as long as one had something good to show them. I often expressed my doubts to you on that score."

And, turning to strategic considerations, Reich argued, "In the meantime, we have discovered that partially disclosing individual experiments or dealing with fragments of the whole involves an enormous risk, namely severe disappointment, which then does more harm than the good that might have been obtained. We have no doubt at all here that bion research cannot be understood, and therefore the practical details of the work cannot be mastered, without one's taking into account *all aspects* of sex-economic theory and the dialectical-materialistic method." And Reich asserted a rule that he stuck with throughout the rest of his career after emigrating to America: "The way I have set things up here with my assistants is that they must carry out accurately and independently the entire set of experiments, starting from the very beginning, before I demonstrate the more advanced, complicated experiments to them. Therefore, it is also impossible to understand the cancer experiment and to carry it out in practice unless, for example, one is firmly convinced from personal observation that (1) the protozoa really are seen to develop from disintegrating grass tissues and (2) in principle precisely the same process is involved in cancer tissue."[63] Announcing the connection between the bion experiments and cancer would bring disaster, Reich said, because there were too many steps in the logical path from the orgasm theory, and connecting the two experimental areas, that most simply could not follow the chain without repeating for themselves all the intervening steps.

It is not difficult to see why many researchers would balk at giving so much time to an enterprise of which they were skeptical. While there is a rational element to Reich's demand for "immanent criticism," it is not difficult to see that most scientists would resist this set of rules and feel it imposed too much on their independence. Also, there was a tension between Reich's caution about exposing his work to irrational, unfair criticism and his nevertheless continuing to engage with scientists he considered open-minded. Like du Teil, another colleague Reich trusted, educator A. S. Neill, pointed out this same contradiction to Reich soon after. This is discussed later in this chapter.

Reich wanted to make a proposal regarding the cancer work to du Teil, if he wanted to involve any third parties: "I am trying as far as possible to devise simple test procedures that are achieved gradually, without any initial relation to the cancer problem, and that thus will automatically open one's eyes to the cancer question. Please let me know whether you are in agreement. I would then arrange the experiments in a logical manner and finally lead up to the cancer experiments. Choosing any other route appears too hazardous to me."[64] Explaining why he was so cautious, Reich told du Teil about his opponents in Oslo making inquiries behind his back to Bonnet, Martiny, and others. He asked du Teil to please let him know if indeed Bonnet, Martiny, or others in France had achieved negative results in their attempts to replicate bion experiments, "So I can see to quashing any rumors that arise. It is better to be aware of an unpleasant truth than to live in a world of illusion." And, finally, Reich requested for du Teil "to enjoin those colleagues who are seriously working on the matter to take whatever steps they can to ensure that we are granted the necessary peace and quiet to continue here. It would be sufficient for one or the other of them merely to confirm that the experiments are being carried out and that it will take a long time to complete them."[65]

Reich did not think all looked black; he closed his letter of April 19 with some good news: numerous positive articles on the bion work had appeared in Scandinavia and in Hungary. In June the director of the University of Pedagogy in Copenhagen planned to deliver a radio lecture about the bion book. Reich congratulated du Teil on replicating the latest type of bion preparation and reported to him, "I have recently obtained the round cells from a Preparation 14 that rested for two weeks. I actually believe that I have succeeded in artificially creating cells of the cancer cell sort; however, I would like to point out one small mistake. Cancer cells are elongated, and only when they are dying do they assume a circular shape and then resemble the cells from Preparation 13."[66] In the midst of the controversy, new experimental work continued in Reich's lab. What were the developments with Bonnet and Martiny that Reich so urgently asked du Teil about?

Bonnet and Debré in Paris: Reenter Kreyberg

The reader will recall that Parisian bacteriologist Henri Bonnet had in August–September 1937 undertaken to replicate key bion experiments in the laboratory of Professor Robert Debré at the University of Paris Medical Faculty, after seeing du Teil demonstrate those experiments in Paris on August 9–10, 1937. Du Teil worked with Bonnet in Paris in early October 1937. Then the outcome of this work went into a peculiar limbo for months, during which time Reich could not get a straight answer from du Teil about what outcome transpired from Bonnet's experiments. During the November 1937 press attacks on Reich by Norwegian psychiatrists, one of Reich's chief financial backers, Lars Christensen, was "bombarded with rumors" about Reich "from an unknown source in Denmark," eventually leading Christensen to withdraw financial support from the bion experiments. In addition, during this time, Odd Havrevold was feeling pressured (probably by Oslo University authorities) to prove that official labs abroad really were testing the validity of Reich's bions.[67] He asked Reich "to name the laboratories in France at which the control studies were being conducted. After some hesitation," Reich "told him the names of the laboratories and gave him Dr. Bonnet's name."[68] Christensen's withdrawal of support occurred in late March or early April 1938, just two weeks or so after the Nazi *Anschluss* of Austria.[69] Reich wrote to du Teil on March 30, concerned about this development:

> The extremely stupid smear campaign conducted against the experiments here has taken on serious proportions. Among other things, Lars Christensen, who was our financial sponsor, is being severely criticized from Copenhagen on the grounds that at the

time I did not have a residency or work permit. Now, under pressure from our friends, Christensen has inquired through the Norwegian embassy in Paris how the experiments are proceeding. All this is just for your information. I must stress once more that it is crucial to our work and our existence that we learn of any positive result that is obtained in France. We will be able to survive the immediate future only if we can counter this abyss of envy and infamy by responding with clear, factual answers based on experiments.[70]

Plan for an International Commission on the Bion Experiments

On April 21, seven weeks into the press firestorm ignited by the publication of Die Bione, Danish plant pathologist Paul Neergaard wrote Reich to suggest that the irrational part of the controversy could be quickly put to rest if an international scientific commission were organized, to publicly observe the experiments and officially pronounce upon their scientific validity: "Would it not be smarter—if at all possible—to invite the indignant scientists to form an investigative committee where they could then have the charge to complete your and du Teil's experiments with you in your laboratory under the strictest of controls (sealing the thermostats at night)? Whatever happens, that could only have a positive outcome: that you welcome skeptics, if they only come, with open arms. The entire bion theory, whether right or wrong, would be seriously tested through this. No matter what happens, I find that such an offer would put you personally in a good position with respect to your adversaries."[71] Neergaard volunteered himself to be a member of such a commission. Reich wrote back on April 23 to say he thought it an excellent idea.[72] At the suggestion of the anatomist Professor Birger Bergersen in Oslo, Reich also wrote to Professor Axel Palmgren at the University of Stockholm, asking him if he would participate.[73] As Reich wrote to du Teil, "Prof. Palmgren is the best anatomist in Scandinavia; also an embryologist. He's been in psychoanalysis for a long time and thinks highly of my work."[74] Reich told Palmgren: "The Press discussion flared up without my doing. I would have preferred to have had a public discussion six months to a year later. However, it can no longer be avoided. In order to put an end to the fruitless and non-objective discussions, I had to make the decision to perform a public demonstration of some of my experiments in the presence of my opponents and with other experts as my witnesses." Reich also asked whether Palmgren knew any Swedish bacteriologist who might be suitable to participate in the commission. Reich anticipated du Teil's fear that he (Reich) would have deliberately precipitated a public controversy; he wrote, "Today again only a short situation-report. I ask you before all else not to believe that I was somehow aggressive and provoked the matter." Reich told du Teil, "It is hard to believe how fantastic and childish people's reactions are to the issue of the origin of life. Obviously, forces and fantasies of which we are totally unaware are behind all this."[75] In his next letter, on April 25, he added that the public demonstration before an international commission "is the only possibility to counter people's . . . mysticism and the expectations that are connected to the expression 'origin of life.'" Reich clearly believed the reasons behind the visceral nature of the emotional storm in the press had little to do with science: "'Bions' themselves do not disturb people, rather the philosophical and emotional significance they have," he remarked.

Reich told du Teil he imagined the public demonstration taking place in June, August, or September 1938.[76] Du Teil replied that for him June and September were impossible but that he would very much like to be present and to participate, so he suggested the time window of July through August 15.[77] Reich wrote back to du Teil on April 28, suggesting a list of possible nominees for the commission:

> In the first glance I thought of as authorities for the control commission and for the witness function:
> Professor Palmgren, anatomist, Stockholm (I have already written to him)
> Dr. Paul Neergaard, plant pathologist, Copenhagen
> Professor Bergersen, anatomist, Oslo
> Would Debré or Cornil come? Travel and accommodation would be paid of course.
> As witnesses could include:
> Professor of Psychology Schjelderup, who has already accepted.
> Sigurd Hoel, writer
> Dr. Ola Raknes, religion, psychology
> Dr. Philipson, doctor, Copenhagen
> Professor Thjøtta, Bacteriologist (whom I reject on a provisional basis for the Commission, for his behavior is not objective.)
> The university here in Oslo has requested that our opponents Kreyberg and Thjøtta be allowed to sit on the commission.

But Kreyberg is absolutely out of the question [*kommt überhaupt nicht in Frage*] because the commission must be objective, which Kreyberg was not.[78]

More Machinations by Kreyberg, Schreiner, and Dysthe

It is clear that Havrevold passed along the information about French labs he learned from Reich because meanwhile a Norwegian named Roald Dysthe traveled to meet Bonnet in Paris and du Teil in Nice, to gather information damaging to Reich by questioning French labs' support for Reich's claims. It is not clear who put Dysthe up to this, though it may have been Schreiner.[79] About Dysthe, after some inquiries Reich could only say, "Dr. Raknes will answer your inquiry concerning Roald Dysthe. I do not know the man. He is supposedly a wealthy and respectable businessman."[80]

By February or March 1938, Kreyberg, at a meeting of the Socialist Physicians Association, "stated that he had been in Paris, had talked to Bonnet in the presence of Debré, and Debré had merely laughed at the whole business." Reich added tartly, "I do not know what Kreyberg said when he was with Bonnet, but quite a bit can be deduced from his behavior here."[81] Reich thought Kreyberg had stooped to lying and elaborate, extremely unprofessional scheming out of a personal vendetta against him. He asked his lawyer, Annaeus Schjødt, to "use all legal means at your disposal" to clarify Kreyberg's machinations in "this matter, which is assuming a criminal character."[82] "Kreyberg rages because he is against the bacterial etiology of cancer," confided Reich in his diary.[83] "A common liar and scoundrel," Reich later described him to du Teil.[84]

While du Teil had been rather evasive since October 1937 about the outcome of Bonnet's experiments, Reich was reasonably confident that Debré and Bonnet had not simply dismissed the bions as laughable, as Kreyberg asserted—particularly because he knew of Van den Areud's being engaged in serious conversation with Debré about the experiments (as discussed in Chapter 3). Reich began to press du Teil for any kind of written response he might have gotten from Bonnet, saying, for example, on April 28, "I ask you to give me an exact report about your conference with Cornil. I discovered here that Martiny and Bonnet were negative, while a third—and perhaps a fourth—person are continuing to work on the issue. It was not a good idea to not share the issue with me then, for I could not refute the local rumors. You obviously had the same problems with the authorities here as I do. However, I would very much like to hear more about it."[85]

Reich was referring to a recent meeting of Dean Schreiner, Dr. Reidar Gording (another medical professor),[86] and Havrevold, at which Dysthe claimed that "Bonnet told him . . . that Debré knew nothing of the bion work and that he, Dr. Bonnet, ought to know, because everything for Professor Debré passed through his hands."[87] Reich was taken aback by this, not having heard from du Teil anything about definitive results—positive, negative, or equivocal—from Bonnet or Martiny. But he knew from Van den Areud's letter that Debré had just asked for more cultures to study.

One of Reich's coworkers became so anxious that he himself wished to hear assurances from Debré and Bonnet's lab in Paris, "to allay his own lack of certainty," as Reich put it.[88] This probably refers to Havrevold, having passed along to Schreiner and Gording the names of the Paris labs. Gording got a letter from Bonnet, declaring that the Paris lab had found entirely negative results in its attempts to replicate Reich's experiments. This occurred, however, at the same time Debré was asking Van den Areud to obtain fresh bion cultures—suggesting that the lab continued serious interest in the bions. What, then, do we have as evidence for what actually took place in those French labs?

Reich wrote to Bonnet, attempting to get a direct answer from him about what actually took place in the Paris lab, but Bonnet did not reply to him until July, and only then after being contacted by Reich's lawyer. He said only, "I did not keep the experimental protocols of the 'bions.' The experiment was made with the samples which Mr. du Teil brought us and they have been conducted according to the technique which you have indicated to him and of which I do not know anything. My part was restricted to the sterilization procedures and to the observation of the tubes thereafter. After three months everything was still sterile in all series." Bonnet said the delay in his reply was because when he first got Reich's letter, he had asked du Teil to refresh his memory about the details of the October 1937 experiments before he wrote back. But he had not heard from du Teil since. Almost as an afterthought, Bonnet added, "PS I can also tell you that the cultures which were shown to me as being cultures of 'bions' contained, without doubt, accidental contaminations."[89]

Du Teil Is Fired from His University Job

If du Teil's caution was to avoid putting any of his French colleagues in an embarrassing situation, his plan backfired because of the Norwegian attack on Reich and the willingness of Kreyberg and Schreiner to get public statements from Paris that would discredit Reich's bion experiments. The Oslo officials' inquiry triggered a chain of events that led to officials at du Teil's university becoming alarmed that, they claimed, "his work with Reich, done privately, had involved the university's name in a public polemic."[90] Du Teil's university felt the publicity made it appear that they had been forced to conceal private work du Teil was doing on their premises. The outcome, as du Teil told Reich on June 5, was, "Consequently—and the decision comes from the Minister—I have been suspended from my post and forced from duty without salary. In any case I will no longer belong to the University Centre as of the end of this month."[91] He urged Reich and his supporters that, whatever they might try to do on his behalf, they do it with absolutely no publicity. Under no circumstances did he want them to publicly defend him. He felt more publicity would make matters infinitely worse and that, given time, he might be able to smooth over the problem and get his job back.

Reich and his Oslo coworkers were horrified at this outcome. Reich wrote back, "I would like to assure you personally, on behalf of myself and also a number of very important friends, that we are entirely at your disposal and you may count on us in all respects, if necessary even for financial support. It is vital for us to maintain solidarity in such a battle."[92] In addition, Reich contacted his lawyer in Oslo to tell him to begin investigating the increasingly sinister behind-the-scenes campaign against him, which he thought now bordered on criminal activity. Reich's lawyer was Annaeus Schjødt—a man famous for accepting unpopular cases in defense of civil liberties. Schjødt had first gotten in contact with Reich and offered his services pro bono, because he felt the treatment of Reich in the Oslo newspapers was so unfair and shameful.[93] Schjødt was still more famous later as chief prosecutor against Vidkun Quisling, the head of the Norwegian fascist government under Nazi occupation.

Despite offering immediate and full support, however, Reich felt that because he and du Teil trusted one another greatly, Reich must tell him, "The situation requires that I speak very openly to you, drawing on all the battle experience and severe disappointments of the past eighteen and more years. Right from the start, I asked and warned you not to rely too much on universities and authorities." If du Teil was irritated at hearing "I told you so," Reich was still more concerned that du Teil seemed unwilling to send him a key piece of evidence needed to fight off the claims of his attackers, that the Paris labs had never produced anything except negative results.

Du Teil had told Reich that Bonnet wrote to him on October 23, 1937, that is, only about three weeks after du Teil had visited Bonnet's lab in Paris and gotten the replication work started there. According to du Teil, this letter "contained confirmatory information about the work in Paris."[94] Reich repeatedly asked du Teil to send him the letter and let him copy it, but du Teil continually stalled and made excuses. Yet du Teil had shown the letter to Roald Dysthe when Dysthe passed through Nice, not realizing Dysthe's intent to harm Reich's work. In fact, Dysthe said he had the impression the letter was a fake because Bonnet's signature was in very small handwriting, while in the letter Bonnet sent to Gording after Dysthe's visit the signature was much larger and in a different hand.[95] Now, when Reich thought their enemies were on the verge of getting the upper hand, he gently chastised du Teil, as noted earlier, saying that denying him access to the details of what went on in Paris rendered him vulnerable to his enemies, able to be surprised by them and unable to refute their claims.

Ola Raknes, a therapist in training and supporter of Reich, wrote to René Allendy in Paris to see whether he could learn anything about what had taken place during the replication attempts in the lab of Debré and Bonnet. Since Allendy had originally interested Guy Van den Areud in the bions, Raknes now asked Allendy if he could find out what became of the additional bion culture Reich had sent because Van den Areud asked for it in his March 8 letter, saying he wanted it because "Professor Debré asks me to ask you if you will kindly send us some more samples of 'bions.'"[96] Raknes politely pleaded, "You would do us a great service if you could obtain from Professor Debré precise information on the following questions: (1) Have the tubes sent by Mr. du Teil to Mr. van den Areud been examined by Professor Debré? If yes, (2) by what procedure were they examined? (3) Can we have a copy of the protocol? (4) Did the test give vague or precise, positive or negative results?"[97]

Meanwhile, on June 21, awaiting the axe to fall, terminating his job, du Teil finally wrote back to directly answer Reich about the October 23, 1937 letter from Bonnet. Yet, despite "directly" answering Reich's request for the letter, he still remained strangely vague, saying,

> It is necessary on the other hand for us to reach an agreement definitely, concerning a document to which you seem to attach an unwarranted importance. This document might produce a negative effect, moreover. By keeping certain documents in my

possession without communicating to anyone in the world, I really acted with responsibility which is not fragmentary but global, and because I am the only one able to judge whether communicating them—even to you—might be useful or harmful. The event that just happened confirms that I was not wrong—since it is a publication from Oslo that set off the press campaign of which you know the unfortunate, if not tragic result. Actually, I alone can judge the repercussions in France of such documents, and it is my imperative duty—even in your interest—to ward off the dangers we would all risk because of the ignorance you have of the reactions in France.[98]

Du Teil would not ever name Bonnet, but he did finally give an (albeit maddeningly vague) account of what had taken place: "Admit for an instant that what they reported to you is exact and accurately interpreted, that is to say that one day a researcher informed me personally and confidentially that he obtained a certain positive result, among several others that were negative, but asked me not to take account of it until he had verified statistically; that subsequently these statistical experiments, if he made them, were negative—and that he then concluded that the sole positive result stemmed from an error, included moreover, in the percentage of accidental errors seen in all laboratories (let's say 2% to clarify matters)."[99] Du Teil seemingly feared that violating confidential conversations with a French colleague was the worst kind of mistake, or breach of trust, possible, particularly if it put Bonnet in the position of appearing to have contradicted himself in what had become a heated international controversy:

> That he [Bonnet] decided then to continue, trusting in his personal discretion, he purely and simply replied "negative," without entering into the details and about what he might consider an error of detail from his laboratory. I sympathize wholeheartedly, you see, with your hypothesis in the most favorable way; and I wonder if you think one can deny such a researcher the right to do as he did—and if you don't think it would be both highly incorrect and useless, even dangerous, to make use of a personal fragmentary confidence made in the course of work to try to make this person appear to contradict himself! Incorrect and even worse, when one knows the risks to which one would expose the

person in question, considering the actual tone of the arguments; as for me, I am incapable of envisaging such a thing."[100]

And du Teil emphasized this was even more explosive for himself personally, now that his job had been taken away, saying revealing the details would be "dangerous, because a new argument might immediately . . . produc[e] new repercussions in France, where I need a few weeks of silence in order to reestablish myself."[101]

Du Teil also wanted to calm Reich down, thinking he had jumped to conclusions, "based on the report of a certain M. Dysthe, who passed by my house in a gust of wind, doing what he could and interpreting as he wished. You say that you had some 'partial' knowledge about a document. You should have said that you had no knowledge of such a document, for I never spoke to you or anyone else about such a document. The awareness of such a document that Dysthe claims to have had is nonexistent, since he gave me his word not to take my dossiers into account, and I fail to understand how he therefore could have referred to it."[102]

It may seem to some that Reich was intrusive, pressuring du Teil so forcefully to share with him a private letter. Yet here, du Teil also seems to reveal his own naïveté about the nasty intentions of Reich's opponents and their willingness to violate academic etiquette and fairness in any way necessary. Perhaps not surprisingly, despite his warm and collegial feelings for du Teil, Reich found this naïveté alarming, nor was Reich likely to feel calmed by du Teil's conclusion, when the French scientist wrote, "I fail to understand, moreover, how one result alone [i.e., the successful one reported in the October 23 letter]—refuted subsequently as they tell me—can be useful and can not, on the contrary, be dangerous—for a case that only statistics can prove. There is the question finally settled, my dear friend."[103]

Before Reich could reply to du Teil, Ola Raknes wrote to the Frenchman, to let him know of his inquiries through Allendy. Raknes had read du Teil's June 21 letter to Reich while at Reich's house discussing how to help du Teil. He told du Teil, "The hypothetical discussion in your letter seems to me to be of very questionable value, at a time when it is facts and nothing but facts that are important." Raknes wondered aloud in his letter whether it might have occurred to du Teil that, in bending over so far backward to be polite and considerate to Bonnet, it might appear to some (such as Raknes himself) that he was putting loyalty to Bonnet—who had no particular commitment or loyalty toward him—ahead of loyalty or consideration toward Reich and his rather dire need while under attack: "Speaking frankly, I think you are wrong not to let Reich see the documents he asked you about—it is neither right nor possible that you take all the responsibility for the outcome of these experiments. I would undoubtedly respect your loyalty toward M. Bonnet or whomever [sic] it is, if he or they make a point of showing the same loyalty towards you and towards Reich. In the present state of affairs I see only a communication to Reich of documents concerning him, and which could be considered a disloyalty. . . . Please excuse my frankness again, and believe, dear Sir, in my very honorable sentiments."[104]

Raknes also told du Teil he had just heard back from Allendy, who had "bothered" Debré on Reich's behalf to inquire about what had become of the additional bion cultures that Van den Areud had supplied him with three months previously. According to Allendy, Debré replied very briefly: "Mr. Debré does not seem to be at all interested in the experiments and there is nothing positive to expect from him."[105] Bonnet and Debré were sufficiently concerned not to be involved in a potentially embarrassing public polemic that Reich, clearly, could not continue to hope for any help from the Paris lab.

Du Teil's reply of June 25, addressed directly to Reich, reemphasized that he felt—even aside from etiquette—that his tentative job situation made it terrifically impolitic for any additional powers-that-be in French academia to be involved in the controversy on his account. First he thanked Reich and his colleagues for having cabled him money, to help him get through the difficult financial crisis from suddenly losing his income. He added that while some at his university had abandoned him, others had taken up his cause and at the very least guaranteed on the strength of his teaching reputation that he would be able to teach at private schools, "so much so that I am now assured of having work, reasonably speaking, beginning August 15th, and certainly much more after October 15th, in private secondary schools." Du Teil also hoped that if he could "succeed in getting off the ground a project for an institute for the teaching of philosophy at the superior level," he would find himself "with more or less equivalent resources by November." The take-home message was, "The most important thing is that I be able to hold on until October and be able to afford the costs of the publicity necessary. . . , and that will be possible with what is left from what you have already sent me. In these circumstances, I think I will even be in a position to reimburse your group monthly, starting next January, for what you sent me, and which, literally, saved my life."[106]

Next, however, he returned to explaining why he was still set on not sharing any communications from other French scientists. So far, he was grateful to be able to say, the controversy had not spread beyond his university. "Of course it is important," he wrote, "more so than ever, in the coming weeks . . . that my name be kept out of this, and that we avoid putting me in a difficult position with regard to new personalities." He elaborated, "Although Raknes's request was addressed very discreetly and by an intermediary (Allendy I think), I am, I confess, upset that Van den Areud and especially Debré were mixed up in this strange business based on absolutely personal letters from Van den Areud to myself. I can see Debré reproaching me [tomorrow] for my indiscretion, and rightly so, and all of that can only sink me in deeper."[107] Again, belaboring to Reich his qualms about being reproached by Debré for inconveniencing him would not have landed on particularly sympathetic ears.

But in addition, du Teil had some news that might well have shocked those in Oslo wondering about his loyalty to the cause of the work he had shared with them. So great was his concern not to have any of his French colleagues drawn into the controversy, he told Reich, "You will forgive me for having had to decide—you understand why—to destroy everything in my files that comes from third parties or is related to them. In this way I will no longer risk seeing documents misinterpreted that were glimpsed indiscreetly, seeing Debré, Bonnet, Decourt, Pretée and all of them mixed up in this together, and seeing once again, moreover, the Sorbonne and the Paris hospitals set up against my indiscretion, while I have had enough from the University of Aix!"[108] Du Teil urged, above all, that "a lot of patience is needed in this business. I already have, even here, around me, a growing circle of [people who are] convinced. People who are more and more specialized will join them, and we will achieve through a fine rain what a thunder strike will never give us."[109] In other words: I told you I knew best how to handle these French matters. Now do you see I was right?

Reich wrote a lengthy reply to du Teil on June 30, but in the end decided not to send it; instead he filed it in his archives. He began with relief and congratulations that du Teil had apparently found a way out of his financial crisis: "I am extremely happy that you have succeeded so quickly in setting your situation straight. We can deduce from this the moral that nothing is ever as bad as it seems [*immer wieder nicht so heiss gegessen wie gekocht wird*]. It is a major benefit that the whole business was not hushed up and that the hullabaloo that arose without any prompting on our part drew the attention of quite a few good people to my work and to the successful control experiments you have carried out."[110]

But Reich took issue with several of du Teil's interpretations of events. First, he said, "You seem to believe that I was the one who unleashed the press campaign and kept on fanning the flames. That is not true. Apart from one basic statement to the press, in which I asked for calm, I have not responded to a single line or a single objection or reproach from my enemies. They have also ridden themselves totally into the ground and exposed themselves to the world."[111] In Reich's opinion, the Norwegian press campaign against his bion work continued still "because some Norwegian scientists and physicians simply could no longer tolerate the disgraceful behavior exhibited by, in particular, a psychopathological, querulous person like Kreyberg. They are now struggling to eradicate the shame that two—no more than that—private persons have inflicted on Norwegian science." Several of Reich's friends and supporters, including Nic Hoel, Ola Raknes, and Odd Havrevold, were actively writing letters to the papers challenging the most outrageous claims of his detractors, as did even the physiologist Wilhelm Hoffmann, who was not part of Reich's circle and disagreed with most of his scientific claims. Reich told du Teil, "I cannot prevent my friends and the others from doing this, nor do I want to. This is how Debré, Bonnet, and others were brought into the discussion, and they are not entirely blameless."[112] In Reich's opinion, du Teil needed to be set straight on two important points: "(1) Bonnet was not in the least entitled to write a private letter to Dr. Gording stating that negative results had been obtained. (2) Bonnet had even less right—unless he was thinking of publishing and justifying his results—to write a letter to Kreyberg in which he repeated what he had said to Gording and also to receive Kreyberg in Paris and report to him on this matter. Kreyberg published Bonnet's letter here in the press."[113]

Reich was concerned because he saw that du Teil still seemed "to be of the opinion that I get into these difficulties for some political reason. I have already written to you once that this is not true. I am not involved in any political activity, I do not belong to any party, and I have always been attacked solely because of my scientific work."[114] If one recalls du Teil's letter to Havrevold in January, six months previously, opining that Reich's difficulties were caused as much by his own personality as by the revolutionary nature of his research, it is clear that Reich was not imagining this attitude on du Teil's part. Reich had lived through many controversies in which opponents behaved unfairly—not least the machinations that led to his expulsion from the International Psychoanalytic Association. As a result, and because of his psychiatric training, he was wary of irrational motivations provoked by direct discussion of sexual matters. Du Teil, on the other hand, was used to easy intellectual collaboration with lots of other academics, without getting into the kind of heated controversy that Reich often found himself in— at least until he was suspended from his job. So it was natural for him to think Reich's personality must be a key irritant that triggered such irrational attacks.

Reich tried one more time to remind du Teil of the real historical reasons for his attitude toward defending his work: that the bion research had a long history in theory and experiment, inseparably linked to his orgasm theory. Du Teil, he pointed out, had not been involved in that long history and had not risked his career and his personal security as Reich had for fifteen years. So while Reich appreciated the importance of professional etiquette, he wrote, "You must concede that I cannot now abandon the orgasm theory merely because someone like Kreyberg or some aged psychiatrist does not like it. . . . Anyone who wishes to espouse the noble bion cause must of necessity also bear the burdens arising from its history—namely, in the field of scientific and experimental sex research. And the word 'sexuality' as well as the scientific work in this field will long remain a stumbling block. We have to reckon with that." At the very least, this exchange (and du Teil's predicament) illuminated the quite different notions that Reich and du Teil had about the obligations incurred by being a coworker. Reich thought the provocative nature of his research warranted a nonnaïve sense of loyalty from a coworker in the face of a hostile society—with the scientists of that society being no exception in their irrational hostility.

Despite du Teil's temporary successes in interesting some French scientists in the bion work, Reich insisted in the same letter, "I have no illusions that a psychiatrist like Scharffenberg, a pathologist like Kreyberg, or a bacteriologist like Bonnet will very soon declare themselves willing to consider the bion experiments in their total context and to verify them using the methods that derive from my sexual theories."[115] Reich felt du Teil's criticism in his previous letter revealed that the French scientist still harbored such illusions. Thus, for emphasis, he added, "I fully share your opinion that we can only make headway step by step; but thunderclaps like those which we have just experienced cannot be avoided, although we must do everything we can not to provoke them. When they are instigated by enemies, however, we have to strike back forcefully and courageously, making no allowances. I fully understand your wish to show consideration for your French academic colleagues, but when I am fighting for my scientific, material, and intellectual existence, I cannot tolerate such consideration."[116]

Reich also reminded du Teil he had sent him an S-bacillus culture and three different bion cell cultures that he considered important discoveries for which he (Reich) was concerned that he maintain control over what was done with them. He asked du Teil to inform him of precisely what had become of those cultures, saying, "If Debré ever had any cultures in his hands that I produced, literally by the sweat of my brow and under the enormous pressure of my financial and social situation, then I don't care how well respected he is as an academic, I have the right to demand from Debré that he tell me what has happened to these cultures."[117] Reich closed by saying he deeply appreciated du Teil's candor with him, and he felt their trust was strong enough that du Teil could reply with equal candor.

Although Reich in the end decided not to send this letter, it is clear that the close working relationship the two men had engaged in for eighteen months never recovered from the strain of this difficult episode. Even had that been possible, by September 1938 and the Munich crisis, du Teil was preparing to mobilize as the French army prepared for possible imminent war with Germany. After Chamberlain's concession to Hitler at Munich and the passing of the imminent crisis, du Teil was still having difficulty recovering his financial stability. He had borrowed heavily for the laboratory equipment needed to control the bion experiments. This seemed prudent when he had a secure post and a well-established professional routine. However, with a wife and children to support and now a new source of employment to get used to, which might require time before his income would build back up to what it had been before losing his university post, the pressure of those earlier debts felt much more worrisome to him. On top of all that, du Teil had had some health problems in August and September. After a letter on September 28 in the midst of the Munich crisis, Reich did not hear from du Teil again until June 1939—and then only a couple of letters were exchanged—despite repeated letters from Reich to inquire about the lack of contact.[118]

Above and beyond du Teil's personal and financial circumstances, it seems likely du Teil would have come away from the shock of losing his job with much greater caution about being involved with Reich. Reich, too, probably thought differently about du Teil's level of naïveté and deference to authorities. But this did not stop Reich from trying to renew the working relationship. But after Reich immigrated to America in August 1939, he never heard from du Teil again (this despite the fact that in the years after World War II, du Teil remained interested in the bions).[119]

Regarding Scharffenberg's claim that du Teil was not actually a tenured professor, the episode of du Teil being fired and having to take a job as a high school teacher is suggestive. Scharffenberg's claim from the spring 1938 debate was that he could not find du Teil listed in *Minerva* (Abteilungen Universitäten und Fachhochschulen, 1937), the annual book of European professors and scientists; he also claimed that du Teil's University Centre in Nice was not a university but a teaching institution for foreign visitors, that is, not an institution where research was carried out. In his response to Scharffenberg in *Arbeiderbladet*, du Teil signed as "Chargé du Service des Facultés," something Scharffenberg noted and took as proof that du Teil was not a professor. One plausible interpretation of this story is that du Teil was probably not a professor. In France, *professeur* is used much more colloquially; any teacher might be called "professor," and du Teil may have used the title in this colloquial sense, thus confusing Reich, who came from a culture where the professorial title carried a great weight. That could explain how du Teil could be fired for being associated with the Reich controversy, and why he would accept a high school teaching position. If he really were a professor with tenure, he could not have been dismissed so easily, and it would have been his obvious right to follow his own research interests, regardless of criticism from colleagues like Bonnet, Debré, and Lapicque. While Scharffenberg used many underhanded tactics and was open to criticism on many points, he may well have been correct on this point of du Teil's formal credentials.[120]

It is worth noting that at the time du Teil was suspended from his job, Reich's deep skepticism about whether the mainstream scientific community could ever accept his bion theory applied to more than just du Teil's efforts, suggesting Reich truly believed that unless a scientist was capable and willing to understand his orgasm theory and its connection to the bion research, there was no hope at all that she could take seriously the bion experiments. A. S. Neill had written Reich to tell him he was trying to interest J. B. S. Haldane and J. D. Bernal in the bions at this time, for example. And on June 4 Reich wrote Neill, warning him against trying to interest Haldane: "I do beg you to be extremely cautious in making any propaganda for the bion book as far as 'authorities in the field' are concerned."[121] Reich told Neill to imagine if he (Reich) sent one of Neill's books emphasizing children's self-regulation in their education to some conservative educational authority, saying that it would achieve exactly the opposite of the effect intended. That is, it would give someone, like Kreyberg, authority to pompously pronounce that that person had examined Reich's work thoroughly and found it to be shoddy.

Neill wrote back in exasperation, saying, "You are really a most difficult man to help. You seem to think that we are all a lot of damn fools. . . . I grant that my ignorance of science is a handicap, but what I can't grasp is this: if a scientist like Haldane cannot read and understand *Die Bione*, who can? Where can one find a scientist who is a specialist in biology and also in psychology?"[122] Neill had been Reich's patient in therapy, which Neill highly valued; nonetheless, he and Reich criticized one another quite frankly over twenty years and remained friends for most of that time. Neill told Reich, "You seem to me to have a phobia that you will always be misunderstood, but why then write books if everyone is to misunderstand them? I feel that one of your main aims now should be to have [Reich's most recent scientific works] translated into English. But as I say, you are so indefinite and fearful of publicity that I don't know how to help you. It may be that living among a crowd of little Norwegian enemies you are out of touch with the English world."[123] Reich, laughing at himself, replied, "Dear Neill! You are right, I am an incorrigible pessimist, since I do not believe in the good will of academic authorities. Still, I do beg you not to let that influence you, and if you can do something toward the publication of *The Bions* or its distribution, I would be grateful to you."[124] Reich, it appears, could see the tension between his own lack of confidence in scientific authorities and, on the other hand, his confidence that there was a public out there that could take his work seriously (and perhaps, eventually, help change the dominant paradigm the scientists defended).

The Aborted International Commission

In late May the process of setting up an international commission was undermined because (as mentioned earlier) the Oslo University Medical Faculty voted not to recommend extending Reich's stay. In June, the medical faculty voted to refuse to appoint a representative to a control commission, even if one were created. This rather extreme move surely reflects the influence of Mohr, Kreyberg, Gording, and others in convincing the faculty that the bion work was "charlatanism" and not worthy of their time or attention. Not incidentally, it undercut Reich's most forceful and promising bid to get the bion experiments treated seriously and fairly by the world of science. The medical faculty offered the following official rationale: they were, they said, "not convinced that the experiments which are to be verified exist in such a form that they can receive serious scientific consideration."[125]

Reich was stymied. As he wrote to the Zurich physicist Emil Walter—one of those who had volunteered to be a commission member—"Now that the Medical Faculty in Oslo has refused to appoint a representative to the control commission it is not possible to carry out the verification procedures this summer."[126] Du Teil, because of the financial crisis precipitated by loss of his job, had to withdraw from participation, at least until his situation stabilized. Reich still hoped a Preparatory Commission could meet, anticipating that somehow the difficulties could eventually be overcome to allow the official commission to witness the experiments. But he said that even that might depend on financial circumstances, since his enemies had been so effective at driving away one of the major donors to Reich's experimental work.

Walter had urged Reich to consider writing up his bion experiments in something more closely resembling the standard format used in scientific papers. For example, in *Die Bione* Reich did not break out a separate "Materials and Methods" section but rather spread the information about techniques used throughout the entire book. Though pessimistic about the commission, Reich still thought it necessary to clarify his scientific stance on such issues, so he took issue with Walter on this point. He wrote, "I have been carrying out scientific research and producing scientific results for the last eighteen years and, as you will probably admit, my work contains some truths and is not without consequence. So far, I have managed to get by without the so-called strictly scientific form, if by this you mean a certain type of convention in publishing scientific papers. Look at any yearbook of official experimental psychology or of standard biology, as I have often done, and you will have to agree with me that at least 90 percent of all the papers could easily have remained unpublished."[127]

Here one might think the problem is Reich's arrogance: his refusal to follow decorum and show proper respect via obediently complying even when he does not like it. Replicating the experiments based on his report is somewhat challenging; hence my discussion earlier of the importance of crucial details in the correspondence and in the "Bion cookbook." But Reich argued cogently that this is only a ruse, that is, that even if he did what they wanted in terms of writing style, his enemies would still find some excuse to shoot him down. In the big picture as Reich sees it, it is clear that his refusal to be bullied—which in his opponents' eyes can only seem hopeless arrogance—is the real issue. In addition, Reich felt this was rooted, for some of his opponents, in their repulsion over his sexual theories. This is discussed further in the following text.

Many scientists who read Reich's publications today, for example, *The Bion Experiments*, react negatively to Reich's narrative style and use of selective examples because they vary so markedly from the more formalized style of scientific writing in reporting research that has today become ubiquitous. Reich's biographer Myron Sharaf pointed out that a more discursive style like Reich's was much more common in scientific writing in the 1930s, and my own reading of scientific publications from the period confirms this.[128] One need only look at two other 1938 books by scientists on origin-of-life research to see that Reich was not so unusual in this regard: Oparin's *The Origin of Life* and physiologist Reinhard Beutner's *Life Beginnings on the Earth*.

Reich appears to decide here that he is not going "to appear in court," as they ask, that is, in a venue where his opponents control all the rules. He has decided that he is going to continue to publish and rely on the court of fact and public opinion. This is a very meaningful precedent for understanding Reich's later confrontation with the U.S. government in 1954, when he wrote a formal response to the FDA complaint and submitted it to the federal judge in lieu of appearing in court in person. In that 1954 document, he urged that a court of law was a completely inappropriate place to judge the validity of scientific work.[129]

Reich told Walter, "You have unfortunately failed to understand that the Medical Faculty's excuse that my work lacks proper form is nothing more than that—an excuse." And he continued,

> To me, "strictly scientific" means describing as accurately as possible, and in accordance with the truth, things one has seen. . . . My experiments with incandescent matter and the autoclavation tests have been described in such detail that any chemist or bacteriologist could imitate them within five minutes. Instead of doing this, people are now hiding behind the objection that the publication of the results is not scientific enough. I must reject any such statement. I was the one who discovered the vesicular nature of boiled matter and of matter heated to incandescence. It should also be the duty of the so-called highly official, authoritarian science to concern itself not only with the proper form of publication, which in any case is a dreadful procedure, but also

with the truth. I can wait, and I am still prepared to show these things to anyone who is really willing to see.[130]

In the end, the International Commission never came to be. And so a crucial opportunity to give the bion experiments a fair scientific hearing, while all the experimental knowledge was still fresh and Reich still alive, was lost.

When the Munich crisis occurred in September 1938, Reich shifted much of his attention to trying to locate labs that would agree to keep bion cultures safely in the event of war. He prepared detailed descriptions of each type of culture and how to maintain it.[131] Although he contacted labs in Sweden, France, and England, there is no evidence that any offered much help.[132] Some in England, for example, were concerned that if war broke out conditions there would be even more chaotic and resources even more stretched than in Norway.

Reich as Oslo Outsider

There are many things that put Reich at a disadvantage with his opponents, all of whom knew one another in the way only "small town" life can lead to.[133] Theodor Thjøtta ran an independent bacteriology institute, but he was also the professor of bacteriology at Oslo University.[134] As mentioned in Chapter 2, Reich had been persuaded by Odd Havrevold to send Preparation 6 bion cultures to Thjøtta for identification in March 1937. Thjøtta replied at first to Havrevold, telling him "these things have been known all along" and "there were spores in all of them." Reich thought Thjøtta's response flippant, saying he could not explain why no growths developed from the individual ingredients of the mixture when each was tested separately for growth.[135] Havrevold, however, grew extremely anxious, perhaps because a senior Oslo professor—the only professor of bacteriology in Norway—was telling him he was working on a project of no scientific value. Reich noted in his diary that he had been noticing an attitude of disapproval from Havrevold toward himself, because of which Havrevold avoided participation and had lost touch with what was actually going on in the experiments.

As a result, Reich noted, Havrevold "ran to the authorities. I could see his fear of being made a fool of through me. If *I* had my doubts, even after months of actually observing things, then how could *he* help being doubtful? . . . He was struggling to prove his maturity and independence—that is, until yesterday, when he told me that he had caught himself in an act of sabotage against me." Reich was shocked. But he could clearly see the difficult position a junior person was in, if he knew his future career in Oslo depended on relations with scientific authorities there.[136] Just a few days later, Reich wrote in private notes, "Havrevold is afraid. [Vivi] Berle [Reich's bacteriology assistant] is insecure. Havrevold is making her insecure. People do not see the big picture; they get hung up on the details!"[137] Reich gathered his group in the laboratory for an evening discussion of the situation. In private he noted, "They're all cowards in the face of authority. I spoke very frankly with them."[138] Reich, as one of the world's most famous psychiatrists, could face up to authorities if he felt they unfairly used their influence. (In a "small town" like Oslo, on the other hand, Reich's own confidence and strong sense of the importance of his contributions were more likely to come across as brash and unseemly—as we have seen already via Hoffmann's reaction to Reich. Hoffmann himself noted that the bion controversy could occupy a small town like Oslo for months, while in Paris it might have blown over in a few days.)[139] For young physicians or lab assistants who saw their entire future careers in Oslo, the risk felt much greater. Similar pressure on Havrevold and on Nic Hoel was applied by Scharffenberg in March 1938 during the press campaign. Schreiner's warnings to Havrevold have also been discussed. In a letter written during the bion experiments, Sigurd Hoel reported that some of his colleagues felt Reich was directing his anger to the wrong target. Hoel told Ellen Siersted, "I have spoken with Reich and confronted him according to what you mentioned in [your previous] letter—that his anger preferably should be directed towards the place whence all the slander and the intrigues are originating."[140]

There are other factors that worked against Reich because of being an outsider. For example, as a stranger in Oslo, Reich may not have known of several key facts about his opponents. He may not have been aware of Scharffenberg's previous sharp statements that Hitler was a psychopath. This would not have been intuitively obvious if Reich knew that Scharffenberg supported social eugenics and sterilization.[141] Similarly, Reich might not necessarily have known (unless he heard it from Hodann) of Mohr's sharp antieugenic stance or of his campaigning for sex education, abortion rights, and birth control reform. In Norway, Reich's professional colleagues had been mostly psychiatrists sympathetic to psychoanalysis. Mohr and Kreyberg attributed much human disease to Mendelian genetics; in that light, Reich's diary entry describing Mohr as a "socialist, genetic biologist" suggests Reich may have guessed mistakenly that Mohr fell prey to the excesses of eugenics.[142] Reich had no colleagues or contacts among the *Drosophila* research community, but left-wing figure in that world H. J. Muller was a known public advocate of eugenics.

It is also worth recalling how common anti-Semitism was at this time, even among many critical of the Nazis (Muller often felt it impacted his career).[143] Also, Reich turned out to be right about many he suspected of being fascists, such as immigration official Leiv Konstad and Klaus G. Hansen, a university chemistry professor who had been a scientific critic of the bions.[144] Reich clearly understood, too, that the Norwegian Labor Party was being cowed and manipulated by Norwegian fascists such as Konstad.[145] Hansen was a member of the Norwegian Nazi Party who helped the German Nazis during the occupation.

Harald Schjelderup could perhaps have set Reich straight about Mohr, but by late 1937, Schjelderup was already distancing himself from Reich. In fact, by 1940, Schjelderup was so angry at Reich that he harshly criticized Reich's ideas about therapy in print in a book he published that year—this despite a sharp warning from Sigurd Hoel, arguing if that Schjelderup went ahead with this attack while Norway was under the Nazi occupation he could betray many other of Reich's former coworkers (Nic Hoel, Raknes, Havrevold, and others), bringing them to the attention of Nazi rulers as left-wing threats.[146] As I will discuss later in this chapter, Reich's left-wing students and coworkers faced very real danger from the Nazi occupation.

Karl Evang was a prominent Norwegian Socialist public health worker whom Reich did know. Evang knew Sigurd and Nic Hoel from Mot Dag quite well and had begun an organization for sex reform and sex education in Oslo in 1929. In 1932 he founded *Tidskrift for Sexuell Upplysning* (Journal for Sexual Enlightenment) in Oslo; simultaneous editions were published in Stockholm and Copenhagen. Evang was the founder and driving force in the Socialist Physicians Association, serving as its chairman from 1931 to 1938. One might have expected Evang, much like Mohr, to be a natural ally to Reich on many fronts.[147] And Evang did sign Hoel's petition on behalf of Reich's residence permit. But Evang strongly disliked Reich. According to Reich, this dated back to 1933; he speculated on possible reasons: "He visited me one day together with Leunbach when I was in Copenhagen. At this occasion we discussed the problem of puberty, and as far as I recall, a rather heated argument developed over his evasion of the problem of sexual embrace during puberty. I think I reproached him for not being more courageous in this respect. There was also the point that I had established sex consultation centers in Austria already in 1928. He seemed rather peeved since he had believed that he was the first one to establish such centers in Oslo [in 1929]. This is only for your information as a possible explanation of his sustained animosity."[148] Other authors have suggested "Reich's aggressive personality" was what was off-putting to potential allies. Whether Reich is correct about the exact reasons for Evang's hostility, it is clear that, among those on the political left with high ideals about sex reform, personal jealousies and hostilities were capable of denying Reich support from some quarters where one might have looked for it. Thus, there were many pitfalls for Reich because of being an outsider, temporarily transplanted into a "small town" setting where all the local scientists knew one another well but where Reich did not know some key things about the background of most of those who became his opponents.

The U.S. State Department Investigation of 1951–1952

During the Food and Drug Administration's investigation of Reich, the U.S. State Department requested the American Embassy in Oslo to interview participants in the 1930s bion controversy, in order to help evaluate the quality and legitimacy of Reich's scientific work.[149] William M. Kerrigan, second secretary of the embassy, conducted interviews in 1952 with a wide range of subjects. Fortunately all the documents pertinent to these interviews have been preserved in U.S. State Department archives and were obtained by Professor Jerome Greenfield in the 1970s using Freedom of Information Act (FOIA) requests.[150] More fortunately still, Kerrigan proved to be a remarkably resourceful and thoughtful interviewer, often cross-checking by asking sources he thought more objective and trustworthy about the reliability of interviewees he thought more likely to be biased. He was particularly diligent in this regard with respect to technical questions about microscopy and microbiology, in which he did not have expertise. These interviews were conducted, it is true, thirteen years after Reich left Norway for the United States. And as Kerrigan himself notes, "The emotional heat which the controversy still appears to generate among many of the participants left the reporting officer [himself] in some doubt as to the objectivity of some of the persons interviewed."[151] However, it is often the case that recollections gathered and intelligently cross-checked at that time are far more historically useful than oral history interviews conducted several decades after the fact.

In particular, Kerrigan observed that it was difficult to accurately assess the degree to which Reich's highly unorthodox views about sex might influence his opponents' views about his biological work; Kerrigan was, however, "struck by the frequency with which Reich's sexual views were referred to by his opponents when their opposition to Reich's biological theories was the matter under discussion."[152]

In attempting to assess the scientific credentials of Kreyberg and Thjøtta, Kerrigan said this was "extremely difficult and possibly of questionable accuracy . . . because there is no precise yardstick by which the attainments of either man can be measured. For example, Professor Kreyberg is the only professor of pathology in the whole of Norway, and Professor Thjøtta is the only professor of bacteriology." Kerrigan noted that Kreyberg had only three published works, two from 1937 and the third from 1947, whereas Thjøtta had "a long list of published works and memberships in several learned societies, including the Society of American Bacteriologists."

Kerrigan decided to consult Professor Harald Sverdrup, director of the Polar Institute of the University of Oslo, because he had "an unassailable position in the scientific world and a reputation for the highest type of intellectual integrity." What Sverdrup told Kerrigan was, he felt, most enlightening: "In commenting on Professor Kreyberg, Professor Sverdrup said that he would accept Professor Kreyberg's scientific opinions only if they were accompanied by corroborative evidence from some other source. He said that Professor Kreyberg tended to be hasty and careless in his judgments and was also a very high strung and emotional person. He stated that Professor Kreyberg tended to take very strong personal likes and dislikes and if, for example, in the Reich case he had taken such a dislike to Reich or his theories, his emotional reactions would override any scientific objectivity he possessed."[153]

On the other hand, about Thjøtta, Sverdrup stated "that he had considerable confidence in his scientific judgments, and said that Thjøtta was not the man to utter any hasty and ill-considered judgments on any scientific questions. He said that Professor Thjøtta was personally a somewhat difficult and abrupt person, who would be very likely to give a more curt and unpleasant answer than the situation might call for, but that basically he would be very careful in expressing his opinions in scientific matters."[154]

Thjøtta had only replied to Reich by phone in 1937, the reader will recall. He took part a year later in the public polemics against the bions in April 1938, saying Reich's bions were "nothing but staphylococci from air infection." Given this, it is interesting that, by his own account when interviewed later, "Professor Thjøtta stated that he never saw Dr. Reich in person, but only had the above described telephone conversation, and possibly one or two more with him. Professor Thjøtta says that he never had any correspondence with Dr. Reich. Professor Thjøtta stated that he did not witness the experiments to which Reich had invited him along with Professor Kreyberg and the other two professors, because he considered that his examination of Reich's cultures was sufficient evidence to show that Reich's theories were nonsensical."[155]

Kerrigan found Sverdrup, thus, a useful gauge to evaluate the claims of Sigurd Hoel and others among Reich's supporters, who thought that Scharffenberg, Kreyberg, and others were incapable of objectivity about Reich's science because they found his unorthodox sexual theories deeply offensive. Kerrigan heard from many that the eighty-two-year-old Scharffenberg had "an irreproachable reputation for personal integrity." But he also noted that "his attacks on Reich were the most violent and sustained of the entire controversy." Having interviewed Hoel as well, Kerrigan noted, "Mr. Hoel described Scharffenberg as being an extremely eccentric man who has refused to marry because, as Scharffenberg once publicly stated, there had been insanity in his family background and he was afraid that the strain of insanity might be transmitted to any possible offspring that he might have. Mr. Hoel obviously considered that Scharffenberg did not have the normal adult attitudes toward sexual matters and his attack on Reich was actually an indirect attempt to cover up for his own personal inadequacies in that direction."[156]

Again, Sverdrup's assessment was informative: he "professed the very highest regard for his [Scharffenberg's] intellectual and moral honesty. I put the question to him as to whether in the Reich case Dr. Scharffenberg's personal ethical attitudes and his attitudes on the subject of sex might not have been deeply offended by Reich's very unorthodox views and that those views might not have furnished the actual basis for Scharffenberg's attacks on Reich. Professor Sverdrup said that in view of Scharffenberg's opinions on sexual and ethical questions it was by no means unlikely that he may have found Reich's sexual ideas deeply disturbing and as a result have used them as a basis for his attacks on Reich, which attacks of course had another ostensible basis."[157]

In his comments evaluating his interview with Otto Mohr, Kerrigan noted, "It will be seen from the foregoing that Professor Mohr has a violently antagonistic attitude toward Reich and his theories." Kerrigan had a clearer sense than with the others he interviewed that Mohr had "a very low opinion of psychoanalysis in general" and a strikingly violent reaction to Reich's sexual theories. Kerrigan's impression is expressed in terms that bear quoting at length:

> Professor Mohr did not describe Reich's [bion] experiments... and in general appeared not too well acquainted with this specific aspect of Reich's activities.... Professor Mohr, in reply to my question [about Reich's sexual theories,] said that Dr. Reich's theories are an offshoot from the Freudian theories about sex and the libido and described Reich's theories as "pornographic nonsense."... It will also be seen that Professor Mohr gave me no specific refutation of any of Reich's ideas, although that of course does not mean that the refutation for those ideas do [sic] not exist. It may be that since the controversy over Dr. Reich occurred 13 to 15 years ago Professor Mohr's memory of the details of Dr. Reich's theories and the refutation of them have been blurred by time. I was particularly impressed by the violence of Professor Mohr's reaction in speaking of Reich's sexual theories. It seemed to me, in the absence of any specific explanation of, and refutation of Dr. Reich's sexual theories, that Professor Mohr may simply have been reacting as many elderly and conservative-minded men might at what he considered to be an unnecessary and uncalled-for discussion of a subject that he, Mohr, considered would be better left undiscussed.
>
> Professor Mohr described Reich as a fraud and a humbug. When I asked Professor Mohr if he felt that Dr. Reich honestly believes his own theories, particularly about the creation of life... Professor Mohr said that, although he himself had never met Dr. Reich, he found it difficult to believe that he could actually believe these things himself.
>
> It will be interesting to determine in the course of this investigation the importance of anti-Semitism in this case. Professor Mohr made such an issue of the fact that he himself was not anti-Semitic, that there may be some possibility that anti-Semitism (Dr. Reich is apparently Jewish) may have played a fairly considerable part in the case here in Norway.[158]

It seems clear from this assessment that Kerrigan was unlikely to be aware of Mohr's (and his wife's) extensive efforts in Norway for reform of sex-education in the schools, and liberalization of abortion laws. (These activities by Mohr do not preclude, of course, the possibility that he could still have reacted viscerally to Reich's particular ideas about adolescent sexuality, compulsive monogamy, or other such topics.)

However, when Kerrigan interviewed psychiatrist Gabriel Langfeldt, another very strident public critic of Reich's bion experiments, he found that Langfeldt was considerably more "approving of" Reich's "psychiatric theories, as distinguished from his theories of the creation of life and the existence of orgone." Langfeldt "felt that Dr. Reich's theories about the importance of sexual attitudes towards mental health had considerable validity, although he felt that Dr. Reich had taken much too one-sided an attitude on the question."[159] Langfeldt spoke similarly about Reich's technique of vegetotherapy. From this Kerrigan concluded, "Professor Langfeldt's attitude toward Dr. Reich's psychiatric theories appear [sic] to confirm my suspicion that Professor Mohr was reacting emotionally rather than scientifically to those same theories in his comments on them to me." Langfeldt, unlike Mohr, "was very definite in saying that he felt very certain that Dr. Reich was completely sincere in thinking that he had discovered how to create life." Like Mohr, he had never met Reich in person.[160] It does seem remarkable, given what a relatively small town Oslo was and how many months the controversy raged in the newspapers and public meetings (although not in the scientific literature), that neither man ever met Reich (or tried to), or that Thjøtta never spoke with Reich except by phone, choosing instead to communicate through the more junior scientist Havrevold. None of them ever looked through Reich's microscopes. Only Kreyberg ever visited Reich's lab, once for one hour. Thjøtta refused to do so, since he was certain from a single sample that Reich's entire scientific project was worthless. The anti-Semitism issue was never conclusively proven or disproven.[161]

Langfeldt, incidentally, took the opportunity of Kerrigan's interview with him to denounce Nic Waal as a Communist, as well as Havrevold. Kerrigan corrects this by noting on the record that the "controlled American source" in Oslo (i.e., an American intelligence operative) "states that she is not now, and never has been a Communist."[162] Kreyberg, along with Gabriel Langfeldt, orchestrated a successful behind-the-scenes campaign in 1950 to sabotage Nic Waal's attempt to get a professorship at Oslo University, because of her continued interest in Reich's work.[163]

The fate of Reich and his bion experiments, then, became entangled in the twists and turns of the politics of science and sex reform in Norway, internal university politics, and the dynamics of the RF "old boy network." And all these entanglements acted against Reich's chances of getting funding or even a further hearing. So quite independently of the scientific merits, Reich was fighting an uphill battle to get his side of the story a fair hearing before the scientific community.

To point out their biases against Reich is not intended to single these men out for having more biases than other scientists (or human beings). Many of them behaved very bravely in the face of Nazi occupation—the details of which Reich, having immigrated to the United States in 1939, was not in a position to know, even after 1945.[164] Many of Reich's students and colleagues showed equal courage: Nic Waal, for example, used her own car to rescue children from Nina Hasvold's Jewish orphanage and drive them to safe houses, on their way to eventual safety in Sweden.[165] In October 1943, Harald Schjelderup was arrested, and he was held at Grini concentration camp for the rest of the war. Arnulf Øverland and Christian August Lange were arrested and spent most of the war in the German concentration camp at Sachsenhausen. Øverland's wife, Margete Aamot, was imprisoned in Ravensbrück.[166] Reich's student Karl Motesiczky (who wrote for SexPol under the pen name Karl Teschitz and participated as a subject in the bioelectric experiments) returned to Vienna in 1938 and helped organize resistance to the Nazis; he died in Auschwitz in 1943 at age thirty-nine.[167] With Reich's help, their colleague, pianist Erna Gál, escaped the Nazis from Vienna to Oslo and then, with help from A. S. Neill, to Britain.[168]

Nils Roll-Hansen describes Mohr as a believer in liberal values, social progress, and the modern welfare state based on science. In his assessment of Norwegian debates about eugenics, Roll-Hansen argues that—particularly for Mohr—the actual facts of the science mattered greatly. Science had as a primary goal understanding what the world is really like; thus science was not *merely* an instrument of political and social forces. When the facts of genetics differed from the wishes of eugenicists, Mohr pointed this out forcefully and helped shape social policies that were, to at least some extent, guided by science in an enlightened way. However, in the story of Reich's bion experiments, one can only be struck by how few of Reich's opponents, including Mohr, ever seriously engaged with Reich's science. Then as now, they judged it immediately to be pseudoscience and therefore not worthy of any of them actually ever taking the trouble to visit Reich's lab.[169] In this regard, whether Reich had an aggressive or defensive personality is beside the point. Kreyberg apparently had such a personality as well, but this was not an obstacle to his opinions being highly respected by Mohr, Schreiner, and most others from Oslo involved in the controversy. The interviews by Kerrigan conducted in 1952 underscore dramatically that none of Reich's opponents had more than a tiny amount of direct knowledge of his work, with the sole exception of Kreyberg, who had a small amount.

Roll-Hansen has also given us an excellent account of how Mohr used his seniority and connections in the *Drosophila* genetics network, and in the international scientific network broadly, to help keep politics out of science to the greatest extent possible, even in highly politicized negotiations such as those surrounding the planned 1937 international genetics congress in Moscow. Says Roll-Hansen, "One purpose of telling this detailed story about the cancellation of the international genetics congress in Moscow is to give a picture of how the international scientific network was able to function even under difficult political circumstances. We see how important the personal relationship between the leading scientists and their direct responsibility for the decision-making process was."[170]

Robert Kohler gives a similar assessment of the *Drosophila* "fly-network" of geneticists (again, including Mohr): "The fly group did not have a monopoly, exactly, but they did have the ability to control any significant new lines of research if they chose to do so. . . . It encouraged generosity and sharing between center and rank and file. It also fostered a kind of benevolent paternalism, whether the fly group consciously intended it or not, and this tendency was reinforced because many of the smaller [players] . . . were fly group alumni, who retained habits of loyalty and deference from their apprenticeship experience."[171]

The flip side of this same coin is the RF "old boy network's" exclusivity to "outsiders," so defined for whatever reasons suited the small handful of influential Haldanes, Mohrs, Dunns, Dobzhanskys and Bernals. Even Kreyberg and Thjøtta had connections to the network already, Thjøtta from being at the Rockefeller Institute in New York City, Kreyberg because of Mohr's patronage. So for this book, one purpose of telling this story is to show, at another time of difficult political circumstances, how important personal relationships between scientists, their paternalism, and their direct responsibility for policing the scientific network can be in declaring a given researcher persona non grata, or in declaring a given body of work "pseudoscience."

Divergence in Life Science Research Agendas

It has been well documented that physicist Warren Weaver, head of the RF Natural Sciences Program for over four decades, had a clear agenda in funding life sciences research that would import the methods of physics and chemistry into biology. Only thus would life sciences progress, Weaver believed. After Watson and Crick he even claimed his selective funding along these lines had been aimed at creating "molecular biology." Whether he actually thought of that term in the 1930s, there is no doubt that Weaver's intervention was a major part of what shifted biology in the direction of explaining living phenomena in terms of macromolecules. Because his selective funding spanned the Depression years, when the RF was practically "the only game in town" for large-scale life sciences funding, the influence of Weaver's vision was greatly magnified. Whereas Staudinger in 1926 may have had to fight the vast majority of his colleagues to believe proteins actually were macromolecules rather than variably structured colloids, by 1945 the ultracentrifuge, the electron microscope, X-ray diffraction, the electrophoresis apparatus, and other tools supported by the RF had won over the vast majority of life scientists to the macromolecular point of view. Thus, mainstream life science took a fairly sharp turn from the late 1930s through the 1940s, away from numerous older research strategies and overwhelmingly toward physical-chemical reductionism of an unprecedented extremity (the Watson-Crick "Central Dogma" era of ca. 1953–1970) and toward macromolecular explanations of living phenomena in place of older (e.g., colloid) explanations. If the 1930s were still an era when many sought a middle way between mechanism and neovitalism, by 1945 vitalism was completely delegitimized and the entire research spectrum moved far over toward Weaver's way of thinking. A research path like Reich's, thus, after 1940–1945 would look as far out of the new mainstream (or further) than vitalism had appeared to be in 1930. This was accentuated still further by the movement of many former physicists into the life sciences—particularly after Hiroshima—bringing with them a view that "we cracked the problem of atomic structure in just two decades; we'll do the same for heredity and other problems biologists have long considered Gordian knots that may take centuries to solve."[172]

I have captured this widening divergence between Reich's research agenda and the mainstream in Figure 6.5. Disparaging evaluations of Reich's science point to the widening gap between his line of research and the mainstream

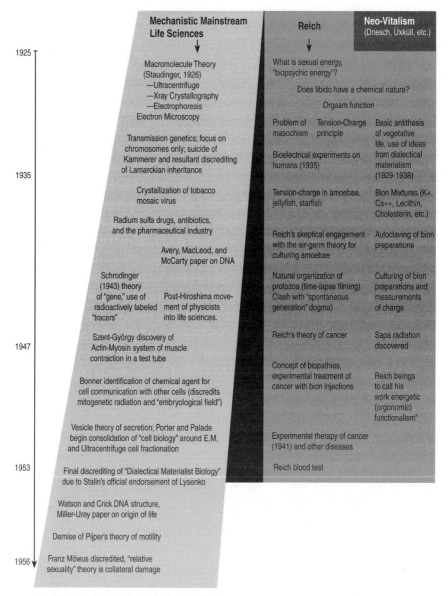

Figure 6.5: Divergence in Life Sciences Research, 1925–1953.

and suggest that means it was he who "veered off" in fruitless, even irrational directions. But that picture of a "duck" can also be seen as a "rabbit." (I am referring to a classic ambiguous form used in psychology to illustrate the Gestalt shift phenomenon.) The same picture could be interpreted to emphasize RF funding and other forces that caused the widening gap by leading the mainstream to veer far off the direction it had been going in through the mid- to late 1920s.

Thus, after Watson and Crick's DNA structure in 1953 solidified the victory of the new molecular biology, it would be almost impossible for a modern biologist to even think in the same language as Reich, or take seriously the questions he was asking—almost entirely from the point of view of an energy principle or bioelectric charge. That gap has only widened more and more rapidly in the years since 1953. In other words, a major part of why Reich's scientific writings sound so foreign to a biologist trained today—like it cannot be understood as biology at all—is because of a historical shift in the direction of the mainstream, rather than (or at least as much) because of a shift in Reich's own direction.

In 2014, it is extremely difficult for a biologist to see what Reich was doing as sensible or connected to any existing (even if minority) research direction. In the 1930s, as I showed in Chapter 1, Reich's research was very clearly seen as rooted in a number of respected (if minority) research traditions. Indeed, the research agenda that sought to identify a specific life energy probably received a far greater death-blow from the success and fame of Watson and Crick, of antibiotics, of unraveling biochemical metabolic pathways—and the sheer productivity and money-making power of these approaches, via pharmaceuticals—than from any decisive experiment (there was none) disproving the existence of such bions and their lifelike properties—or even from discrediting because of the later U.S. government case against Reich.

7

SAPA BIONS AND REICH'S DEPARTURE FOR THE UNITED STATES

After the intense press controversy died down in the fall of 1938, Reich tried to keep his work out of public view, and he pursued ever more actively a visa to immigrate to the United States. Nonetheless, he continued experimenting, particularly pursuing his cancer research. Reich believed he discovered a new form of specific, biological energy in a particular type of bion culture—SAPA bions—in February and March 1939. This is the final major episode—and in many ways the most astonishing, or strangest—in the bion experiments prior to Reich's departure for America, where this story ends. Reich published none of this work until July 1939, on the very eve of his departure for the United States, and even then only in a very preliminary form. But even though these experiments were not part of the Norwegian press controversy, their key role—especially these final SAPA bion experiments—in all Reich's later studies of orgone energy means these need to be explicated in some detail.

SAPA Bions and Their Radiation

In November 1938, Reich received a letter from a Dutch physicist, Willem Frederik Bon, who had read *Die Bione* and found Reich's experiments fascinating (see Figure 7.1). Bon said he had no doubt Reich's explanation of the bions as transitional steps to the origin of life was correct; however, he noted, "Whether the world is ripe for your discovery I can not judge. There are in the history of science many examples of new theories confirmed by lots of evidence that were simply denied." Most of all, Bon wanted to know whether, when bions form, "radiation is emitted and whether absorption of radiation is necessary for the emergence of bions."[1] Bon believed that life was a radiation phenomenon, but he had had little success in convincing

Figure 7.1: Willem Bon (at left) in his lab, Amersfoort, 1939. Courtesy of Wilhelm Reich Infant Trust.

physicist colleagues of his idea. As I mentioned in Chapter 5, Reich had been speculating about the possible existence of radiation, "life rays," and similar ideas for almost three years at that point, since the bioelectric experiments. However, since he had as yet no concrete evidence—no direct detection of radiation in bion experiments—he replied to Bon in the negative.

On Sunday, January 15, 1939, Reich was demonstrating the incandescent soot experiment to a Danish therapist who had studied with him, Tage Philipson, MD. By mistake, one of the lab assistants took sand from a jar on the laboratory shelf rather than carbon, heated it to incandescence, and plunged it into the usual liquid culture medium Reich used, 50 percent 0.1 N KCl solution and 50 percent nutrient bouillon. The sand was from Nevlunghavn, on the coast of Norway south of Oslo.[2] Bions were produced, fairly large ones that were much less motile than soot bions. Since they had the "packet" shape of other "PA" bions, Reich called them SAPA ("sand packet") bions. Reich found these bions were culturable; he repeated the experiment with sand and found he got the same bions in five out of eight attempts (thus, six out of nine total).[3] The broth culture produced by directly adding the incandescent sand was incubated at 37°C for five days, to watch for possible growth of contaminants. After five days, the liquid culture was reinoculated on Reich's solid egg medium IV. "After 48 hours," Reich reported, "the medium in question displayed an intense yellow, creamy growth. A microscopic examination showed bions of the packet amoeba type."[4] These bions differed from the otherwise common packet amoeba and bions, however, because of their different size (about 10–15 µm in diameter) and a relative immobility when compared to other cultures. The cultures grew well when reinoculated onto blood agar or simple agar, not only on the initial egg medium. He found the cultures fascinating and studied them under the microscope for several weeks.[5]

By Friday, February 3, for the first time in a laboratory notebook he made a brief entry about SAPA bions that notes "Strahlung?" [radiation].[6] That is, he saw a very bright field around the SAPA bions and several other phenomena that made him think they might be emitting radiation (see Figure 7.2). For example, he noted, "I had the impression that I could see unusual light phenomena when using a dark field condenser (Reichert and Co.) at a magnification of 2000×–3000×. Under one specific dark field condenser setting, one could see peculiar streaks of light similar to rays, sometimes appearing as continuous lines, which regularly appeared and seemed to originate from especially bright, glowing places inside the bions." Still more attention-getting, Reich said, was that "these radiations moved in rhythm with the bion movements, and their position could not be influenced by repositioning the reflecting mirror. A gradual reduction of the microscope's light completely weakened the radiation; however, the tiniest ray of light caused them to light up again. Likewise, it is noted that, with correct light and dark field condenser settings, the bions were filled with light and dark lines at regular intervals."[7] Despite Bon's suggestion a few months previously, Reich said at first it did not occur to him that radiation might be involved. However, after four weeks of looking at SAPA bions every day under the microscope, Reich's eyes began to hurt. "They felt tight, I could hardly use the microscope because they were so sensitive to light, and my eye pressure was increased. An ophthalmologist in Oslo diagnosed conjunctivitis and ordered rest. For several days, I wore black-lensed safety goggles until my eyes improved."[8]

Reich said he tried looking at the bions only with one eye, through a monocular microscope. "After about 1.5 hours, that *one* eye began to hurt and to

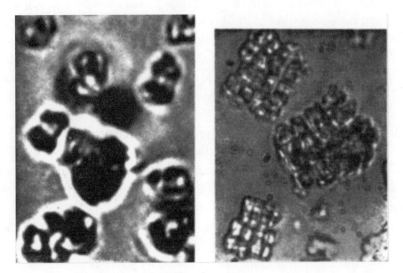

Figure 7.2: SAPA bions, magnified 4000×. Courtesy of Wilhelm Reich Infant Trust.

display the same symptoms. Following extended periods of microscopic observation, previously unknown and strange looking, deep-blue images appeared until the eye left the eyepiece."[9] By Thursday, February 16, Reich wrote excitedly in his diary that he was certain he was dealing with radiation and, thus, *"A completely new world!"*[10] On February 17, Odd Havrevold was studying SAPA bions microscopically and also complained of eye pain similar to Reich's.[11] Several lines of subjective evidence seemed to reinforce the perception of some kind of rays from the bions with biological effects. Reich noted that people whose hand palms have a high electrical charge (he could measure this with the equipment from the bioelectric experiments) "perceive a subtle prickling in their palms and sometimes warmth when exposed to SAPA cultures in agar tubes held steadily at a distance of 2–5 mm for 2–5 minutes. The same reaction results from smearing an eyelet [a small amount in a wire inoculation loop] of the culture on a quartz specimen slide and placing on the skin." Reich was not satisfied because of the element of variability, saying, "This experiment is unreliable because results depend on the electrical charge of each individual's skin and on each subject's ability to perceive the prickle."[12]

Objectively, however, a distinct reddening of the skin usually occurred within five to ten minutes, if a small amount of SAPA bion culture was placed on a quartz microscope slide, then that slide placed upon bare skin. Quartz was used because, unlike glass, it is transparent to radiation in the ultraviolet wavelength range.[13] Reich tried this because he had in mind the model of mitogenetic radiation (MR), which also passed through quartz but not glass. When the SAPA bions were observed on a microscope slide in the vicinity of T-bacilli or cancer cells, they paralyzed and killed these objects much more quickly than previous PA bion types had done (see Figure 5.3).[14] Reich said that when bion cultures were placed on top of (or even near) new, unexposed photographic plates still in their cardboard box and inside a paper wrapper, the plates became fogged. The rays emitted by the cultures, then, could penetrate cardboard and paper, suggesting they were more powerful than alpha particles.

As a result of all this new evidence, Reich asked Tage Philipson, his student in Copenhagen, if he would contact Niels Bohr and ask "whether he might be interested in carrying out the necessary physical investigations" to identify and characterize the radiation. He also asked Philipson to see if he could find a specialist eye doctor in Copenhagen who knew "what danger exists for the retina when it is exposed to radiation from radium-active substances."[15] That same day, he also wrote to physicist Willem Bon in Amersfoort, Holland, asking if he would be willing to help with physical investigations of the radiation and asking if it could be done with "the utmost discretion."[16]

Reich hoped the laboratory workers might be somewhat shielded from rays with powerful biological effects, by keeping the SAPA culture tubes and plates inside a Faraday cage, because photographic plates fogged less or not at all when protected by a Faraday cage with the cultures outside it.[17] Invented in 1836 by British physicist and chemist Michael Faraday, the Faraday cage is a grounded metal enclosure—usually comprising mesh walls, but sometimes solid—that blocks out electromagnetic energy from the outside to ensure the enclosure is free of electrical charge. Inside the cage, electroscopes are used to confirm the absence of any influx of electricity. Reich used Faraday cages of various sizes, including one large enough to sit and work in, 2.5 meters in length and width (not unlike the first "cage" experiment of Faraday himself that involved a foil-lined room). Reich built the large Faraday cage in the basement of his house—where these experiments were taking place. Thus, observations in a dark room could be combined with observing the effect of a Faraday cage. To isolate SAPA phenomena from any electrical energy for study of its properties, Reich placed test tubes and Petri dishes of SAPA cultures *inside* Faraday cages for microscopic work and experimentation. Working inside the large Faraday cage, he noticed that "tingling on the palm of a hand can also be felt at a distance of approx. 3 millimeters from the copper wall of a Faraday cage after it has had many SAPA cultures placed within it and left for several hours, or, better still, several days."[18]

Some microscopists might give a different interpretation of the visual effect that Reich later called the "energy field" around the SAPA bions, which he also described around other bions and around red blood cells.[19] Bacteriologists might interpret this as a capsule (often polysaccharide material) or "array envelope" around what they would interpret as bacterial cells.[20] (This could perhaps be tested by analyzing its chemical makeup.) Traditional microscopists would probably interpret this ring of light around the edge of a tiny object as a diffraction ring or as refraction through the cytoplasm/fluid content of a living object, which would differ from that of its surroundings.[21] Reich points out that when using good apochromatic lenses, this field is blue (around red blood cells as well), more or less strongly blue in parallel with the electrical charge and culturability of the bions. He also notes (and uses as a criterion in his Reich Blood Test) that red blood cells from one patient to another (and of the same patient in varying states of health or illness) vary dramatically in the width of this "field" around themselves.[22] That argues for the phenomenon not being mere diffraction. Traditional microscopists might respond that red blood cells could differ in such properties as their refractive index compared to that of the physiological saline solution in which Reich observed them.[23] This dispute might require some careful experimental work to resolve. And even if Reich turned out wrong to think this visual effect was an energy field, with SAPA cultures the remaining, more objective effects such as skin reddening and electroscope effects still remain to be explained. At this point in the experiments, all the other biological and physical effects of the radiation the SAPA bions emanated persuaded Reich that the light around the bions was a "field" of the energy they were emanating. And as described here, he was able to photograph that energy through the microscope using darkfield techniques, independent of this "field" phenomenon.

Reich also turned to using zinc sulfide discs (which luminesced when struck by radioactive emissions) and darkroom observations, as developed by Rutherford and others thirty to thirty-five years earlier for the study of alpha particles and which had since become standard techniques for studying radiation emitted by radioactive substances.[24] Havrevold, and later numerous others, "without knowing anything beforehand," confirmed the majority of Reich's observations in the darkroom.[25] Philipson advised Reich that, unfortunately, Niels Bohr was currently in America and unable to conduct the physical studies Reich wanted. He told Reich, however, that an assistant at Bohr's institute who was "very skeptical" was willing "to investigate it to the extent possible."[26]

Reich, his self-confidence badly damaged by the recent Norwegian press campaign, was extremely cautious about the possibility of another episode like he had had with Kreyberg or Thjøtta, where a "skeptical" observer used such an opportunity to publicly dismiss the work without giving it a serious or fair examination.[27] He wrote Philipson that before he would send samples to the physics institute, he wanted to be sure they really would do serious measurements and take into account all he had done so far, which made clear the phenomenon was real.[28] In the end, this never came to pass with Bohr's institute. Reich turned increasingly for advice to Bon in Holland, a physicist he knew from prior correspondence who took the bion experiments seriously. Reich was impressed that Bon had even foreseen the possibility of a bion culture that gave off radiation. He described these experiments and darkroom observations in detail to Bon in a letter of February 27 as well as in a report he wrote on March 10. He summarized these subjective observations thus:

> In total darkness, preferably inside a Faraday cage, one can observe [around the bion cultures] a diffuse, pale, bluish glow that is boundless and unstable (like patches of fog). After about half an hour (when one is accustomed to the darkness and can perceive the glow), putting on a pair of black-tinted glasses will decrease the intensity of the pale radiance. On the other hand, the glow on the copper walls of the cage and on metal objects can be seen more clearly when the glasses are removed.
>
> Wearing the glasses and staring at a metal wall in total darkness for an extended period of time, one sees deep-blue-violet points of radiation rhythmically emerge. This phenomenon can be intensified by holding up a sheet coated with zinc sulfide.
>
> The same happens when one holds a strong magnifying glass between one's eyes and the shiny glow, adjusting the distance

until the shiny glow appears denser. The violet points of light can then be clearly observed, especially if the magnifier is connected to a focusing screen at the correct distance.[29]

Reich described in detail several other groups of subjective phenomena: "If two people sit in the dark [in an enclosure with SAPA cultures], the second person, totally without prompting, will report that the same pale glow appears and is denser around the white lab coat, the head, and the hands, etc. It was possible to reach for the glowing objects." He noted that by repeatedly putting on and taking off the black-tinted glasses, one could become convinced of the accuracy of the observations. They became less intense when viewed through dark glasses. "One must differentiate between a general pale glow and the deep-blue-violet radiance," warned Reich. One afternoon, Reich said, when he had been sitting in the dark room continuously for five hours with the SAPA bions, the fingers and palms of his hands glowed. "This shimmer diminished after a few days. However, the palm of my left hand retained a feeling of warmth accompanied by tingling, alternating whiteness and redness, and a bright glow visible in the dark room exactly on the spot where I had regularly held the test tubes containing the preparations during subjective tests of the emissions."[30] Eventually that spot on his hand became inflamed and painful.

It bears repeating that Reich said these visual impressions became clear only after the observer first sat in total darkness for half an hour to allow the eyes to acclimate and that he asked others to describe what they saw without any prior prompting or description of what he had seen. But if a magnifying glass made the effects appear brighter and dark glasses made them dimmer, and if different observers saw similar visual impressions, Reich felt this was strong presumptive evidence that the phenomena were outside the eye and not some subjective impression within the eye.

If an observer sat for one-half to one hour in the basement Faraday cage, "the air was perceived to be 'dense,' 'heavy,' 'oppressive,' depending on the individual. If one remained in the Faraday cage too long—two to three hours—one felt compelled to leave it again. Once outside of the room, one could breathe a sigh of relief. If one is exposed to air influenced by the preparations for longer periods of time, one becomes tired and listless. Fresh air quickly dispels such conditions." This left one feeing fresh and strong, Reich said. He pointed out that this impression, though subjective, was also reported by other people, and that "mice and guinea-pigs clearly reacted with unrest when placed in the Faraday cage [containing the SAPA cultures] at a distance of ten to twenty centimeters from the wall."[31]

Finally, Reich said that examining the SAPA-bions under a monocular microscope for longer than about thirty minutes at high magnification "(again varying with the individual), results in pain in that one eye but not in the other one. Bluish afterimages appear. Rest and dark glasses relieve the sensitivity to light, the eye pressure, and relieve the conjunctivitis. Thus, the SAPA-bions should be examined under a microscope very cautiously."[32] This caution was necessary, he noted, as long as the nature and intensity of the radiation remained unknown. The connection between SAPA radiation and the bluish and blue-violet visual impressions in the darkroom observations led Reich to recall that some of his previous bion cultures had appeared to have a blue color, when viewed microscopically using apochromatic lenses.

Reich became fearful for the safety of himself and his assistants. For more than twenty-five years it had been known that X-rays caused cancer and other harmful biological effects. H. J. Muller in 1927 had shown they could cause gene mutations.[33] Ghostly looking emanations around his own body in the darkroom observations were alarming, he noted in his diary on Monday, March 6. He immediately telephoned Bon in Amersfoort to ask what further precautions, if any, he could take.[34] Bon was concerned and suggested precautions, but it was not obvious how to do so while continuing to investigate the SAPA bion cultures.[35] By March 10, Reich had implemented the following list of protective measures:

a. All of the cultures are located in the Faraday cage to limit the effect of the radiation. The cage is in the basement; however, it is connected to the living area through a radiator system.
b. There is continuous ventilation in the basement and living areas to counteract any radiation that might come through.
c. Microscopic observations are held to a minimum.
d. The possibility of obtaining equipment for eye protection was raised with a firm specializing in such equipment.
e. Lab coats and clothing are immediately exchanged and aired following contact with the preparations.
f. To start with, consideration is being given to using asbestos, lead-lined protective aprons, and radium protective gloves when working.

g. A radium-lead container was procured for the preparations. However, it is not certain whether this container will provide enough protection given the intensity of the radiation.[36]

In addition, Reich had injected SAPA bions into seven experimental mice on February 24, and after two weeks he could report that so far the mice had suffered no ill effects. Reich thought this hopeful but cautioned that one must wait longer before passing judgment.[37]

Out of this concern, and despite his caution about "Kreyberg-type" dismissals of his work that kept him from sending samples to the Bohr institute, Reich approached the radium physicist at the Radium Hospital in Oslo, Dr. N. H. Moxnes. When Moxnes held a SAPA culture up to his electroscope (designed for testing radium, via ionization produced by its emissions), there was no reaction. He immediately declared, "There was no radiation." Reich objected that since "his electroscope was designed for radium only, . . . the negative result meant only that there was no *radium activity*, but not that there was no radiation at all." Moxnes remained firm in his initial claim, however.[38] (Within a few weeks, Reich concluded that this new form of radiation, orgone, affected an electroscope only by way of insulating substances—"rubber gloves, paper, cotton, cellulose and other organic substances"—that had been charged with it. This will be discussed further later in this chapter.)[39] Moxnes's behavior, despite all the other phenomena Reich had observed, seemed to Reich to confirm his worst fears about the inability to get a fair hearing for a "fantastic"-sounding phenomenon like "radiating microbes."[40] This took place at a time when many scientists recalled the overblown claims of Butler Burke's "radiobes" and were aware of recent work claiming to discredit the phenomenon of MR. More on this follows.

Bon had written telling Reich he would be willing to help characterize the radiation, but saying that since he worked in a small lab, the "absolute discretion" Reich requested might not be possible.[41] By March 11, Reich wrote that he was willing to relinquish his demand for absolute confidentiality about the experiments, saying, "It makes no sense, and it is also impossible because I have to constantly discuss the subject with optical factories and many different other people." He sent Bon SAPA cultures and the broth-KCl medium and a sample of the original sand so that Bon could replicate creating the bions himself.[42]

Meanwhile, Reich had learned more about the radiation from the SAPA bions. Not only could it penetrate cardboard or thin wooden boards to fog film, including X-ray film. When Reich put the cultures near X-ray films, even if the films had lead coverings, the radiation penetrated any available crack. "The most remarkable thing is that the unknown factor X travels around corners, that is, passes through cracks to reach the films wrapped in lead. Probably it does not pass through the lead itself," said Reich.[43] When the SAPA cultures had been in one room for several weeks, Reich's oscillograph—in that same room—"ceased functioning. Some batteries in the room lost more of their charge than is usual under normal circumstances—the negative pole lost more power than the positive." And numerous metallic objects in the room near the cultures (such as scissors, pincers, and needles) became magnetized.[44] From this host of observations, Reich could assert confidently to Bon that the radiation was real. And their growth in culture through successive generations proved the SAPA bions were living. However, Reich added, "I am extremely worried that, since conventional scientific theory makes a sharp distinction between life and nonlife, the fact that living organisms were obtained from lifeless minerals and that these living organisms are radioactive will be thought shocking. . . . I do not know how to get around this problem. I do not wish for a second time to become the target of another highly upsetting campaign against my work of the kind that raged for a whole year in Norway after the publication of my bion book."[45] Reich told Bon he would be extremely grateful for any help in solving this conundrum.

Bon replied with some good news: his "former mentor, Professor Doctor [J. M. M.] Smits, director of the chemistry-physics radiation department at the University of Amsterdam ha[d] offered usage of his laboratory to determine the bion radiation type. He also guaranteed me the utmost discretion."[46] Bon wanted to know the exact composition and the precise instructions for creating the bions, in order to rule out whether the radiation might be emitted by one of the constituent substances rather than only by the bions after their creation. Given all the phenomena Reich had described, Bon remarked, "There must be immense quantities of energy involved!"[47] Reich invited Bon to come to Oslo to study creation of the bions and culturing techniques at first hand; Bon telegraphed that, regrettably, he could not come.[48]

Reich had subjected SAPA bion cultures to fifteen-volt pulses for an hour to study the effect on the radiation, if any. He called this type of culture exposed to electrical pulses SAPA I (or "Evin"). The bions of this type were larger than in previous SAPA cultures. A second variant culture type had bions with "vesicular vibrating 'sheaths'" around them. Reich called these SAPA II (or "Lorin") bions.[49] Reich named the two types after his daughters, Eva and Lore. He sent Bon one of each type, along with instructions for how to maintain the cultures. They were unstable and had to be reinoculated onto nutrient agar or blood agar every three to four days. After inoculation, they needed to be kept ten to twenty hours in a 37°C incubator while growing. Then they could be kept at room temperature. If exposed to excessive heat, the bions disintegrated into tiny particles.[50] "Occasionally, they decompose resulting in tissue shrinkage and the appearance of bacteria rods observable under a microscope." These degenerated cultures should not be used, Reich said, since there were already too many unknowns about the SAPA radiation.[51] Reich said if one of Bon's cultures got ruined, he would gladly send another; also that "rhythmic electrical charge pulses from about 10 volts enlarge the structures (in KCl) and charge them." The bion cultures, when healthy, were always an intense yellow color, Reich said, and they should be used for radiation studies only if they had this color.[52] Several cultures prepared simultaneously and kept together in the same room for at least twenty-four hours made the radiation effects more prominent and easier to study, said Reich. Another indication of healthy cultures was that, when observed in darkfield at 300–400×, they should "glow brightly . . . and when observed in brightfield at 2000× [they should] stretch or swell when in a KCl solution."[53]

Following up on his observation about the large amount of energy the SAPA radiation must have to cause the effects Reich described, Bon wrote to warn that the greatest danger was the possibility that the bions were emitting gamma rays, since "those rays are what burn the skin and internal organs, and can cause cancer if the body is exposed to them too long." The bright fluorescing of the bions need not indicate gamma radiation, Bon thought, but was more likely to indicate alpha or beta rays, as in radium. However, he recommended extreme caution against the possibility that gamma rays might also be emitted. Inhaling a substance that emits gamma rays could lead to terrible health effects. Thus, the bions should be kept in tightly closed glass containers, probably shielded with lead, he advised. One-centimeter-thick lead shielding was probably enough, but since the true intensity was not yet known, as much as ten centimeters of lead shielding might prove necessary. In addition, the lab should be very well ventilated.[54] "It is absolutely possible, though, that we are dealing with something completely unknown here," Bon added, remarking that quantitative and qualitative studies of an unknown radiation were likely to be "a time-consuming ordeal." He thought it well worth the effort, however, and encouraged Reich: "I am convinced that an exceptionally important detail has been discovered. I am glad that I called your attention to the possibility of radiation in my first letter, perhaps even helped to inspire you with this small nudge towards discovering the presence of radiation." Bon hoped they might jointly publish whatever final conclusion they arrived at.[55]

Reich sent Bon a detailed report on March 10, summarizing all the observations—both subjective and objective—made on the radiation to date and some preliminary conclusions. In carrying out experiments with photographic or X-ray film plates, Reich now warned that one must use control plates from the same series as those to be exposed to SAPA radiation, but the unexposed control plates must not have even been kept in the same room with the SAPA cultures. "The experiment showed that, since emanation apparently occurs, even such plates that are in the same room with the preparation located some distance away are exposed. A roll of x-ray film intended for the control experiment and coincidently left in the room was consequently no longer usable. Control plates should be stored away from the trial location, such as at the place where they were procured," explained Reich.[56] Reich gave very detailed instructions for equipment and methods needed to photograph the SAPA bion rays directly through a microscope in darkfield. (This entire report is posted on the website accompanying this book, http://wilhelmreichbiologist.org/, for those who wish to know the details.)

The precise instructions for each observation are given in this report, including, for example, how to carry out the skin-reddening reaction: "After a few minutes, between 5 and 10 minutes depending on the individual, it begins to tingle. The skin reddens after 5–15 more minutes of exposure exactly in the form of the culture's shape on the object slide. Sometimes the edges are red and the center is whitish (central anemia). It is of utmost importance to patiently wait until the reaction appears. Vasomotoric and easily excitable individuals react quicker and more noticeably than people whose skin carries a poor electrical charge."[57] Once again, as in so many procedures in different parts of the story of the bion experiments, patiently and carefully watching for an extended time is crucial, as well as paying attention to the vegetative differences between experimental subjects and having the equipment to measure such differences.[58] Reich saw this as a fascinating subject of research by itself, but he realized it was problematic. In addition to experiments with film, electroscopes, and other equipment, he told Bon he was "working on a method using a special microscope outfitted with a special device to intensify the radiation to the point where those who do not wish to say yes, must say yes [i.e., the phenomenon was of such clarity that anybody could see it unambiguously]. You realize what a major role the subjective source of error plays in assessment and observation."[59]

After noticing the magnetization of metal objects, Reich sought to test whether the SAPA radiation would affect a compass needle. Direct contact of the needle with the bion tubes did not produce clear results. However, Reich saw that the radiation had a strong attraction for iron, nickel, and other metals. If he kept SAPA cultures in the Faraday cage for at least twenty-four hours, then a compass needle brought twenty to thirty centimeters from the inner copper wall of the cage would become agitated. The cage was grounded, so it had no field of its own; however, the walls of the cage acted as though they were charged by the SAPA cultures. Reich realized he needed to carry out control experiments in the cage with no SAPA cultures in it. But for the time being, he felt it was the only safe place to keep the cultures for any length of time. As he expressed it, "Shielding has not yet been determined. A Faraday cage appears to be the best means of deflecting rays towards grounded metal; however, it does not stop emanation."[60]

In his report of March 10, Reich noted that an electroscope did register effects from the cultures, but this was still confusing. Reich felt from the magnetic effects as well that the radiation was in some way connected with electromagnetic phenomena. The reader will recall from the incandescent carbon experiments that Reich believed the energy that gave bions their electrical charge when they formed was energy released when the bonds that held solid matter together were broken by heating to incandescence (or more gradually, by swelling up and breaking down in fluid). In his report he said, "The creation of SAPA-bions from Norwegian sea sand and their functions align with the theories of physics regarding matter waves and particles of energy and with the basic theory of bions that life is created from matter through swelling and electrical charge."[61] Solid sand, he thought, was held together internally by energy of cohesion that could be released by heating. In some sense, he thought the original source of the energy trapped in the sand's internal bonds was energy from the sun and the formation of the solar system, an idea inspired by his reading of Arrhenius's *Worlds in the Making* in August 1938.[62]

Experiments with the Electroscope: "Orgone" Radiation

Reich began by the end of February to experiment with electroscopes, trying to solve the puzzle of why an energy connected in so many ways to electromagnetic energy still had produced no reaction at Moxnes's radium electroscope. Within two weeks Reich had concluded "it has been established that insulating materials such as porcelain, wool, rubber, and glass have an extraordinary affinity for the radiation. Metals—copper, iron, etc.—re-emit the radiation in greater intensity. As I have already written to tell you, the self-destruction of the bion cultures is most likely a form of reflected radiation from the surrounding metals. The problem of screening has still not been solved."[63] (As a result, Reich warned Bon not to keep the cultures anywhere near metals.)[64] By March 17 (outside the laboratory, the German invasion of Czechoslovakia had just occurred a day earlier), he had coined the name "orgone" radiation, to indicate that the radiation was absorbed by "organic" (insulating) materials and that this discovery grew out of his research originally into the orgasm formula.[65] By Tuesday, March 28, Reich's logic moved a step further. In a diary entry that day, Reich first explicitly concludes that orgone is the specific biological energy—the idea suggested by Kammerer and others twenty years previously: "Orgone is a type of energy that is the opposite of electricity; it is the specific form of biological energy. In keeping with the orgasm theory, which equates the sexual and the vegetative, it must at the same time be the specific sexual energy, orgasm energy."[66] What, then, were the results of the experiments that led Reich to conclude this energy had fundamentally different properties from magnetism or electricity (or even bioelectricity)?

At first, as in so many experiments, the results were confusing. Reich told Bon, "All the phenomena are completely new and unknown to me. I have no idea if they are known to anybody else. But I think it possible that the radiation emitted by the SAPA bions, which does not directly influence the electroscope, might have something to do with the dielectric of poor conductors."[67] But soon a pattern was apparent. While he was convinced it was not electricity or magnetism, the radiation nonetheless "influences matter in such a way that phenomena resembling magnetism and electricity manifest themselves at the electroscope." First, metallic materials in contact with (or near) SAPA cultures for any length of time were influenced in such a way that they attracted the north pole of sensitive magnetic needles such as compass needles. By contrast with metals, Reich found, "Materials that are otherwise used to provide electrical insulation, such as rubber (gloves), hard rubber, glass, etc., are influenced in such a way that the leaves of the electroscope move even when the objects are held far away from them, and in certain positions this material *attracts* the leaves."[68]

One of the most important experiments was to place thin rubber gloves (or soft rubber in any other form) in the same metal cage with "fresh and intact" SAPA bion cultures for several hours. As the control, similar pieces of the same kind of rubber were exposed "to freely moving air in the shade," but *not* in a closed space to which moving air had no access. The control gloves gave no reaction at all when brought near an electroscope, or even in direct contact with it. The gloves that had been in contact with the SAPA bion cultures, by contrast, caused a very marked deflection of the electroscope leaves. This deflection was caused by direct contact with the electroscope or even just by bringing the gloves within a few centimeters of it. As Reich put it, "there is no doubt that when, for example, a pair of thick rubber gloves that have been exposed for a long time to the preparation are brought up to the protective glass cover surrounding the leaf electroscope, the same effect is produced as if one had placed an electrically charged condenser plate in the vicinity. The leaf of the electroscope is rapidly attracted to the side of the glass wall and remains stuck there."[69] Neither the control gloves nor the SAPA-treated gloves had ever been rubbed or subjected to any friction.

If the two gloves were exchanged—taking those that had been in the cage and putting them in the open, shaded air, and vice versa—their reaction at the electroscope is reversed, as one would expect if the source of the charging ability was contact with the SAPA cultures. Either thorough airing or immersing in water of the previously charged gloves removed their charge completely.[70] Reich felt he now understood the negative reaction at Moxnes's radium electroscope: the highly charged bion cultures could produce a deflection, but only if the charge was conducted to the electroscope via charged insulating substances, not by the cultures themselves directly brought near the electroscope.

In a second experiment, Reich placed the same rubber gloves in bright sunlight for a period of time and found that was also sufficient to charge them so they produced a deflection of the electroscope on approaching it, or upon contact with its charging plate. Control gloves kept in shady, moving air—or immersed in water—produced no reaction at the electroscope. This was suggested to Reich by his hypothesis that the energy liberated from the sand and radiated by the SAPA bions might be identical to solar energy. The results were consistent with that hypothesis. After a preliminary trial of this experiment, Reich wrote Bon to ask if he could replicate it. Reich wanted to do numerous replications himself, but as he told Bon, "I have been unable to repeat the test since last week because the sky is always clouded over, but I am too impatient."[71]

Reich explained to Bon, "When exposed to fresh air under a cloudy sky, rubber reacts in the known manner, i.e., it emits orgone and therefore does not produce any deflection of the electroscope. On the other hand, a rubber glove or a hard rubber rod that has definitely discharged all its SAPA orgone but has been exposed for fifteen minutes or more to bright sunshine produces a deflection on the electroscope. A control glove placed in the shade produces no such deflection." Reich was excited about this new finding: "I was a little giddy when I noticed this reaction occurring several times one after the other, but I will repeat it as often as possible so that even the remotest doubt is eliminated. I was not yet able to verify whether the 'solar orgone' behaves like SAPA orgone in relation to friction electricity. You will understand that I immediately connected the phenomenon with my earlier experiments on the human body, in particular the erogenous zones."[72]

In a third experiment, Reich placed the same type of rubber gloves, which had previously given no reaction at the electroscope, on the bare skin of, for example, the abdomen of an experimental subject, carefully avoiding any friction. After about five to ten minutes of contact, the rubber produced a similar reaction at the electroscope to what had been seen in the two preceding experiments. Again, control gloves that had been kept only in fresh, moving air in the shade produced no reaction. Again, if the experimental and control gloves were exchanged, the results were reversed. As Reich had told Bon, this experiment was clearly suggested to Reich by his earlier results in the bioelectric experiments, using many of the same subjects who had participated in those experiments.

In this case, there was an additional, fascinating result: "Persons with poor respiration and diminished vegetative excitability cause only a slight deflection of the electroscope leaves or none at all." This was completely parallel with the findings from the bioelectric experiments, where Reich knew well the relative lively variations of electrical skin charge (and respiration) of the experimental subjects. In the earlier experiments, the most extremely diminished reactions were from catatonic patents at Dikemark Asylum, which in this case Reich could not include. This variability led Reich to believe that the SAPA bion radiation (and solar energy) might be closely related to, or perhaps identical with, the moving bioelectric energy he had found in humans and—in the bion experiments—in at least some kinds of microorganisms.

As a follow-up, Reich reported that "if one has the experimental subjects who react poorly, breathe more fully, then the reaction occasionally succeeds." Subjects "with strong vegetative excitability give strong and quick reactions," he emphasized, very similar to the bioelectric experiments in 1935.[73]

Another unexpected phenomenon was that if an electroscope was previously charged using electricity, then the presence of an orgone charged object would make it slowly *lose* the previous electrical charge.[74] By March 18, Reich reported to Bon, "So far, there is absolutely no doubt that some type of radiation exists. It can best be observed when the preparation is allowed to act on the lens system of a microscope, and a good matte screen, possibly reinforced with a fluorescent agent, is held continuously at the focal point of the ocular."[75] All of these results, taken together—the darkroom observations, magnetizing metal objects, effects on electroscopes, photographic and X-ray film plates, visualization of the energy via a microscope, and, not least, the biological effects such as reddening of the skin—led Reich to conclude that the energy he was measuring was not electricity or magnetism, but a new form of energy that was interchangeable with electricity and magnetism under the right conditions. He also concluded this energy was specific to living organisms (and the sun); thus, he reasoned, what had manifested itself in the bioelectric fluctuations of potential at the skin surface in earlier experiments—that is, the fundamental energy moving in emotions, which itself triggered such electrical manifestations—was orgone energy.

Concerns about Profiteering

Reich raised a new issue with Bon, saying, "In the last management meeting at the Institute, we addressed the question of how we should handle the SAPA bions from a purely business point of view. It is clear that, with full recognition of the existing radiation by several physics institutions, we will be dealing with assets of gigantic proportions. Please do not lose sight of the Union Miniere du Haut-Katanga for Radium Production."[76] Here Reich was following upon the concern he had communicated to du Teil in December 1937, about monopolies or unscrupulous capitalists attempting to hijack his discovery and control it for profit first, with medical necessity a secondary concern (assuming they failed to squelch it entirely, to protect the existing radium monopoly). Reich emphasized that his research group had neither the possibility nor the desire to profit materially for themselves from the new discovery. But they wished that "under no circumstances may the SAPA bions fall subject to some monopoly or to the patent constraints of some individual."

With this in view, they decided that "the best option is to quickly allow distribution of the preparations as widely as possible, so that a private, monopolizing exploitation is no longer possible. A provision should also be included that gives the inventor of such material the right to veto any patent claims to the preparation."[77]

Meanwhile, Reich apprised Bon of some new experiments, injecting cancerous mice with SAPA bions to see if they had any potentially medically beneficial effects. "Please be very careful," Reich said. "A test conducted on mice has resulted in the death of one animal with precancerous neoplasms in the liver and a change in the blood that very strongly indicates carbon monoxide–hemoglobin. This has not yet been verified. (Admittedly, a very large injection was given, corresponding to ten liters in the case of a human being)."[78]

Bon responded to Reich's latest experiments in a letter of March 23, saying he thought the SAPA rays must be a combination of visible light wavelengths and long-wave ultraviolet radiation. He reasoned, "The fact that 'orgone' beams are refracted by typical glass lenses is certain proof that at least a part of these rays are pure and simple beams of light. The fact that one can even see them without needing a zinc sulfide screen alone is proof enough. That one can see them better using a fluorescent screen in the ocular shows that they are, in addition, *long-wave UV radiation*."[79] Bon added, "Because UV radiation can set electrons free in metal, etc., the latter has a *positive* charge when exposed to light. Thus, it is nothing special that cultures that emit UV rays have an effect on an electroscope. I cannot yet say for certain whether things are going according to regulation or whether what we have here is entirely something new." Bon thought that the phenomena of charging insulators such as glass and rubber could be understood as the "charge being stored in the insulating materials in the form of ions." When a condenser is disassembled, for example, he said, the glass (dielectric, or insulating) layer retains the charge.[80]

Sarcinae *Redux*

Reich's description of "creamy, yellow cultures" of SAPA bions is quite similar in terms of their macroscopic appearance to his earlier description of "packet-amoeba" PA bion cultures.[81] Perhaps not surprisingly, then, Bon was advised by bacteriologists he asked that the cultures looked to them like *Sarcina lutea*. For anyone familiar with the packet amoeba cultures, Reich had pointed out that the SAPA bions were much less motile and significantly larger than the earlier packet amoebae when viewed under the microscope. While both types stained gram positive, the "tetrad"-shaped packets of SAPA bions were much more rigid (though *Sarcinae* tetrads were more rigid still). Reich described the color of packet amoebae under the microscope as "intense yellow to dark yellow"; SAPA bions by contrast he described as "intensely golden-yellow."[82]

Bon wrote on March 23, telling Reich that the bacteriologist's comment had made his mentor anxious about the possibility that the bions were an accidental contaminant: "The director of the Institute for Inorganic Chemistry, professor Smits, told us that he insists that we repeat your basic experiment: beach sand, bouillon, 0.1N KCl, and this without any possibility of contamination. Although I, myself, am skeptical and believe that this will not be so easily done, we still must begin, for it is nice to remain on friendly terms with one's old professor."[83] Bon was aware of Reich's warning that the bions were fairly easy to create, but then getting the first culture to succeed was quite difficult. Reich wrote back quickly, to emphasize the point: "The culture experiment is difficult. The bion solution must be inoculated onto the special egg-medium at exactly the right moment, determined partially by gut-feeling and partially through experience. Personally, many culture experiments of mine failed, as I mentioned." Reich added, "The infection argument that the bacteriologists are constantly bringing up is just as bothersome as it is incorrect because the experiment, as described, reveals the process without further ado. The labels that the bacteriologists have for the structures do not say much."[84] Reich observed that the terminology of the bacteriologists contained assumptions that raised the question: "The structures known to the bacteriologists must also have a source. After all, control experiments with air-borne infections in the laboratory have shown that a common air-borne infection looks totally different. The 'germ' argument also can be easily refuted by the question of how the germs originated."[85]

Not wanting Bon to be too discouraged by this conundrum, Reich explicated at some length: "Under certain conditions, one can observe the development of matter resembling spores in crystal, coal, iron, etc. I know that I am not telling you anything new, but I have made it a habit of having the important arguments ready when dealing with bacteriologists. If you wish to go to the trouble of producing the preparations yourself, I would suggest that you prepare, besides sand, blood-charcoal (carbo sanguinis), soft iron, etc. using exactly the same procedure and follow the development; only then is one convinced beyond a doubt of the swelling and formation of vesicles in the materials." Reich reminded Bon, "You yourself expressed doubt whether the present-day scientific world would be willing to accept our data. I do not believe that it is spite or simple conservatism; rather, that we are dealing with an unexpressed fear of the process by which inorganic matter becomes organic, a psychological defensive attitude."[86]

Reich warned that vesicular disintegration of the sand and bion formation fails "when the incandescent heating process is not thorough enough and the swelling does not work long enough," but he added that if Bon's experiment with culturing SAPA bions failed, Bon could still forge ahead with the physical studies on the radiation from the cultures he (Reich) could supply. He thought Bon's guesses about the radiation as long-wave ultraviolet sounded similar to his own hunches, adding, "If solar energy froze in the earth's material eons ago, then it is not surprising that this energy could be set free again by blasting the material [i.e., heating it to incandescence] and that we are dealing with simple solar energy. Your detailed comments were very valuable because, although I am very familiar with theoretical physics, my knowledge of applied physics is very lacking."[87]

This situation must have reminded Reich of (as he saw it) du Teil's naïveté about the irrational behavior his work seemed to provoke in opponents— a situation Reich thought du Teil was still in denial about (and at least partly blamed Reich for), even after du Teil lost his job. After a few more days of thinking about the possibility of a similar catastrophe for Bon, Reich wrote on March 30 and put the issue very frankly to the Dutch physicist, asking if Bon was sure he knew what he was getting into:

> I would like to say a few words on an extremely important point that has so far played a very decisive role in my twenty years of scientific activity. I have hesitated for a long time to write you about it because I did not know how much knowledge you have of my work. I was afraid for the promising scientific relationship between us, but on the advice of a scientific friend here in Oslo I have decided to talk about the matter. My work has a major shortcoming. Its results seem simple and obvious, and they are very attractive to people of various professions. But the world as it is constituted today places extraordinary dangers, in particular of a personal kind, in the way of this work. From defamation through sabotage all the way to direct personal threats, these dangers come in all shapes and sizes. Over and over again it hap-

pens that a scientific worker embraces these questions with enthusiasm, totally unaware of the threat posed by the forces of mysticism. When the danger becomes acute and has to be faced, most people tend to fail, not because they are indecent, but simply because they have no practice in such matters and because as a rule scientific work does not have to be performed in such a dangerous environment. I must therefore warn you of this. Sooner or later there will be trouble from one side or another. There is enormous irrational fear about discovering the origin of life; therefore it would be a good idea to consider at this stage whether, in view of the dangers, you would prefer for the time being to carry out the work privately and not have to be ready to convince somebody now or in the very near future, because our enemies do not make rational decisions. You know that I very much need your scientific assistance. Please write and tell me honestly where you stand.[88]

With this letter, he sent Bon a report on the Norwegian press campaign of the previous year. If one thinks—as many of Reich's opponents claimed—that his bion work was attacked because it was scientifically unsound, then Reich here would sound quite paranoid. In that interpretation, du Teil lost his job because of internal political issues within French academic circles that were incidental to the scientific evidence. If one thinks—as Reich did—that sabotaging du Teil's job was a not-so-incidental outcome of a Richard Nixon–style "dirty tricks" campaign that would stop at nothing to silence his bion work, then Reich's caution here could sound well warranted and his frank warnings to a younger colleague could sound very considerate. When one, as a historian, concludes the outcome was underdetermined by the experimental evidence—even though not necessarily accepting the degree to which Reich attributed irrational, emotional, largely unconscious motives to his opponents—then examining a possible role for "externalist" factors does not seem out of place. Competition for a fixed pool of Rockefeller Foundation (RF) research grant money exemplifies this, for instance, among other factors I discussed in the Chapter 6.

Bon seemed to appreciate Reich's frankness on this issue, since, as he related, he, too, had long had difficulties with colleagues. From the very beginning of his studies in 1921, Bon said, "I have primarily attempted to prove my hypothesis: 'Life expresses itself as radiation.' I have spoken and *argued* a lot with my friends, teachers, and colleagues, but the most important result was simply that they considered me *a good-natured fool.* In these past years, it appeared to me that a better atmosphere developed here in Holland; at least I noticed a somewhat friendlier response to my ideas. However, I am fully aware that it is best to complete experiments concerning SAPA radiation in private."[89] Bon went on to recount a recent "incident": "The Director of the University of Amsterdam's Bacteriological Laboratory read your bion book and stated that it was insane. He has, in spite of this, given permission to grow bion cultures (quietly) in his laboratory!!! I do not care at all about that man's opinion. I will continue to complete the experiments, but will no longer speak to others about them, because there is absolutely no point in doing so." Bon agreed with Reich that "there really seems to be a certain fear concerning the question of Life." For the time being, Bon said, he felt the question of the origin of life had "been answered through the discovery of bions," so that he would "now focus solely on the question if and how the SAPA bions radiate."[90]

Yet Bon did become rattled in his confidence by the bacteriologist's claim of "contaminant *Sarcinae*" and his former mentor's anxiety over that possibility. He wrote Reich on May 12, after weeks of trying to take radiation measurements on the SAPA cultures Reich had sent him but having had no success at all. "Unfortunately, I must inform you that the SAPA cultures we were sent have resulted in completely negative radiation results. We could not determine any visible radiation, no skin irritation, and no effect on photographic materials. My co-workers are completely convinced that we are not dealing with radiating bions, rather that the cultures contain perfectly normal types of *Sarcina*." It is only rational for a young scientist to think carefully about getting too involved in something highly controversial before being better established in his career—particularly if opponents raise objections from a field (bacteriology) that he (a physicist) knows little about. Bon told Reich that, at least for the time being, he must "suspend analysis of these so strongly negative results." However, he immediately also stated, "I am busy trying to find other co-workers because I cannot accept that we are dealing with an illusion here."[91]

Reich had told Bon repeatedly, "It is a good idea to make an assortment of agar plate cultures and to have them in one stack if one wishes to have easy access to the phenomena."[92] Bon does not say whether he ever grew up enough SAPA cultures to follow Reich's repeated advice; it seems unlikely, since his wording suggests he only used the original two or three cultures

Reich sent. This may have been a reason he could not measure any radiation phenomena.

Reich wrote back to say he was happy Bon was in the end not shaken into dismissing the whole matter as illusory. "What was I supposed to do," Reich asked, "with a system of scientific thought that has to believe at all costs that, four times in a row, I had cultured something from sand that by chance were *Sarcinae* from the air—four times in a row and each time in pure culture! In fact, I learned that this type of bacterium is extremely difficult to find, and occurs only rarely."[93] Reich is not even mentioning the fact that the sand was heated to incandescence—a greater guarantee of sterility is hard to imagine—and that the culture media were all autoclaved by standard procedures and had shown no trace of contamination before the sterile transfer of the heated sand into them. It is not impossible that *Sarcina lutea* existed in Reich's lab as a contaminant. But that his sterile technique was faulty so many times in succession is quite unlikely. Alternatively, as in the previous round of discussion of this question, even if the organisms did correspond to the bacteriologists' *Sarcina lutea*—"Oh, that's just *Sarcina* . . ."—to assume that *must* mean they were an accidental contaminant from without (because they could not have arisen endogenously) was, of course, to beg the very question at issue.[94]

Reich seems to have believed that this episode may have undermined Bon's confidence more than Bon was admitting (or perhaps was aware of). In a later document, Reich wrote, "Several months later, I sent our friend in Mexico, Mr. [Alfonso] Herrera, likewise a culture sample and received the notification that it must surely be an airborne infection because he had only found 'sarcinae.' Such incidents occurred so often in my contact with biologists, bacteriologists, and physicists that the reason had to be an underlying, shared misconception of the new facts. Those types of misconceptions are embarrassing, cloud friendships, and easily leave the impression behind that one succumbed to a deception oneself. A friend does not hold such a 'misconception' against you, he understands. Such things occur easily. What one is supposed to do in such cases is to simply understand and give up the investigation."[95] First, Reich emphasized, the two forms could be distinguished morphologically, despite a certain amount of resemblance. "Sarcinae consist of four vesicles that form a square," whereas SAPA bions "consist of irregular clusters with *blue* vesicles."[96] And as noted earlier, the irregular SAPA clusters had a more fluid shape, while the Sarcina tetrads were rigid and immobile. However, Reich said, the logic of a large number of other experiments made it quite clear that even if these organisms *were Sarcinae*, then they were not "just *Sarcinae*." That is, those organisms gave off a previously unknown form of radiation of which microbiologists had no idea—and could kill cancer cells—whatever name one gave them. The killing of cancer cells, fogging of photographic plates, and numerous other lines of evidence had not been published by Reich, so, he said, it was natural that many investigators would not be aware of the striking evidence for why SAPA bions were not "just *Sarcinae*." Most of all, originating as a pure culture from ocean sand heated to incandescence was not in the least a property of the *Sarcinae* the bacteriologists thought they knew.[97]

In addition, Reich said, "In the scientific world, words that had once merely been coined from objects or processes come to take on an independent existence. The expression 'staphylococcus' denotes a certain sized coccus organized in clusters and with a gram positive color reaction. The word tells nothing about the origin or the function of the structure."[98] He expanded on this to say, "Likewise the word 'sarcinae' denotes only another form and explains nothing. I witnessed how these forms *developed* through their different stages in matter. For me, the blue colored gram staining only told me that bions behave *like* staphylococci. They also appeared to indicate how staphylococci and sarcinae *develop*, yet the name told nothing about their origin. For the friend or the opponent who was unfamiliar with the bion process, and who had also maintained the necessary [sterility] precautions, the similarity between bions and other known forms seemed to speak against the bions. It often appeared to me that my friends and my opponents were both disappointed when the bion experiments did not result in 'totally new,' 'never seen before' structures."[99] Thus, Reich thought, those observers missed the main point that bions might well be related to known bacteria, but their origin and properties might suggest that bacteriologists' ideas about those bacteria could be totally wrong. It was not just a pleomorphist claim that "bions" could turn into "bacteria" and vice versa. Rather, Reich believed the bion experiments proved a more fundamental claim: that bacteriology might—because it overlooked natural processes of microbes originating from nonliving matter—fundamentally misunderstand what those microbes are, what they are capable of, and how they might originate.

Reich and Herrera might put different interpretations on bions; nonetheless, they both seemed to agree on this basic critique of a "standard bacteriology" account that amounted to a "non-explanation" of bions.[100] When Herrera learned Reich was leaving Norway to come to America, he eagerly invited

Reich to visit him in Mexico City. In the event, Reich was never able to do so; Herrera died in 1942.[101] Reich did send Herrera, from New York, a culture of t-bacilli to examine. This resulted in another episode of confusion over bacteriological concepts. Because Herrera's microscope could not achieve magnification of greater than 1500× in bright field, he could not see the tiny t-bacilli and focused instead on some much larger, rod-shaped bacteria in the culture. This led him to conclude the culture contained only contaminants from air infection. Reich later wrote,

> I explained in a letter to Herrera that the putrefactive bacteria evolve from infection and also through degeneration. It happened again and again in my exchange with amicable minded researchers that they highly regarded my theoretical position, my microscope photography caused wonder and approval, but at the same time misunderstandings developed when my friends could not separate themselves from the catchphrases and terminology of mechanism.
>
> Today the meaning of autogenous putrefaction [i.e., endogenous infection] in cancer is verified and established through multiple experiences. Had I relented to the incorrect hypothesis that the only origin of putrefactive bacteria was "from the air," then I would have never been able to make cancer comprehensible. However, my microscopic investigations, film images, and other means of observation showed me again and again that tissue degenerated into sepsis rods and the sepsis rods into t-cells. Even "in the air," sepsis rods could not form by any other means than the decomposition of organic matter.[102]

Orgone or Electricity? Electroscope Experiments, Part 2

On Sunday, April 2, Bon wrote Reich to say he had looked carefully over the various experiments Reich described with rubber gloves at the electroscope. Bon suggested that many of the experiments could be explained as well by assuming the rubber gloves had somehow acquired a negative electrical charge from the radiation given off by the bion cultures, rather than needing to assume a new and different kind of radiation. "In my opinion, there is no question that these experiments prove that bion radiation gives the glove a negative charge. I do not believe that we are dealing here with a specific orgone radiation, rather 'only' with an electrical one," Bon said.

However, he emphasized that, even if this were so, "that is important enough, for such an effect can only be accomplished with an extremely high level of energy."[103] Reich replied ten days later, saying, "I was extremely pleased in all respects by your letter of 2 April because you are obviously closely following the line of my bion research and vegetotheory, . . . and finally because you were able to confirm my electroscope experiments. We will no doubt very soon reach agreement on the question of whether the radiation is a kind of negative electricity or in fact, as I believe, a new kind of energy. The experiments will probably make clear which is the case." Reich said that, as far as he understood, the electricity of glass charged by friction was positive, and the electricity of amber charged by friction was negative.[104] Now upon reading Bon's letter he seems to have realized glass was an inorganic substance, while amber—fossilized tree resin—was an organic one. He speculated, "It might in fact be that negative electricity exists along with a form of energy that has something important to do with the function of life. All this is very uncertain, but for the time being I do not wish to abandon my hypothesis (at least not until it has been proved totally unusable and false) that a specific biological energy exists that is similar to electrical energy. Interpreted in this way, it would not initially contradict the existence of negative electricity." Thus, Reich concluded, "If you are in agreement, each of us should continue to work with our own theories and we will let the experiments decide."[105]

Reich then went on to describe a variety of new experimental findings; for example, a photographic plate had been exposed to bright sunshine for ten minutes as a control and was uniformly gray. But, said Reich,

> the second one, on which you will see straight black parallel streaks, was also exposed for ten minutes to the bright sun but had been covered over by a horseshoe-shaped magnet. According to common photochemical knowledge, the part of the plate that had been covered by the horseshoe-shaped magnet should have been significantly *lighter* in the negative than the remaining non-covered part, and in the positive it should have been darker in contrast to the bright field that corresponds to the part exposed to the strong sunlight. To my astonishment the reverse effect was obtained. The area on the plate that was covered with the magnet is *darker*. This contradicts photochemical experience, but it cor-

responds entirely to the experience I have had with the orgone—namely, that the radiation has a particular affinity to metals and rubber. A parallel test conducted with rubber yielded exactly the same result as that obtained with the magnet. I must therefore assume that rubber and the magnet both take up solar orgone radiation and pass it on in amplified form to the photographic plate.[106]

Reich asked Bon to please repeat this experiment and inform him of the results. At the end of this April 12 letter, Reich informed Bon that he would be moving his laboratory to New York in the next couple of months and asked whether Bon's lab could care for six other important bion cultures, to protect them in the event of war.

Bon continued to get negative results on the radiation from the cultures Reich had sent and was baffled. In future letters, Reich described a host of small, technical details that mattered quite a bit for him in getting various experiments to produce the results he was reporting. Thus, for the SAPA bions and their radiation effects—much as with the original bion experiments—close study of the original correspondence is necessary in order to genuinely carry out replication. For example, in a letter of May 24, Reich noted that only after he had mailed a previous description of new experimental results did he realize he had left out many such seemingly minor but decisive details. When observing the radiation under the microscope and attempting to photograph it, for example, because it was in the ultraviolet range, Reich used "quartz glass lenses. The condenser is also made of quartz glass. I allow the camera chamber of the microscope to 'absorb' the emanation for a few hours, because then it shows up more clearly. I can only say that once the phenomenon has been correctly observed, one never forgets it, because it becomes so typical."[107] It bears repeating that Reich said he himself saw ambiguous results until he adopted such methodological changes. Regarding his darkroom observations—perhaps the most striking and counterintuitive phenomena he reports—it is perhaps worth reproducing an extended description Reich gave of the important technical details involved:

> When I set about making the first observations in the darkroom, everything went wrong until I rid myself of a false expectation. I had assumed that the rays were similar to radium radiation, that they would produce bright flashes when I directed them onto the

fluorescent screen. I had not anticipated that the rays would be simply visible. However, no rapid "lightning flashes" were to be seen. Since the only preparations I took into the darkroom were those I had previously checked out via the skin reaction and the sensation of prickling in the vicinity of the skin (palm of the hand or cheek, also tongue), I was very disturbed when, for a long time, I did not see the expected flashes of radiation. I was not aware of what I now see all the time in the darkroom. On the day I telephoned you, the phenomenon had become startlingly clear for the first time. It is quite different from what I had originally expected. I will try to describe it to you. Once the brightness images begin to fade in the darkroom, the room at first starts to turn dark again. In the control room the darkness increases and nothing can be seen. In the room containing the preparations, on the other hand, it starts to get significantly *lighter* after about half an hour. Instead of seeing nothing but blackness, one is aware of a strange *gray-blue* color. What one sees is diffuse but not uniform. It is like a surging and seething, in which the gray-blue color alternatingly becomes more and then less intense. It looks like fumes, fog, or something similar. I have experienced it myself, and most test persons hesitate a long time before describing what they have perceived. Some of them say it is "ghostly." As time goes by, one believes that one is seeing blue "afterimages." But they are not afterimages; instead, very fine deep blue-violet points of light are emitted from the copper walls that are resting against brick walls. They are quite different from "flashes." It looks as if they were slowly floating in the air, moving through arcs or in an undulating fashion. They fly or float slowly through the room. If a fluorescent screen is positioned on the north wall, the surging clouds and the deep blue floating points shimmer as if bundled or clustered together. The motion is not constant but occurs in a very slow rhythm. The same thing can be observed when a photographic plate that has been exposed to orgone for several days is developed. The plate emits a deep blue-violet light in the developer. A pale grayish-blue halo appears more and more clearly around hands, white coats, cotton, and porcelain, but it is diffuse and without distinct boundaries. Several persons who had no idea what was going on drew my attention to the phenomenon with shouts of astonishment. "Your hands are lit up." "Your head is." Havrevold once compared it to the northern lights. On the fluorescent screen one sees what looks like billowing clouds.

I have been working constantly on the technical problem of how to make these phenomena stand out so clearly that even the most unwilling person will see them immediately. As yet, I have not progressed very far, but at least I am trying.[108]

Reich told Bon he considered it very important to help the Dutch physicist replicate in exact detail for the sake of controls. But he added, "You are also performing control tests to confirm my findings, but I am less interested in that aspect. For me, your collaboration is not a 'control' procedure, but instead I regard it as essential assistance from a fellow researcher who can help explain these peculiar phenomena. I cannot spend all that much time on the purely physicochemical tests because I am very busy with my clinical work."[109]

In Bon's last letter, dated May 30, 1939, he expressed thanks for all the detailed information about technique, since he was still experiencing negative results with most of the findings, with one possible exception: *"I could only perhaps verify one thing, namely I noticed an effect on my eye. Because I could not replicate this, I regarded it as an optical illusion."*[110] Bon wondered whether, indeed, the cultures had somehow deteriorated in shipment from Oslo or after their arrival in Holland. He was still very much committed, though with a skeptical mind, to solving the riddle: "It speaks for itself that we must seek a means of demonstrating the phenomenon so clearly that at least the somewhat reasonably logical [people] can be convinced right now. I count myself as one of them!" Bon closed, however, by saying that he must temporarily suspend his work on the SAPA problem "because I have problems of a social kind that need all of my time, but I have every intent of focusing on these intensely important experiments as soon as it is possible."[111] Reich wrote again with more technical details on June 1 and on June 8; however, by this time his lab had largely been packed up for shipment to New York. Thus he was unable to do new experiments. The disruption of contact had causes at both ends, but it is unclear why, despite Reich's continued attempts to reestablish contact and despite the courageous, defiant note in Bon's letters of May 12 and May 30, Bon never wrote again.

Loss of Contact with Bon; Humidity as the Cause of Negative Results?

In early June, Reich's laboratory had been shipped to New York, accompanied by his assistant Gertrud Gaasland Brandt, so he could not continue experimenting until the lab was up and functioning in America. That was not until September 1939, because his own departure was delayed bureaucratically until mid-August. World War II broke out just days after he arrived in New York.

During Reich's voyage across the Atlantic from August 18 to August 28, 1939, he continued to do electroscope measurements in the ocean air, which led to a surprising discovery. The "static electricity"—or as Reich now thought them, orgone—phenomena were measurable at the electroscope until August 23. Then, on August 24, "I got no more reactions south of 57 degrees northern latitude and 25 degrees western longitude. *The high amount of moisture in the air blanketed all phenomena.* I first relocated the radiation phenomena at the electroscope on 16 September, in New York, when the relative humidity sank below 50%."[112] Reich concluded that this might well explain Bon's negative findings about most of the SAPA radiation phenomena, even in the cultures Reich had sent him: "*The humidity in Oslo is relatively low, in Holland high.* This appeared to me to explain the negative results well. Since 1939 in New York, I have lost the phenomenon in the summer and found them again in the fall."[113] Reich reported these findings to Bon from New York, but never received an answer. In the meantime, all the turmoil brought on by the war intervened.

With the Nazi occupation of the Netherlands in the spring of 1940, Bon was no longer in any position to continue the experimental work even if he wanted to—much as with the occupation of Norway, had there been any scientists there who wanted to pursue Reich's bion work further.[114] One prominent scientist, microbiologist Albert J. Kluyver, who lived through the war in the Netherlands, writing in November 1945, described the war years there (I quote at length to capture more fully the dramatic scene):

> I feel the more so [grateful, compared to what others endured], because my laboratory remained practically intact, and because Delft is probably the least damaged town in Holland. At your next visit you will find the town almost unchanged; the only difference is perhaps that owing to the lack of fuel many

trees have been cut down, but happily those bordering the main canals have been saved, so that the inner town did not lose its picturesqueness.

We went through a well organized famine, and I lost considerably in weight, but six months after our liberation these troubles are nearly forgotten. However, our work has been badly disorganized. Already in February 1943 all students had to leave, and so did my assistants soon afterwards. In September 1944 the military situation led to an almost complete lack of fuel; electricity and gas practically dropped out which gave a final blow to our activity. Moreover, there was the German manhunting, and through my window I saw ships with slaves passing, a sight not unworthy of antiquity. It was quite terrible. In the last year the Germans have partly destroyed our harbors, looted our factories (both for raw materials, and what is worse for machinery) and blown up the larger bridges.

This means, of course, that we are a very poor people now, our export capacity has nearly gone. Although final recovery will be slow, some real progress has been made in the months since our liberation. The inundated areas have been nearly reclaimed, traffic is coming up again, but since nearly all railway carriages had been stolen and our liberators do not allow us to fetch back our properties from Germany, it is still on a very reduced scale as compared with normal times.

Our very restricted amount of foreign currency has to be reserved for the purchase of locomotives, building materials, etc., so that it will take a long time before we will be able to afford the subscription to foreign periodicals, and the purchase of scientific books.[115]

By the end of the war, it is not surprising, then, if Bon's interest was forced to move on to much more practical matters for an extended time. He did actively continue scientific work and began some popular science writing.[116] There is no evidence in Reich's archives of any further contact in either direction after Reich's last letter to Bon shortly after his arrival in New York. Similar to the situation with du Teil, all the disruptions of the war contributed to cutting off contact. It is also possible that Bon, like du Teil, had been naïve initially about the degree of trouble he could get into for collaborating with Reich on such highly controversial experiments and was frightened when senior colleagues created difficulties for him as a result. If this is so, the distractions of the war may have provided an opportunity to reconsider whether they wished to continue a collaboration that was risky for their careers.

Whatever the reasons, given that du Teil and Bon (and to a lesser extent Paul Neergaard) were practically the only scientists (aside from Reich's students) who ever made a serious attempt to learn Reich's techniques and actually replicate any of the bion experiments, the wartime disruption was extremely unfortunate for Reich's chances of getting a hearing. And, as I discussed in Chapter 6, by the postwar period research in the life sciences had "moved on" to a great extent, committing itself henceforth more and more exclusively to what became the reductionist, macromolecule-based "molecular biology" and leaving behind any remnant research programs committed to colloid chemistry, or to an "energy principle" in the sense of Reich's work, of MR, or of "field theory." The "break" caused in research by the war made what was a growing trend of change in life sciences look more in retrospect like a complete epistemological break with many older research traditions in which Reich was rooted.

Scientific Reactions to "Life Rays" from SAPA Bions

In America, Reich quickly became more absorbed than ever in experiments using SAPA bion injections as a possible cancer therapy. Over time, nonetheless, he felt more than ever the need for collaboration with a physicist, to help him understand the physical implications of orgone radiation. After about fifteen months in America, he made contact with Albert Einstein and visited him in Princeton, New Jersey, to demonstrate some of his experimental work on orgone energy. Einstein was at least temporarily quite intrigued by Reich's work, though he then went silent on the subject. Reich's side of the exchange of letters between the two has been published, in the volume titled *American Odyssey*. However, as this discussion concentrated mostly on the orgone energy accumulator that Reich invented in late 1940 rather than on SAPA bions, a detailed explication of their exchange is beyond the scope of this book.[117]

In 1941 Julius Weinberger, an engineer in the License Division at RCA Laboratories, interested in Reich's work on orgone energy, on his own initiative wrote to Warren Weaver at the RF about Reich's work with orgone radiation. Weaver expressed extreme skepticism, saying that "the number of mysterious and poorly defined 'radiations' is discouragingly large." In the wake of N-rays, J-rays, and radiobes, the sciences were only too aware of rays that turned out to be illusory. Weaver strongly suspected that any test of Reich's SAPA bions by a qualified person would be "completely negative." Weaver said that if Reich would not be willing to try such a test he would have little interest.[118] He then suggested that the best person to talk with would be a Professor Loofbourow at MIT. Weinberger forwarded this letter to Reich, who responded to Weinberger that he would be more than willing to make such demonstrations in his lab to all interested observers. Reich said he found Weaver's tone arrogant, and he recalled the foundation's abrupt refusal to fund him after first encouraging him to apply in 1937, so that he had "been 'my own Rockefeller Foundation' for about fifteen years."[119] Since the RF had already known about his work for several years and had told him in 1938 they no longer had any interest, Reich felt that Weaver was already prejudiced against any possibility of accepting something as outlandish as orgone energy.[120] Weinberger subsequently wrote to Weaver that Reich would be willing to have tests. But nothing came of this, which confirmed Reich in his views about the "establishment."

Reich's psychoanalytic insight amplified his view of the role of the irrational among scientists, but Reich was hardly alone in this view. As bacteriologist Otto Rahn had already observed in 1933, about the reaction to Gurwitsch's MR discovery, "Fundamentally new scientific discoveries are by no means readily accepted by fellow scientists. They are, as a group, quite conservative, and intolerant toward radically new ideas. . . . Mitogenetic radiation, proved by hundreds of onion root experiments and tens of thousands of yeast experiments, . . . is still doubted by many biologists. Two things seem to have made the biologist suspicious. He does not like the conception of radiating organisms. It smacks of the popular conception of radiating personality, miraculous healing powers and similar mystic superstitions. Further, some biologists did not obtain the same results when they repeated the work of Russian, German, French or Dutch workers."[121]

I began a discussion of this topic—the idea of "life rays," particularly of MR, in Chapter 5. By 1941 when Weaver was writing, most scientists had become convinced that MR was illusory. But how did this happen? Since one in such an influential position was ready immediately to lump SAPA radiation in with MR, it will be useful for us to look a bit more closely at the process by which MR came to be viewed by most (not all) scientists as discredited.

In September 1935, Alexander Hollaender, one of the early researchers in the United States to study MR, completed a long review article, "The Problem of Mitogenetic Rays," published in 1936.[122] He was beginning to see problems with a lot of the MR work but still basically believed that the phenomenon was real. Hollaender stated that the "politics of appearances" is just as important or more so in the skeptical reaction of most scientists to MR than the data themselves.[123] Hollaender lamented, "It is doubly unfortunate that the problem has attracted some workers who, apparently, see in the problem only an opportunity to deal in the spectacular. In some respects Gurwitsch, himself, is to blame for this. He tends very pronouncedly to accept the work of investigators whose data agree with his theories and to reject almost entirely criticism of a contradictory nature. The result has been that a large number of scientific workers have become prejudiced against the problem; and the problem has not received the consideration which the work of a man of Gurwitsch's integrity deserves."[124]

Hollaender noted, however, that many labs were now reporting an inability to confirm MR effects. Hollaender thought that increasing skepticism was also (perhaps as much as because of evidence) a result of issues related to the nature of doing science. For example, he noted, "A comparative statistical evaluation of all experiments verifying the effect and of those yielding negative results would obviously be unfair because of the general disinclination to publish negative results, in the feeling that these may be due to faulty technique." In other cases, it appeared that negative results might be a result of changes made in the technique by some subsequent researchers. They believed they were "improving" on the original technique but might be inadvertently altering or overlooking some key element. This reminds one of du Teil and Neergaard having difficulties replicating Reich's bion experiments—difficulties that were eventually found to be a result of very specific issues of technique, as simple a matter as using red gelatin as opposed to colorless gelatin, or having inclined binocular tubes and apochromatic lenses on one's microscope. Many times, such details are realized to be important in close collaboration or correspondence only when a scientist makes sincere attempts at replication and inquires conscientiously of the original scientist about very fine points of methodological detail.

Then, too, often the original scientist is reticent to publish key details at first out of concern to keep control of priority for the discovery, future patent possibilities, and other aspects of the work, so one often finds the answers in private correspondence rather than in the published "Materials and Methods" sections of articles. Gurwitsch had such reticence, and so did Reich—at least until he should be given credit for the full import of his discovery, as he saw it, and prevent its exploitation for profit. So Hollaender complains that some of the skepticism over MR "can be understood easily since Gurwitsch has not published a clear-cut description of the methods he has found most successful. If the methods are such that each laboratory must largely develop its own procedure, Gurwitsch himself should direct attention to this point, but this he has not done."[125]

It is worth repeating that Reich *did* publish many of these caveats in *Die Bione*. Even so, most readers ignored them, for example, the caveat about needing the same optics. With the exceptions of Reich's demonstration in Fischer's lab and of Kreyberg briefly visiting Reich's lab to watch bions being made for one hour, none of Reich's critics ever seriously tried to replicate the most important details or to study any bion preparations in the living state. By the time Reich found the SAPA bions in 1939, he was deeply reticent to publish at all in Norway, where most of the scientific community had become hopelessly prejudiced that all bion work was worthless at best, "charlatanism" at worst. Neither du Teil nor Bon was able to visit Reich's lab to learn the technique directly, as du Teil had learned so many of the previous techniques. Perhaps it is not surprising, then, that of all Reich's bion types, SAPA bions have proven most difficult for subsequent researchers to replicate.[126] (A recent, noteworthy exception is discussed in the Epilogue.)

An additional problem Hollaender raised about MR was that very low-intensity radiations were often "almost outside the range of our most sensitive physical instruments," so that living organisms had to be used as "biological detectors." A special instrument, a modified Geiger-Müller counter, had been designed, he said, but it picked up the rays "from relatively few senders." He concluded, "Thus, it seems that with any investigation in this field one must count, in the main, on a biological detection. That biological detectors may be more sensitive than the most sensitive physical instruments will not surprise the biologist, who realizes that many biological materials are influenced by factors which are as yet imperceptible with our most sensitive physical and chemical tools."[127] Yet this introduces the variability—often emphasized by those with medical training—for which living things are notorious, and which physical scientists so abhor. If one takes Reich's SAPA bion radiation seriously, many of these issues are potential problems to be faced. Reich himself reported that he obtained the SAPA bions only five times out of eight trials, for instance, though using the same procedure and materials.

Hollaender discussed technical issues of this sort in his 1936 article as well. When yeast was used as the detector organism, he said,

> to be able to use any yeast technique successfully (that is, in obtaining positive results) one must first have acquired, it is said, some facility in handling this material. Still, each investigator in any other laboratory must work out his own special procedure. There have been times, too, when investigators usually "successful" with the yeast technique have reported unsuccessful experiments. However, such incidents have occurred at times when several or all workers in the particular laboratory were unable to get positive results, and the conclusion was reached that "something had gone wrong with the yeast." It is unfortunate that those working with the yeast cultures in this field have not emphasized the point referred to; the descriptions as published read as if it were possible to repeat the work precisely as in the case of any other standard experiment. A more cautious approach to the problem should at least have been suggested.[128]

By June 30, 1937, in a major turnaround, Hollaender and a colleague, Walter Claus, had completed a much larger range of experiments at the University of Wisconsin, funded by the RF and the National Research Council (NRC). This led them to conclude that MR had not ever been demonstrated to be a real effect. When these experiments were published as a bulletin of the NRC, belief in the reality of MR began very quickly to decline among biologists.[129] By the 1940s MR was being used as a textbook case of experimenters seeing what they hoped to see because of an insufficiently statistical approach.

As I stated in Chapter 5, physics historian Spencer Weart is correct that a lot of mystical ideas about "life rays" and "death rays" abounded and recognition of this compromised scientists' objectivity to varying degrees with "radiobes," "N-rays," and other such reported phenomena. There are other such stories—for example, "J-Rays"—and physicist Kurt Stern has used MR as an object lesson about misplaced scientific authority.[130] Weart points out that many of those ideas (including life rays and death rays) were bound up in the popular imagination with sex, for example, "electric beds guaranteed to lead to conception." In addition, premature publicity, sensationalist reporting, and controversy in the popular press often undermine the chances for a given set of experiments to be treated fairly by the scientific community. In 1952, in his autobiography, Otto Rahn states that, despite the problems caused to his career by newspaper sensationalism around MR, "I am still convinced that thorough scientific work, unhampered by newspaper writers and uninformed supervisors, will reveal the existence of biological radiations."[131] The sensationalism of the heated press campaign over Reich's bion experiments was even greater than that surrounding MR, and—particularly since the press, in addition, drew in his sex research and his highly unconventional ideas about sex—this surely did not help Reich's chances of getting a fair hearing.

Because the public has deep, unconscious archetypal myths about sex energy, does this automatically mean that any attempt to study such a connection by definition cannot be scientific? Reich and his ideas about orgone energy are rather easily cast without much examination into this "epistemological wastebasket," as Warren Weaver's response, based on very little knowledge, suggests. I have shown, however, that Reich's sexual theories may well have provoked a striking lack of objectivity among his critics with regard to the bion experiments. Might this not suggest that with regard to SAPA bions and the radiation phenomena they exhibit, it is an open question whether the subjectivity resides with Reich or with his opponents?

So What Does It Matter?

During its legal prosecution of Reich in the 1950s, the Food and Drug Administration (FDA) hired scientists to help it prove Reich's accumulator was worthless and that orgone energy did not exist. I put it this way deliberately, because it was clear from the outset the FDA believed Reich's research was a sham, merely a "front" for a "sex racket." And the FDA conveyed this to researchers such as physicist Kurt Lion at MIT, who was asked to test the physical properties of the accumulator. A detailed scientific critique has shown these flawed experiments cannot be considered legitimately to have replicated Reich's experiments.[132] Lion "was called upon to prove that the box was just a box and that Dr. Reich was a fraud," as Lion's son recalled in a later letter. Lion thus neglected to include crucial controls for relative humidity and other meteorological conditions that Reich's experiments had shown could at times lead to negative results.[133] Another critical analysis showed similar methodological errors in the FDA's attempts to test the biological effects of the orgone energy accumulator, errors which reflected bias and lack of detailed knowledge of Reich's published experiments.[134]

The FDA did not attempt any replication at all of bion experiments—including the SAPA bions and their radiation, the key experiment in which Reich had reported the very discovery of orgone energy and the first study of its properties and its biological effects. Yet since Reich's laboratory notebooks indicate the bion experiments actually did take place and that most of the phenomena are reported there as they came to light, it seems to me likely that Reich actually saw and measured something real.

Many who are interested in Reich's work and who have some scientific training have since repeated some parts of the bion experiments and reported seeing much the same things Reich reported.[135] In one study, on the migration of bions in an electrical field, for example, the authors replicated Reich's electrical examination of bions and confirmed his finding that they are electrically charged. They were also able to quantify this charge by measuring migration velocity, and found that migration velocity declines over time and can be increased by autoclavation. This work has been dismissed or overlooked by the scientific community because it is only being done by Reichians, mostly scientific amateurs. Among such "Reichians," the most difficult experiments to replicate have been important ones: growing bions in culture through successive generations and producing SAPA bions. I have suggested throughout this narrative numerous potentially critical technical issues that could be the reason for such difficulties. Regarding the tautological critique of "Reichians," it is relevant to repeat the way I posed the question previously: does one speak of "Einsteinians" or, rather, of students of relativity theory? In the end, it is the evidence that matters.

I have also shown here that the response in Norway to Reich's experiments was largely based on negative reactions to his sexual theories, his politics, and other issues; that none of his opponents at the time ever made a serious attempt to replicate his bion experiments; and that of the scientists who did to some extent do so, du Teil confirmed most of Reich's experimental observations. (Bon never traveled to Reich's lab, nor did Neergaard learn the techniques firsthand as du Teil did.) The results that Reich reports appear to be largely substantiated by the records to be found in his laboratory notebooks, even though these are not as detailed as one might wish by more recent "materials and methods" standards. Thus, this study concludes that Reich's bion experiments never got a fair hearing based on the evidence.[136]

Reich's lab notebooks are not complete enough to make as many or as detailed reconstructions of the chain of logic between the experiments and the development of new theories as was done by Geison for Pasteur's work. Some important exceptions I discuss, however, are the detailed descriptions of control experiments used, for example, in "Preparation 6," to isolate the role of each ingredient in a bion mixture; or in the incandescent carbon experiment, to exclude air infection as a possibility and thus undergird Reich's hypothesis that the microbes were developing from the mixture itself. Even clearer examples are the incandescent carbon experiments and how Reich's studies on staphylococci and streptococci from an osteomyelitis patient—at first just to compare those organisms with his bions—led Reich to begin thinking of disease processes as endogenous infection processes.

Historians of science might find the various twists and turns of this story interesting in their own right, particularly for the light they shed (often by contrast) on other developments in the context of biology and medicine in the 1930s. But scientists might well ask, so why does Reich's work matter, particularly since mainstream life sciences research went off in very different directions after this period and gave us the fascinating new tools and discoveries of molecular biology? There are a great many possible ways to answer this question. Were Reich's T-bacilli related to prions, for instance? Or were they cancer viruses (at a time when even Rous sarcoma virus was still not accepted as real)? And if so, could the discrediting of Reich's work have delayed recognition of one or both of those pathogenic agents longer than necessary?

In light microscopy and study of materials in the living state, Shinya Inoue not long after this story proved that crucial features of mitosis, especially the "spindle," could be understood only by pressing the limits of both these areas with polarized light microscopy. Could Reich be right—notwithstanding the aid later given by Pap smears on fixed, stained cells—that advance warning of cancer or other diseases could be improved by examining secretions in the living state and often at magnifications well above the range usually used in research laboratories? Perhaps most of all, if Reich's theory of the endogenous origins of many cancers is correct to any extent, many thousands of lives are at stake in finding an answer to that question based on careful scientific research thoroughly grounded in the technical details Reich claimed were crucial to replicating his work. Yet because Reich's cancer theory has been "black-boxed" as a textbook case of "pseudoscience" or outright fraud and relegated to derisive narratives about the "orgone box" or the "orgasmatron," such a serious scientific reexamination has been prevented for sixty years.[137]

Reich could, of course, be mistaken in his interpretation of his bions and on their connection with cancer and other diseases, on some or on all counts.

The results of his experiments could mean instead, as P. and A. Correa argue, that microbes such as *Myxosarcina* (in ocean sand) and many others can survive previously unimagined temperatures—such as being heated to incandescence—and then can immediately thrive in fully active condition afterward, as well as be grown through serial transfers in pure culture. This would represent a major discovery all by itself, if it were the correct interpretation of Reich's results. Or as Young and Martel argue, Reich's "bions" could be nonliving, yet capable of replication in culture and actively involved in health and disease. Both of these cases are discussed further in the Epilogue. However, if Reich were right about even one of his main interpretations, he could have been on to something important about the genesis of cancer, *or* about orgone energy (the two, though related, do not necessarily stand or fall together, as this history has shown). Or Reich could be on to some important insight about bioelectrical pulsation being crucial to life and therefore to understanding the origin of life. If any one of these novel ideas were to be verified, it would be a major contribution, let alone if more than one of them were validated. The case cannot be finally resolved until careful, thorough replication experiments are carried out—such as never were done by the FDA. In addition, such a resolution might shed light on why the FDA felt so provoked or threatened by Reich's scientific work that it was felt necessary to literally burn his books and journals, tons of literature having been shoveled into the flames—surely an unusual measure in America in 1956. Thus careful, well-informed attempts at replication would be very much worthwhile.

EPILOGUE

Recent Discussion of Reductionism in Life Sciences

Is this story of purely historical interest? Beginning in the 1930s came Warren Weaver's major drive to fund a "molecular biology," at a time when Rockefeller Foundation funding was almost the only game in town during the Depression years for any large-scale life sciences research. This reenergizing of physical-chemical reductionism in biology surged still further with the migration of quite a few powerful figures from physics into life sciences in the wake of World War II and the bombing of Hiroshima. The renewed reductionist tide reached a record high water mark with the Watson, Crick, and Franklin DNA structure, the Urey-Miller origin-of-life experiment (both in the spring of 1953), and the cracking of the genetic code. Indeed, to judge by the vast bulk of popular science writing, even of the caliber of the *New York Times*, one might get the impression that any and all previous doubts about reductionism were finally laid to rest, never to revive as serious science. It all seems just so much mystical "vitalism" now. Even E. O. Wilson's scathing critique of the excesses of Jim Watson seems basically an apologia for why population and behavioral life sciences simply had to face the need to become reductionist or die.

It ain't necessarily so. A small but steadily growing number of prestigious biological researchers, particularly in microbiology and cell biology, have been sounding an alarm, saying that reductionism has once again presumptuously overreached. From a tiny trickle of such voices a generation or more ago (imagine the opening passage of Bedrich Smetana's orchestral piece "Vltava/ Die Moldau"), this outpouring has swelled to a robust current, descending from the ethereal highlands toward the broad plains of daily research. Here I will mention only some of the more noteworthy recent books on the subject, as well as some critics further toward the fringes.[1] More than just criticizing biological or genetic determinism in behavior studies, or invoking epigenetics and "evo-devo" to debunk "master molecules" narratives about DNA, these authors claim that the reductionist program exemplified by Watson and Crick's "central dogma" fails fundamentally to answer some of the most basic questions about what life is, how cells function, and what drives evolution. There is "big trouble in biology," opined microbiologist Lynn Margulis.[2] Epigenetics and symbiogenesis provide a dramatically revised view of evolution, according to James A. Shapiro.[3] And all these authors agree, to varying degrees, although the points of disagreement among them are as enlightening as their shared views.

There are numerous wellsprings of opposition to the strong program of molecular reductionism epitomized by Watson and Crick. Many of them are discussed in these works: from neurobiology and embryology to physiology and cell biology. Some noteworthy early works in this vein since 1953 include Chambers and Chambers's *Explorations into the Nature of the Living Cell*, Paul Weiss's *The Science of Life*, Hans Selye's *In Vivo*, Gilbert Ling's *A Physical Theory of the Living State*, and Harold Hillman's *Certainty and Uncertainty in Biochemical Techniques*.[4] The noted zoologist Marston Bates was impressed by the recent breakthroughs of Watson and Crick; nonetheless, in his popular 1960 book, *The Forest and the Sea*, he opined that, rather than the molecule, the level of the whole organism ("the rabbit and the raspberry bush") must remain the primary level at which a biologist must understand her subject.[5]

For anyone who has ever worked in a biology lab and seen "theory-laden observation" in practice, while the researchers still steadfastly refuse to admit that is what they are doing, physiologist Stephen Rothman's *Lessons from the Living Cell* is extremely informative. Rothman insists that molecular biologists and electron microscopists have mistakenly come to believe that checking their models against the final arbiter of actual living cells is no longer necessary. This can often lead, says Rothman, to partially or completely false views of how cell processes, such as secretion, actually work.[6] Harold Hillman and Gerald Pollack argue this position even more forcefully (and both give much credit to Gilbert Ling), suggesting that many of the structures seen in electron microscopy (EM) or inferred from biochemistry are not what they seem. Pollack says that supposed "pump" and "channel" proteins in cell membranes may have fundamentally different functions. Hillman claims that most structures seen in EM, including endoplasmic reticulum, lysosomes, and golgi complex, either are totally artifactual or are so distorted from their living state that their appearance in EM photos is more deceiving than informative.

While less sharply critical of his mainstream colleagues for their reductionist worldview, microbiologist Franklin Harold nonetheless says, "I hold the mildly heretical view . . . that the genetic paradigm as it stands is insufficient, incomplete and fundamentally misleading. Briefly, biological organization is made up of multiple layers, which span the range from molecules to cells. Genes do, of course, specify order at the level of molecules and of supramolecular complexes that arise by self-assembly, such as ribosomes. But molecular structures do not suffice to specify cellular structure, for cells do not arise by self-assembly of their molecular constituents. Instead, cells grow."[7] Harold wishes to distance himself from those with more drastically revisionist views such as Pollack, or even more so, Hillman or Ling. Still, he grants (and this is more than most cell biologists are willing to do) that while the existing membrane theory "is as close to the truth as we can come" and seems unlikely to be overturned, nonetheless "it is important to keep the possibility open." In explaining why, Harold concisely sums up one of the main points of this review: "Ling, Pollack and their cohorts do represent a heresy that continues to smolder on the margins of cell biology. Heretics should not be ignored: they question what no one else doubts, draw attention to contradictions that were swept under the carpet, and spotlight the peculiarities of science as a way of knowing. . . . However fallacious their ideas, by forcing us to reexamine periodically what we hold to be self-evident, Ling and his supporters do science a service."[8] So these heretics fall on a spectrum: one important distinction between these authors is the extent to which they think the results of modern biochemistry and EM are fundamentally misleading as these techniques have been used to date (Ling and Hillman); that these results have a lot of important value but are misleading with respect to at least some major cell processes (Pollack and Rothman); or that the molecular and EM results are basically reliable but only capable of explaining events at the molecular level, without understanding other forces that come into play at the level of the whole cell (Harold).

Though Harold attempts to keep the more radical critics at arm's length, even with his restrained criticism it is noteworthy that, in his own words, his "holistic position remains altogether marginal" among his colleagues in cell biology, microbiology, and molecular biology, so much so that Jan Sapp recently counted Harold in a list of what Sapp called "cytoplasmic heretics."[9] And those whose more revisionist views Harold says deserve our attention are for the most part ignored outright by mainstream molecular and cell biology. The intolerance of reductionist molecular biology for these points of view (Harold is the exception that proves the rule) itself points up one of the most important issues discussed by the authors reviewed here. Rothman puts it best, in a helpful discussion of the many and varied meanings of "reductionism," when he observes,

> In the broadest sense, science and science studied from the reductionist perspective have been synonymous since the time of Newton. From this standpoint, to call yourself a *reductionist* is simply to say that you are a scientist, and to talk of a "reductionist science" is redundant, or of a "nonreductionist science" oxymoronic. Thus it should not be surprising that those who have questioned the reductionist philosophy and research program have from time to time been viewed as "antiscientific," even if they are scientists themselves.
>
> In biology such skeptics have traditionally been greeted with the charge that they are *vitalists*. . . . Vitalism was rejected [by the early twentieth century] with such force that to be called a *vitalist* was the most opprobrious of labels. It designated an individual as being ignorant of modern science and as holding antiscientific attitudes and even magical beliefs.[10]

I have argued elsewhere that this was the state of affairs in biology already by 1949 or 1950.[11] And it has become even more so today; the most common reception of the works under review here has been to suggest that even asking such questions about the reductionist approach, or to suggest (either on technical grounds or in principle) that reductionism alone cannot fully explain every living process means the writers must be vitalistic. Indeed, one reviewer of Pollack's book said its critical tone was reminiscent of creationist writings and its style "almost anti-scientific." Thus, it seems the excesses of reductionism may be very much alive in some quarters in biology, as well as in medicine.[12] So the attempts to solve this same problem in the 1930s may have relevance for today after all.

Recent Experimental Work Relevant to Reich's Bions

As indicated in several places earlier in this book, for instance, in the concluding section of Chapter 7, despite the eclipse of experimental work caused by the "pseudoscience" narrative, at least a certain amount of work relevant to the bion experiments has recently been carried out. Given much more sophisticated imaging (video microscopy rather than time-lapse photography, etc.), computerized data analysis, and other sophisticated capabilities with today's equipment, should it not be possible to clarify some of Reich's findings or to falsify them? The work that has been done is highly suggestive on this count.

The first issue to note is that some researchers have reported great difficulty publishing anything in a mainstream scientific journal if it in any way directly mentions Reich or his bion experiments by name. Physicist Adolph Smith had learned from previous papers that a paper would be published in *Perspectives in Biology and Medicine* only if all reference to Reich was removed.[13] Another researcher was told this directly by the editor of a prominent scientific journal but consented to my reporting this only if I did not identify any member of that research team in print. This gives an idea of the difficulty created in the sciences once a body of work has been subjected to the kind of slanderous rumor campaign widespread about Reich for the past sixty years. The Wilhelm Reich Infant Trust was preparing a grant application to fund work replicating some of Reich's cancer experiments when Reich's book *The Cancer Biopathy* (one of the titles burned by the Food and Drug Administration) was brought back into print in 1973. Mary Boyd Higgins, the trustee, asked advice from publisher Roger Straus. Straus made some inquiries; his reply suggested the climate of scientific politics was a major barrier:

> I have two very close pals who are cellular physiologists and have been very much supported for a number of years (although they are both quite young) by various foundation grants in their field, and off the record please, they write me as follows:
>
> "We read the grant application you sent and I'm afraid it's so far out that no foundation with a good scientific advisory board would touch it. A lot of very competent scientists (i.e. Pasteur and others of that calibre) have investigated the question of the origin of cells in decaying straw infusions and have concluded from their experiments that cells only come from pre-existing cells. I think most biologists would feel quite sure that Reich was observing first disintegration processes and then the growth of new cells from pre-existing ones contaminating his preparations. The evidence on this question is strong enough that the people who generated this application are being religious rather than scientific in weighing Reich's observations more heavily

than the enormous body of contradictory evidence which has formed scientific thought on this subject over the last 100 years. I think their only hope is to find a foundation that actually goes in for supporting borderline schemes, which, human nature being as varied as it is, must exist somewhere out there, we feel sure."[14]

It is telling that, even in expressing this opinion to Straus, these physiologists wished to remain off the record.

Grad and Činátl

Bernard Grad as a young biologist studied with Reich at his lab in Maine from 1949 through the early 1950s. In replicating one of Reich's experiments, he produced some unusual structures much larger than most bions, 6–20 µm in diameter (he called them "spore-like" structures or "primordial forms"). Their large size, more typical of eukaryotic cells, was part of what attracted Grad's attention.[15] Reich had not seen such forms previously and asked Grad to try to find out what had produced them. After Reich's death, Grad returned to those experiments only several years afterward, eventually concluding by 1968 that calcium carbonate ($CaCO_3$) was essential for the structures to form (though these experiments were not published until many years later).[16] When the $CaCO_3$ first formed by mixing the ingredients, Grad noted that the expected hexagonal or orthorhombic crystals "were either not present, or present in small numbers," so that the majority of the $CaCO_3$ was present in some other form, allowing creation of the primordial forms. Over the course of a few hours, however, the $CaCO_3$ transformed into hexagonal crystals. By the end of twenty-four hours, "the hexagonal crystals were in the great majority with only an occasional primordial form being observed."[17] Grad continued varying the mixtures used to produce these forms over several decades. X-ray diffraction analysis revealed that the $CaCO_3$ in the primordial forms was in the form of calcite. When the forms were prepared in the presence of humic acid, pectin, albumen, RNA, or DNA, Grad reports that they took on interesting forms—in the case of albumen, for example, structures were produced resembling the Haversian systems in bone. RNA added to the $CaCO_3$ mixture produced structures that looked like a nucleus. Added DNA produced forms resembling a syncytium, a thin sheet of cells from a tissue. In the sense of earlier "cell-model experiments," Grad saw these experiments as conveying something about combinations of chemicals that imitate, or perhaps actually create the form of living systems.

From 1966 to 1969, Prague biologist Jaroslav Činátl reported findings in many ways similar to Grad's, but Činátl apparently had been working with no knowledge of Grad's work. Činátl described "calcareous forms" (i.e., forms with a high calcium content) that were created from alkalinization of salt solutions, of tissue culture media, and in biological fluids and he interpreted them as prebiological structures. He observed that "in salt solutions and culture media they arise by alkalinization of concentrated solutions and are of inorganic nature," whereas "structures in sera and other biological fluids are formed only after long-term incubation at 37°C in a sterile isotonic milieu; besides salts they contain organic substances such as lipids, polysaccharides and proteins. On sterile serum agar plates these structures may be induced by nuclei of dead cells, nucleic acids and some enzymes." Činátl thought it unfortunate that some investigators (including Lepeschinskaya) had until quite recently mistakenly regarded these structures as living organisms. His argument, by contrast, was that they should be regarded as "on the borderline between living and non-living nature" and he considered them "structures with a subvital character." This is quite similar to Reich's earlier understanding of bions. Činátl also felt that "these multimolecular complexes with a high calcium content are of significance also for the solving of practical biological problems concerning the deposition of calcium."[18] He lamented the fact that science usually recognizes only a binary distinction between life and nonlife, since that prevents recognition that an intermediate state does exist. He called the "calcareous structures" subvital because they exhibited "states resembling passaging, propagation, ageing and adaptation, degeneration, disintegration and purposeful reaction"; because they were highly variable; and because "their growth is based on intussusception," rather than on apposition, by which nonliving crystals grow.[19] Despite the appearance in the modern world of "a sharp boundary between the living and nonliving," Činátl argued that from an evolutionary viewpoint, there must once have been "a time when there existed side-by-side the non-living world and subvital structures. These quickly disappeared after the emergence of the first organisms which were better adapted for existence and used up for themselves all the material available for the continued growth and development of subvital structures."[20] Since that time early in the history of life on Earth, he argued that these subvital structures can be created and endure long only in sterile media. Nonetheless, the calcareous structures might then be models that could teach us much about the early stages of the origin of life, reasoned Činátl. His research was discussed further in 1972 by Adolph Smith and Dean Kenyon.[21]

Pollack and Coworkers

Much more recently, Gerald Pollack and coworkers in 2009 carried further Pollack's previously published argument that the properties of interfacial water (water adjacent to hydrophilic surfaces) are key to living processes. In this recent paper, they argue that such properties may also be key to understanding the origin of life. Water that is found surrounding particles in the gel state—including in the coacervates discussed by Bungenberg de Jong and later Oparin—exists in this structured state in which it causes charge separation and accumulation of negative charge. Pollack and coworkers assert that the energy thus produced could have driven coalescence of molecules such as simple sugars or short-chain amino acids. Either radiant energy from the sun or heat from hydrothermal vents could contribute to this process, they argue, much as heat causes amino acids to spontaneously assemble into "thermal proteins."[22] Since water still has this characteristic property when in interfacial condition, they further suggest that this mechanism might make ongoing origination of life under present-Earth conditions possible. This is a claim Reich made, yet it is still today considered so impossible by most as to render them skeptical of the judgment of anyone who would put the claim forward. About the simple gel or coacervate droplets, Pollack and coworkers say, "Quite remarkable is the extent to which such entities can resemble cells. Gels/coascervates [sic] can have negative electrical potentials just like cells; they absorb energy from the environment just like cells; and, like cells, their charged constituents move, here driven by energy from the environment. These constituents, including both water and dissolved/suspended substances, can be ordered, much like the cell. And they can also grow like the cell . . . so long as environmental energy is present to drive the expansion. Hence, these primitive gels/coascervates are more cell-like than one might initially suppose. The line of distinction between gels/coascervates and cells is somewhat blurred."[23]

Young, Martel, and Coworkers

Another recent researcher, John D. Young, earned fame in the 1980s for work that discovered the main pathways used by lymphocytes to kill tumor- and virus-infected cells, as well as how immune cells lead to programmed cell death (apoptosis) in general. In the past decade, he and several coworkers—most prominently Jan Martel—have investigated mineralo-organic nanoparticles (NPs) that share many properties in common with Reich's bions. Young and Martel even call the particles they study "bions" (though giving no indication of awareness of Reich's previous work), reporting that these can be created in a laboratory but have also "been observed in various body fluids . . . they assume marked morphological *pleomorphism*, with the spherical NPs flattening out to become spindles, platelets and films. These granular shapes are present throughout nature."[24] Carbonate hydroxyapatite minerals were the minerals they examined most thoroughly. Their structures share a great deal with the "calcareous structures" of Činátl and the $CaCO_3$ structures of Grad. For a time after they first formed, as in Grad's structures, the minerals were not in the form of large, immobile crystals, but over time more and more of the minerals converted to crystalline form, removing most of the lifelike properties of the mineral NPs.

Young, Martel, and their team state that they initially became interested in these NPs because a vast literature in the past twenty years describes them as the smallest living symbionts and claims they are involved in a wide range of disease processes, from Alzheimer's disease, arthritis, and atherosclerosis to kidney disease, cancer, and co-infection with HIV, among others. They conclude that their bions "mimic" many life processes of microorganisms and may well be involved in disease processes (especially those involving calcium metabolism in the body). Nevertheless, in contrast to much of that previous literature, they conclude their bions are *not* living organisms. They argue for recognition of such biomineral NPs as a distinct class of structure: involved in many living processes and capable of many characteristics of life, such as ability to be subcultured through many successive generations in fresh, sterile culture media—but not themselves alive. This is quite similar to Činátl's argument. Young, Martel, and their team argue that this interpretation of such structures also makes more sense as a way of understanding the meaning of much previous research, which has viewed the biomimetic properties of such NPs as proof that they represent life. For example, they argue that the putative "fossilized Martian nanobacteria" reported in 1996 by a NASA research team in Mars meteorite ALH84001 are much better explained as the kind of nonliving NPs they call bions.[25] By the same logic, Young and Martel's evidence would suggest a similar argument about Reich's original bions: that he did see novel structures with many lifelike properties, not merely buffeted by Brownian movement, seventy years before they did, but was merely mistaken in thinking they really were living or partially living. This is a fascinating possible way to reinterpret Reich's findings in light of very recent research.

Young and Martel's evidence does not seem anywhere to provide conclusive or compelling reason that their bions must be viewed as completely nonliving, however, rather than interpreting them as transitional stages between nonliving and living, the way Reich interpreted his bions. A paradigm that sees present-day origin of life in nature as impossible would predispose one to see Young and Martel's explanation as more compelling. But for one like Reich, Činátl—or now Gerald Pollack as well—who does not necessarily accept that paradigm as a given, then the interpretation that the bions are alive (or genuinely transitional between life and nonlife) seems to be a potentially valid alternative explanation to seeing growth, division, subculturing, and other such processes as merely "mimicking" living processes. Pollack's team poses this question rather directly as a question still open for discussion. They ask, if we are in the realm of such close mimicking of biological processes, then might not formation of such particles in fact *be* the kind of process by which the transition from nonliving to living matter occurred, and could still be occurring today?

Some argue that a microscopic structure cannot qualify as "living" unless it contains nucleic acids or some other high-fidelity replication apparatus capable of transmitting genetic information. But in the origin-of-life research community there have long been other researchers who argue that a membrane-enclosed structure capable of metabolism probably constituted the first "protocell," with a replication apparatus probably being added later in life's development. H. J. Muller had certainly made the "gene first" argument as early as 1926, while Oparin's coacervate models for protocells suggested a "metabolism first" approach. Reich had very little to say about genes, except to deplore their excessive invocation without evidence in Nazi "race biology." (The role of DNA in genes was, of course, not established until Avery, MacLeod, and McCarty's 1944 paper.)[26] But Reich's theoretical "basic antithesis" approach to understanding the nature of living matter would certainly make him much more sympathetic to "metabolism first" approaches in any case. So he did not lose much sleep over the fact that genes were unlikely to exist in bions from incandescent carbon, iron, or sand. Many origin-of-life researchers today still favor "metabolism first."[27]

How are the NPs of Young and colleagues related to human disease? These researchers claim to have disproven the idea put forward by earlier workers, that the NPs are infectious agents that directly cause disease. Instead, they argue, "mineralo-organic NPs such as the bions presented here do not represent an infectious cause of disease but may instead be part of a physiological cycle that regulates the function, transport and disposal of mineral ions in the body." By this interpretation, accumulation of excess nanoparticles could sometimes occur, with pathological results "when calcium homeostasis is disturbed and when clearance mechanisms are overwhelmed." This would often affect the kidneys and arteries. These researchers cite work by A. M. Gatti and colleagues to also suggest "that accumulation of mineral NPs in the body may induce chronic inflammation and even cancer in genetically-susceptible individuals."[28] Young, Martel, and colleagues suggest that, because they demonstrate macrophages can phagocytize excess calcium bions in vitro, they may be cleared in the body by the liver and spleen.[29] One cannot help but recall Reich's hypothesis that when cancer tumors bionously disintegrated (for example, in SAPA bion injection experiments), the immediate cause of death for the patient was overwhelming of the kidneys, liver, and other organs as they attempted to rid the body of the excess detritus.

Young, Martel, and colleagues also argue that their "bions" show several shared characteristics with prions, the particles that, when misfolded, are associated with BSE or "mad cow disease." Just as misfolded prions can reproduce themselves "when seeded into a solution containing sufficient amounts of the normal cellular prion protein," their bions can reproduce themselves when seeded into a solution with an adequate concentration of precipitating ions. In addition, just as prion protein is present during normal metabolic processes and leads to pathology only if it undergoes a conformational change, so their bions "may form in vivo from dissolved ions, proteins and organic molecules due to various factors . . . [such as] when there is excess of precipitating ions in the body fluid or cell milieu." However, unlike misfolded prions, which can actually trigger pathological folding of other prions and thus act as infectious agents of disease, "in this sense, . . . bions can not be deemed as infectious agents per se but they may certainly result in stone-related pathologies when they accumulate in the body."[30]

Paulo and Alexandra Correa

Recent work related to Reich's SAPA bions and published in 2010 by P. and A. Correa in Canada creates a situation in many ways similar to that created by the work of Young and colleagues. They validate most of Reich's observations, but they frequently differ in the interpretation of those findings. In this case, the researchers are aware of Reich's earlier work and are explicitly attempting to replicate Reich's key findings in experiments with numerous bion types, including SAPA bions. They recognize that earlier, out-of-hand dismissals of Reich's experiments—by merely *assuming* his sterile technique was consistently faulty—are not legitimate. First, they say, "Reich must be taken seriously with respect to his procedures and observations. These, one can, of course, confirm for oneself in the laboratory and by replicating his work. . . . One can certainly admit that Reich might have made mistakes, that here and there the sterility of his technique or those of his assistants . . . might not have been the best, or what, it turns out, is actually required to kill prokaryotic spores. But these claims must be based on solid facts, new or well founded facts, that can explain Reich's own results."[31]

The Correas argue, however, that Reich's categories "bions" and "PA bions" lumped together a wide range of very different kinds of structures. In some cases they suggest ultra-heat-resistant living structures, while in the case of bions from disintegrating plant matter they are convinced what Reich saw (and they, too, observed) were symbionts such as mitochondria, chloroplasts, or plastids from the disintegrating plant cells. They make an important point: Reich justified defining such a wide range of objects as bions because they were all freed from other matter by heating and swelling and all showed motility. (Retrospectively, after the discovery of orgone energy, Reich also said they were all energy vesicles freed from decomposing matter.) If Reich had at his disposal more recent knowledge from serial endosymbiosis theory, about what a wide range of endosymbionts a plant cell contains, perhaps he would judge those objects as more different than alike, though of course there is no way of knowing. Ivan Wallin was suggesting an early version of endosymbiosis theory during the time of Reich's bion experiments—though highly marginalized for the idea—but there is no evidence Reich was aware of such thinking.

Still, after replicating Reich's experiments, the Correas assert that Reich lumped together as "PA bions" a range of objects that they conclude are quite diverse. In disintegrating plant tissue, they conclude Reich was seeing mitochondria, chloroplasts, and plastids. They seem disposed to this interpretation because they can then make sense of these bions under the rubric of Margulis's serial endosymbiosis theory. Those entities, once released by disintegration, could resume some of the characteristics of previously free-living microbes. This approach might also offer another way to understand Reich's observations of such bions clumping together to form eukaryotic protists, they argue, "inasmuch as it includes indeed consideration of the possible role that mitochondrial or chloroplast precursors or 'cousins' (such as *Paracoccus denitrificans*, *Alcaligenes denitrificans*, various streptococcus, or certain purple nonsulfur bacteria and *Cyanobacteria*) might have in establishing cell-free symbiotic multiplicities that may actually be involved in cellular biopoiesis, specifically in the 'evolutionary' creation or modification of eukaryotic cells."[32] This is a fascinating possible interpretation. They do not, however, offer DNA sequencing or any other concrete evidence for why the objects should be considered mitochondria, plastids, and such rather than "bions," that is, transitional units between living and nonliving. So—much as with Young and Martel—their preference for this interpretation seems to be because it fits within their paradigm while still accounting for the fact that they found most of Reich's observations replicable. They are pushing Margulis's idea of symbiogenesis further than she did, perhaps, in suggesting it could lead to biopoiesis: a "natural organization of protozoa" or other eukaryotic cells, but via symbiogenesis, not via abiogenesis.

When repeating bion experiments with charcoal, sand, or other materials heated to incandescence, the Correas find structures much like Reich's "packet bions." They refer to these, instead, as "sarcinoid . . . encapsulated coccoid packets."[33] They believe what Reich actually found in these incandescence experiments could be microbial spores or "thermally super-resistant endosymbionts" with previously unimaginable heat-resistance—enough to survive heating to incandescence in a Bunsen burner flame. They suggest that "if a thermally resistant spore-based explanation is not eliminated by experimental studies, the conclusion [Reich's] of biopoiesis cannot effectively be validated."[34] Given discoveries since about 1977 about microbial extremophiles (many of which are now recognized as Archaea), for example, living at deep-sea hydrothermal vents, this seems a more pressing demand. We now know of microbes that thrive well above the boiling point of water, under many atmospheres of pressure, and in other extreme conditions. Furthermore, the Correas find, using 0.45 μm filters, that most of the ingredients for Reich's bion mixtures contain tiny, 0.4 μm cocci and/or spores. Only with repeated autoclavations (four times sequentially, with 24–36 hours in between) at pressures up to 27 psi for times up to forty minutes could they finally eliminate all these cocci and/or spores from the ingredients. Since such multiple autoclavations were not Reich's standard procedure, they suggest he might well have been seeing these grow in his bion ingredients not heated to outright incandescence. This is a substantial challenge to Reich's interpretation that must be tested in any further attempts to replicate the bion experiments.

There remains, however, Reich's point that the bions appear, full sized at 1–2 μm and above, *in very large numbers, immediately,* after the preparation is made. This seems to strain belief that all of them could be produced within five minutes from a few tiny cells of 0.4 μm that survived his autoclaving and dry sterilizing. The Correas' suggestion seems still further strained when soot (at least one of the ingredients) was heated to incandescence, as seen in Reich's film of fresh soot bions. The Correas also counter their own suggestion, by demonstrating that only a very small number of known non-spore-forming bacterial types have the characteristics of PA bions and none of those has heat-resistance to survive the procedures Reich used.

In the 1930s Reich's assertion that no microbe or spore could survive incandescent heat for more than a few seconds would have been uncontroversial. Of course—though the Correas do not emphasize this—if the spore idea is merely "not eliminated" by replication experiments, then neither does that logically falsify the possibility of biopoiesis, that is, Reich's interpretation of the observed growth after such extreme heat.

The Correas conclude that there are basically only two possible legitimate interpretations of Reich's repeatable observations: either Reich's results were "evidence for an 'accelerated biopoiesis'" or they were "evidence for the ubiquitous existence of very small spores of unidentified species of spore-forming sarcina."[35]

As the Correas formulate their inquiry, they state that the first question they need to resolve is whether "Reich did, or did not, in fact, discover a methodical series of chemical and energy-injecting steps capable of inducing or catalyzing the spontaneous organization of matter . . . into simple, non-nucleated protocells capable of DNA replication and cellular reproduction (fission)."[36] As I discussed with respect to Young and Martel, however, this is an ahistorical misreading of Reich that raises a central question. Reich never mentioned DNA and did not suppose that bions reproducing themselves in culture by fission necessarily implied they had DNA, or that any DNA they might contain necessarily replicated itself as part of their reproduction. The Correas here are conflating replication of DNA with multiplication or reproduction. As Freeman Dyson has pointed out, this kind of conflation mistakenly muddies the waters of the debate over "genes first" versus "metabolism first" in the origin of life. If one *defines* reproduction, even of protocells, to be a process that requires high-fidelity replication of some information molecule, then one is defining away the possibility that "metabolism first" protocells might have reproduced for a long time by growth and fission (as they do in Young and Martel's experiments), before any kind of high-fidelity *replication* mechanism like nucleic acids existed. To impose "capability of DNA replication" upon Reich's bions is to insist on a trait he did not necessarily believe they had.

Having made this conflation, the Correas compound it in their analysis of Reich's high-temperature sterilization procedures. Since no known nucleic acid polymers can withstand these temperatures, then is it really possible, they wonder, to interpret Reich's experiments as abiogenesis from scratch of "fully formed, energy metabolizing, DNA-carrying, self-replicating prokaryotic cells"?[37] At the risk of redundancy, Reich did not claim bions had these features. He saw them as transitional structures between nonliving and living, not fully formed cells. If we call them in today's language "protocells," their key features were metabolic and their reproduction was not said to be "replication," despite the fact that in pure culture transfers they remained true to type (excluding pleomorphic changes that result from changed media or other environmental conditions). Indeed, Young and Martel's experiments show that reproduction, without genetic replication, is possible in such structures, without needing to view them as bacterial contaminants.

Ten years after the bion experiments and in light of his later theories about orgone energy, Reich wrote a passage in the introduction to his 1948 book *The Cancer Biopathy*, which could be interpreted (the Correas do read it thus) to mean he was reconsidering whether *Staphylococci* were, as fully formed cells, totally different in kind from bions. He wrote, "The bacteriologist, for instance, sees the staphylococcus as a static formation, spherical or oval in shape, about 0.8 micron [µm] in size, reacting with a bluish coloration to Gram stain, and arranged in clusters. These characteristics are important for orgone biophysics, but are not the essentials. The name itself says nothing about the origin, function, and position of the blue coccus in nature. What the bacteriologist calls 'staphylococcus' is, for orgone physics, a small *energy vesicle* in the process of degeneration. Orgone biophysics investigates the origin of the staphylococcus from other forms of life and follows its transformations . . . and produces it experimentally through degenerative processes in bions, cells, etc."[38] If this passage does indeed indicate that Reich by 1948 accepted that bions were in some sense identical to staphylococci—fully functional cells with DNA—then this would lead to a need to consider whether PA bions do have DNA and whether it is the same in sequence as the DNA in staphylococci. (In 1948, such a question was not obvious; even in the wake of the 1944 paper by Avery and colleagues, many biologists still held the belief that proteins were the genetic material.) This could overrule many of the points made in previous paragraphs, about nucleic acids not necessarily being relevant if the bions are regarded as transitional stages rather than as fully formed living cells. It is not clear whether Reich saw these implications or pursued them very far in his thinking.[39] By 1948 most of his experimental work had moved on to other topics.

The Correas consider separately Reich's SAPA bions. They assert that Reich's bions were "bioluminescent, whether spontaneously or upon stimulation," rather than using Reich's language that the bions emit radiation. They also assert, though without any specific evidence, that "it is our view that in darkfield he mistook the refractile capsule as evidence for a proximal energy field radiating from the SAPA bions."[40] The Correas do, however, successfully produce organisms using Reich's procedure that are an extremely good match for his SAPA bions. They conclude that a spore-forming *Sarcina ventriculi* or, more likely, marine cyanobacterial *Myxosarcina* (or their "baeocytes") are the most likely identity of SAPA bions. They suggest that these "baeocytes" or resting cells of *Myxosarcina*—partially protected from the heat of incandescence by being trapped within crystalline quartz or igneous rock—could be released by the heating of those crystals and then begin to grow. The only alternative to assuming such previously unheard-of heat resistance, they argue, is to accept Reich's view that SAPA bions are "microbes that were created *de novo* in the laboratory by an abiogenic process."[41] This is a fascinating new finding that could open up new insights into the SAPA research.

In their description of their SAPA bion experiments, the Correas do not discuss any attempt to replicate Reich's most startling findings: the emission of radiation by these bions, or its characterization, to see if it differs from electromagnetic radiation in the ways Reich found. They barely mention the radiation, other than the rather brief, dismissive comments in the previous paragraph or mentioning that the organisms they produced that so much resemble Reich's SAPA bions were "highly refractile."[42] This is not to belittle the enormous amount of work this team accomplished, or their remarkable success in replicating the production of SAPA bions. But given that they produced microbes agreeing so closely with Reich's description of SAPA bions, and that the radiation was one of Reich's key arguments for why these bions—despite any morphological similarities—must represent something quite different from the bacteriologists' *Sarcinae* or *Myxosarcinae*, it is baffling that this team makes no comment at all on this issue.

Glossary

Reich's Bion Preparation Types

Dramatis Personae

Chronology

Periodicals Important in the Norwegian Press Campaign

Archives Consulted

Notes

Acknowledgments

Index

GLOSSARY

N.B.: For the most part, I use Reich's terminology in describing his experiments since it is so conceptually central to understanding both his choice of experimental approach and his interpretation of the results. This is not tantamount to endorsing Reich's terms and concepts (as I make clear in the Epilogue), but it is helpful in giving an account that takes Reich's laboratory science claims seriously—precisely what I argue has rarely been done until now by either scientists or historians of biology.

Basic antithesis of vegetative life: Reich's theory that the contrast between sexuality or pleasure versus anxiety is an expression of the most basic level of living functioning. The expansion versus contraction of the autonomic (vegetative) nervous system (or of only the protoplasm itself in simpler organisms like the amoeba) that corresponds to these emotions is, Reich thought, the basic pulsation that determines matter is alive. According to Reich's experiments, movement of bioelectric charge from the core of the organism out to the periphery characterizes the expansion and is perceived as pleasure, while movement of that same charge from the periphery in toward the core (the autonomic ganglia) is what constitutes anxiety.

Bionous disintegration: Matter breaking down into bions, either via heating and swelling or by disintegration of dying plant or animal tissue.

Bions: Reich's term for microscopic vesicles, in the size range of bacteria, from about 0.25 to 5 μm in diameter, that he found in disintegrating plant and animal tissue as well as produced by chemical mixtures of certain kinds. They had numerous lifelike properties (internal pulsation, growth and division, electrical charge, accepted biological stains, and cultivability on sterile media through successive generations in pure culture). Because of this, Reich thought they were transitional stages between nonliving and living matter, which could lead to the development of living organisms such as amoebae or, under anaerobic conditions, to protozoal cancer cells.

GLOSSARY

Brownian movement: A jittery, random movement on the spot, visible in tiny bacteria-sized objects because of their bombardment unevenly by random movements of molecules of the fluid in which the objects are immersed. Numerous critics tried to use this to explain that the movement of Reich's bions was not lifelike but a result of this purely physical process. Reich countered by pointing out that at magnifications of 2000× and above, one could clearly see that the bions were pulsating internally (as well as experiencing Brownian movement from without).

Endogenous infection: The long-standing belief among some medical researchers (Reich became one of these) that some kinds of diseases can be caused by microbes generated within the sick organism itself, because of its debilitated condition. John Hughes Bennett and Antoine Béchamp in the nineteenth century called the particles into which unhealthy tissues disintegrate "molecules" or "microzymas," respectively. Lionel Beale called them "bioplasts." They argued those could become agents of disease—some thought, for example, that this was how tuberculosis bacteria originated in a debilitated patient. Reich thought similarly about bionous disintegration of tissues in an unhealthy animal (especially in conditions of poor respiration and, thus, oxygen deficit). Mainstream medicine today considers this idea largely discredited.

Orgone energy: A type of energy Reich first studied as radiation emanating from SAPA bion cultures in the spring of 1939. A series of experiments, he said, convinced him its properties differed from electromagnetic radiation. For example, it could affect an electroscope but only via charging of some insulating material (rubber, wood, fiberglass, etc.), not if the SAPA culture plate was brought directly up to the electroscope itself. Reich derived the name "orgone" from the fact that the energy could charge organic materials and the fact that its discovery grew out of his long series of investigations into the function of the orgasm. He later came to believe it was the specific biological energy conceptualized earlier by biologists like Paul Kammerer.

SAPA bions: A type of bion culture created in January 1939 by heating sea sand to incandescence and then plunging it into a mixture of 50:50 bouillon:KCl solution. These bions could be grown through many generations in pure culture and emitted a radiation capable of strong biological effects (reddening the skin, irritating the optic nerve, etc.) as well as fogging photographic film and other physical effects.

Simultaneous identity and antithesis: Reich's conceptual approach to phenomena that appear superficially to be opposites, for example, love and hate, expansion and contraction, and parasympathetic and sympathetic. For Reich, it made sense not just to see two such "opposites" as dialectically related in some way, like the "thesis → antithesis" of the Hegelian dialectic. He felt one could usually understand the relationship between the opposites best if one could understand them as simultaneously rooted in a deeper underlying unity. In that sense, Reich said, they could be seen as integrated parts that made sense in terms of the functioning of a larger whole system. Thus, the contraction functions of the sympathetic autonomic nerves are not merely antagonistic to the expansion functions of the parasympathetic. Since both are part of the underlying unity of the autonomic nervous system, their functions at a deeper level are complementary and coordinated to produce the unified function of *pulsation* of the autonomic nervous system and the entire organism. Similarly, calcium ion has a dehydrating, contracting effect on tissues; potassium ion a hydrating, swelling effect. But the two together in a membrane-enclosed structure can contribute to its having the lifelike property of pulsation.

t-bacilli: Originally called S-bacilli by Reich, because he first found this type of lancet-shaped bion in sarcoma tissue from a cancer patient. "T-bacilli" was later coined to indicate "Tod" (the German word for death). This type of bion was much smaller than other types (0.2–0.5 μm in length) and stained gram-negative. Reich found it from many different types of putrefying proteinaceous matter, as well as in carbon-bion preparations. Injected into healthy animals in large quantities, it could induce cancer and other destructive, infiltrating growths. It was also produced in large amounts in the final stages of cancer as the tumors putrefied and disintegrated; Reich was convinced that such a flooding of the organism with t-bacilli was often the immediate cause of death.

REICH'S BION PREPARATION TYPES

Preparation 1 = bions from swelling earth.

Preparation 2 = bions from swelling coal.

Preparation 3 = bions from swelling soot.

Preparation 5 = bions from swelling grass.

Preparation 6 = PA (gram +) from lecithin-cholesterin, egg white, and other ingredients, as described in Chapter 2. One unusual variant was yellow packet-amoebae.

Preparation 8 = natural organization of amoebae from moss (see film script 2 Nov. 1936).

Preparation 10 = human blood → PA amoeboid bions.

10e = charred blood or soot → add to autoclaved Preparation 13, add 1 cc blood serum or sterile egg white → BLUKO → round cells (gram +) and interpenetrating soot and blood bions.

Preparation 11 = blood + coal experiment (BECP, p. 17). Blood spread flat on a petri dish and dried out sterile for 24h in incubator, then sterile blood charcoal spread over it, left for 24h. Then add 5–10cc of B + K → PA bions + t-bacilli [according to 1 Oct. 1938 protocol Preparation 11, "This culture originated just once from Preparation 11 on egg-medium IV. It grows at body-temperature only on egg-medium IV, 7.4 pH."].

Preparation 12 = BOPA. (gram +) Blood from a 40-year-old woman autoclaved → 12/I produced SEKO I cells (gram -).

SEKO II: "Originated from the tenth Preparation 12. Macroscopically it is orange-red, creamy, grows on egg-medium IV at room temperature, usually with a whitish border. It grows shallowly. Microscopically it consists at first of large vesicular bions, which one can see organize together in cell formations under the influence of KCl."

Preparation 13 = HAPA. (gram +) Same, from a man of 36, autoclaved ("rose culture"); these produced BLUKO I cells.

Preparation 14 = cancer model experiment: egg medium being penetrated by charcoal bions (BECP photos 24, 25).

Cyankali expt. = t-bacilli (unlike other bacteria) are not killed by KCN, but actually strengthened by it.

Preparation 15ce = egg medium with KOH added. (Bion lab ntbk 3, ms. pp. 94–95, entry of 31 May 1938 contains the formula for preparing these bions.)

Preparation 17 = iron bions.

SEPA I: SEPA-bions originated from "human arm vein blood. . . . The blood was centrifuged in sterile tubes. An egg-medium IV was inoculated from that serum after 24 hours and kept flat in an incubator. A fine, grainy growth was spread around and after a further 24 hours transferred to another egg-medium that was left standing at room temperature. After 6 days, a thick, light-yellow, creamy growth had formed that could be successfully carried over onto agar and egg-medium at both room temperature as well as incubator temperature."

SAPA = PA bions produced from sea sand heated to incandescence, added to 50:50 bouillon:KCl.

DRAMATIS PERSONAE

Christensen, Lars (1884–1965), was a Norwegian shipowner and whaling magnate with a keen interest in the exploration of Antarctica. He became a supporter—including a major financial supporter—of Reich's work when Reich moved to Norway.

Debré, Robert (1882–1978), was a professor of bacteriology at the University of Paris Medical School. In his laboratory under the supervision of his assistant Henri Bonnet, attempts to replicate some of Reich's bion experiments took place from fall of 1937 to spring of 1938 with ambiguous results.

du Teil, Roger (1886–1974), was a professor of natural philosophy at Mediterranean University Center in Nice; he carried out control experiments on bions during 1937–1938. He volunteered to be part of an international commission that would examine the validity of the experiments in the summer of 1938. This commission did not convene.

Fischer, Albert (1891–1956), was a cell-and-tissue-culture specialist at the Rockefeller Institute and the Biological Institute of the Carlsberg Foundation, both in Copenhagen. Fritz Lipmann worked in Fischer's lab there from 1932 to 1939. After initially expressing interest in Reich's experiments and seeing Reich demonstrate creating Preparation 6 bions, Fischer soon afterward ridiculed the bion experiments and invented claims such as that Reich demanded ridiculously high magnifications.

Havrevold, Odd W. (1901–1991), was an Oslo neurologist and endocrinologist who was very interested in psychoanalysis and became a student and patient of Reich's. He was one of Reich's main assistants in his laboratory experiments, particularly in tissue identification and animal experiments with bion injections.

Hoel, Sigurd (1890–1960), novelist; was Norway's most renowned literary figure from 1930 to 1960. He was a patient of Reich's for therapy and became a close personal friend and supporter of Reich's work in Norway. Hoel organized a petition signed by many of the most prominent intellectuals in Norway, defending Reich's right to continue his research, after Reich was viciously attacked in a campaign in Norwegian newspapers beginning in March 1938.

Kreyberg, Leiv (1896–1984), was Norway's only cancer researcher; he worked on the inheritance of cancer susceptibility in tar-painted mice and worked at Oslo Radium Hospital. Kreyberg became a professor of pathology at Oslo University in December 1937. Reich approached him in September 1937 for help in obtaining blood from cancer patients at the Radium Hospital. Their relations, which were friendly at first, quickly soured when Kreyberg attacked Reich's bion work shortly after Mohr (Kreyberg's patron for the university professorship and dean there) declared the bion experiments were scientific nonsense.

Langfeldt, Einar (1884–1966), was an Oslo University physiologist. Along with O. L. Mohr and R. Nicolaysen, he told a Rockefeller Foundation officer he thought Reich was a scientific "charlatan."

Langfeldt, Gabriel (1895–1983), was an Oslo psychiatrist and a strident critic of Reich.

Leunbach, Jonathan Hoegh "Joyce" (1884–1955), was a Danish physician, sex-educator, advocate of liberalizing abortion laws, and staunch supporter of Reich. Leunbach was prominent in the World League for Sexual Reform.

Mohr, Otto Lous (1886–1967), was a professor of genetics at Oslo University and rector from 1945 to 1951. Early on in the bion experiments without having witnessed any of the work, Mohr in September 1937 declared Reich's work nonsense because it appeared to revive claims of spontaneous generation. Mohr consistently thereafter opposed renewal of Reich's visitors permit in Norway, claiming the experimental work had no scientific merit. Kreyberg, a protégé of Mohr's, became a much more vocal public critic of Reich's experiments.

Neergaard, Paul (1907–1987), was a Danish plant pathologist who replicated some of Reich's experiments but encountered numerous technical problems, particularly because of insufficient microscope equipment. He proposed the idea of an international commission in 1938 to resolve the bion controversy. Brother of Reich's student, sociologist Jørgen Neergaard.

Nicolaysen, Ragnar (1902–1986), was an Oslo University assistant to E. Langfeldt and was later appointed professor of nutrition studies with help from Mohr. Told a Rockefeller Foundation officer he thought Reich was a "charlatan."

Nicolaysen Hoel Waal, Caroline "Nic" (1905–1960), was a psychiatrist and specialist in children's mental health. Married to Sigurd Hoel, 1929–1937; to Wessel

Waal, 1939–1960. She was a student of Reich's; after his departure from Norway she became skeptical about his subsequent experimental work on orgone energy.

Øverland, Arnulf (1889–1968), was one of Norway's most famous poets, a Communist and student of Reich's and in therapy with a trainee of Reich's. He wrote famous antifascist poems such as "You Must Not Sleep!" Imprisoned by Nazis at Sachsenhausen; upon his return after World War II, he was made poet laureate of Norway.

Reich, Wilhelm (1897–1957), was a psychoanalytic student of Freud and founder of "body therapy." He eventually moved into laboratory research to understand the energy behind neurotic symptoms, which he understood to be Freud's "libido." That research produced the bioelectric experiments and from 1936 to 1939, the bion experiments. In February through March 1939, Reich thought a new form of energy, specific life energy, was experimentally demonstrated in SAPA bion cultures. Because of the intense controversy generated in Norway by his bion experiments and because of the approaching war, Reich immigrated to the United States in August 1939.

Scharffenberg, Johann (1869–1965), was a Norwegian psychiatrist, eugenics advocate, and public polemicist. One of Reich's most outspoken critics in the Norwegian newspaper campaign of 1938.

Schjelderup, Harald (1895–1974), was a professor of psychology at the University of Oslo. He was initially responsible for obtaining permission for Dr. Reich to lecture to the students of the University of Oslo and to be granted lab space in the Psychological Institute for his bioelectric experiments. "It does not appear that Professor Scheldrup [sic] at any time supported Dr. Reich's biological theories," stated William Kerrigan in a 1952 interview with Schjelderup.

Stenersen, Rolf (1889–1978), was a Norwegian track and field athlete (1920 Olympics) and later a stock broker and writer in Norway and a patient of Reich. He defended Reich during a newspaper smear campaign and donated substantial amounts to support Reich's research.

Thjøtta, Theodor (1885–1955), was a professor of bacteriology and director of the Bacteriological Institute of the University of Oslo. Professor Thjøtta and Professor Kreyberg were Dr. Reich's principal scientific opponents on his biological experiments in Norway.

Walter, Emil (1897–1984), was a Zurich physicist and sociologist. He agreed to participate in a 1938 international panel to evaluate bion experiments.

CHRONOLOGY

1929: Reich's paper "Dialectical Materialism and Psychoanalysis" is published.

1930: Reich moves from Vienna to Berlin.

1932: Reich's article "The Masochistic Character" is published; Reich has a confrontation with Freud over its publication.

1933

Reich publishes *Character Analysis*.

April: Reich relocates to Copenhagen, in exile from Nazis.

September: Reich publishes *The Mass Psychology of Fascism*.

1934

Reich publishes "The Orgasm as an Electrophysiological Discharge" and "The Basic Antithesis of Vegetative Life."

August: The International Psychoanalytic Association holds its conference in Lucerne, at which Reich is expelled from the association.

October–November: Reich moves to Oslo; begins planning the bioelectric experiments on skin potential; completes "Psychic Contact and Vegetative Streaming."

1935

February: Reich begins bioelectric experiments (continuing through November 1936).

1936

Oparin's *The Origin of Life* is published in Russian and German.

April–June: Reich conducts experiments on "natural organization of protozoa," earth bions, and "pseudo-amoebae."

September: Reich begins experiments on *Staphylococci* and *Streptococci*; begins to guess at cancer cell origin.

November 16: Rockefeller Foundation officer visits the lab of Albert Fischer in Copenhagen.

November 20: Reich attends cancer conference, speaks with Blumenthal; is struck by how in a film living cancer cells appear exactly as he had intuited them previously.

December: Reich begins experiments attempting to grow bions in culture.

December 6: Reich demonstrates Preparation 6 bions at Fischer's lab in Copenhagen; positive Giemsa stain result.

1937

January: Reich sends bion preparations, along with a report, to French Academy of Sciences.

January 12–13: Rockefeller Foundation officer H. M. Miller visits Oslo.

February 7: Du Teil has begun control experiments; reports on bion work to Nice Natural Philosophy Society. By this time, Reich has succeeded in culturing Preparation 6 bions.

Late February: Rockefeller Foundation officer Tracy B. Kittredge interviews Reich in Oslo and encourages him to apply for a research grant.

March 4: Reich applies to Rockefeller Foundation for funding. Kittredge inquires of O. Mohr, E. Langfeldt, and R. Nicolaysen opinion of Reich's bions and is told they consider Reich a "charlatan." Rockefeller Foundation turns down Reich's request.

May 18–24: Reich prepares bions from coal, soot heated to incandescence.

July 27–August 5: Du Teil visits Reich's lab in Oslo; learns all bion techniques firsthand; leaves for Paris with films and Reichert microscope for demonstrations there.

August: Reich discovers S-bacillus from sarcoma tissue (later called t-bacillus).

August 10: Du Teil demonstrates bion preparations; shows films to French scientists in Paris.

September 9: Du Teil develops h-tube and Sy-Clos control experiments.

September 19: Reich publishes "Dialectical Materialism in Biological Research," which initiates a round of critical articles in the Norwegian press.

September 21: Mohr criticizes Reich's bion experiments in article in *Tidens Tegn* without having seen anything; Kreyberg visits Reich's Oslo lab, sees preparation of fresh carbon (soot) bions.

September 22: *Aftenposten,* the main conservative paper, carries an article with a warning about Brownian movement; *Tidens Tegn* carries "Vistemann på Kampen" cartoon.

September 23: *Tidens Tegn* carries Reich's letter in response to its September 21 article, asking that the papers contact him first before reporting on scientific research.

Late September: Reich develops theory of connection between carbon (e.g., from tar or tobacco smoke), tissue asphyxiation, and anaerobic cancer cell respiration.

October: Kreyberg blocks Reich from access to fresh cancer tissue from Oslo Radium Hospital for his research.

November: Beginning with I. Nissen and J. Scharffenberg, press campaign begins against Reich's therapy methods.

1938

The first English translation of Oparin's *Origin of Life* is published.

January: French Academy acknowledges bions sent a year earlier; remarks upon their still-active motility.

Early February: Reich begins "cancer model experiment," Preparation 14; begins negotiations with Lapicque about publication of bions in *Comptes Rendus* of French Academy.

Late February: *Die Bione* is published; Lapicque refuses to publish results of bion experiments unless Reich deletes success of culturing experiments and includes Lapicque's "Brownian movement" explanation of bion motility.

March: Norwegian press campaign against Reich's bion experiments is reignited, beginning with Johann Scharffenberg. Sigurd Hoel sponsors letter, signed by many Norwegian luminaries, urging Reich's right to experiment and stay in Norway be extended.

May 18–19: Rockefeller Foundation officer W. E. Tisdale visits Oslo; Mohr requests a large sum.

June: Crisis ensues over du Teil's dismissal from his job; International Commission to observe bion experiments is canceled.

August 28: Bernal responds to *Die Bione*; about this time Bernal completes *The Social Function of Science*.

September: Munich crisis; du Teil and Reich have their last significant communication. Reich attempts to get foreign labs to keep bion cultures safe in the event of war.

October–December: Reich sends thin sections of mice tumors induced by t-bacilli injections to pathology labs for classification of tumor types. His own classifications are largely confirmed.

1939

January 15: SAPA bions are created in Reich's lab; by early February, Reich has developed severe conjunctivitis from observing them.

Late February–early June: Reich has an exchange with W. F. Bon over experiments to characterize SAPA bion radiation.

June: Reich has his last contact with Roger du Teil and W. F. Bon.

August 18: Reich departs Oslo for America on the *Stavangerfjord*.

August 18–28: Reich observes on his voyage to America that relative humidity greater than 50 percent renders all orgone phenomena very hard to detect.

August 28: Reich arrives in New York to take up residence in the United States and to teach at the New School for Social Research.

September 25–November 10: Reich and Herrera exchange letters about t-bacilli.

PERIODICALS IMPORTANT IN THE NORWEGIAN PRESS CAMPAIGN

Aftenposten = conservative

Arbeiderbladet = socialist, organ of Norwegian Labor Party

Dagbladet = liberal

Morgenposten = conservative

Tidens Tegn = conservative, supportive of fascism (see dispute on this point, *IJSO* 2: 190 [1943])

ARCHIVES CONSULTED

American Philosophical Society, (Bentley Glass, L. C. Dunn, Curt Stern papers), abbreviated APS
American Society for Microbiology Archives, University of Maryland Baltimore County, Catonsville, MD
British Psychoanalytic Society Archives, London
Robert Brown papers, British Library and British Museum of Natural History
Jerome Greenfield papers, Taminent Library and Robert F. Wagner Labor Archives, New York University
Sol Kramer papers, University of Florida, Gainesville, FL
Otto Lous Mohr papers, Institute of Medical Genetics, Oslo University
Stuart Mudd Medical Microbiology Archive, University of Pennsylvania
Hermann J. Muller papers, Lilly Library, Indiana University, Bloomington, IN
National Library, Oslo, Norway
National Library of Medicine, Bethesda, MD (Aurora Karrer papers, FDA papers on Reich case)
Leo Raditsa papers, Harvard University Archives
Wilhelm Reich Archives/Orgone Institute Archives, Countway Library of Medicine, Boston, MA, abbreviated RA
Wilhelm Reich Museum, Rangeley, ME (including Reich's personal library, his scientific equipment, and microphotos donated by Charles Oller, MD, in 1980)
Rockefeller Foundation Archives, Rockefeller Archive Center, Tarrytown, NY
Staatsarkiv, Oslo, Norway
Personal papers lent by Bernard Grad, Morton Herskowitz, and Adolph Smith

The following laboratory notebooks in the Reich Archives were consulted in this research (including my abbreviated citation names for some of them). They are usually loose-leaf files of note sheets previously bound in loose-leaf notebooks:

1. Orgone Institute (OI) box 5, small black notebook, 1934 or 1935 through 15 June 1936.
2. OI box 5, notebook 3, "Zur Theorie der psychische-physische-funktional Identität, Sexökonomie."

3. OI box 6, "Versuche" (bioelectric experiments notebook), 14 Oct. 1934 through March 1936 (1st ser. Mar. 1935–Feb. 1936; 2nd ser. begins Feb. 1936).
4. OI box 6, "Pathologie und Krankengeschichten," e.g., 22 Oct. 1936 through 24 Nov. 1936.
5. OI box 6, "Entwicklung der Bionenforschung," Apr. 1936 through 4 Jan. 1937 (hereafter bion lab notebook 1).
6. OI box 7, "Geschichte der Erforschung die Bione," 9 Jan. 1937 through 7 Aug. 1937 (bion lab notebook 2).
7. OI box 8, "Kultur Journal," e.g., 23 Feb. 1937, 16 Aug. 1937.
8. OI box 8, "Research Notebook, 1937–38," 14 Apr. 1937 through 18 May 1938 and 18 Sept. 1939.
9. OI box 7, "Journal, Odd Wenesland," Wenesland's own separate lab notebook.
10. OI box 7, "Notes by a Laboratory Worker, 1938" (bion cookbook).
11. OI box 7, "Basic Research III" (bion lab notebook 3), 10 Aug. 1937 through 28 Oct. 1938 and a few more entries through 31 Oct. 1939.
12. OI box 8, "Präparatherstellung und Protokoll" and "Research Preparations, 1937–39" folders.
13. OI box 11, black notebook titled "Geschichte der Krebsforschung Nov. 1936–Nov. 1938."
14. OI box 11, second cancer notebook: "Bion und Krebsforschung 29.X.38–15.XII.39."
15. OI box 13, "Research Notebook, 1939" (Orgone Research lab ntbk. 1; contains most details of SAPA bion experiments).

All Reich–du Teil letters are in Reich Archives, correspondence box 10, du Teil folders.
All Reich–Paul Neergaard letters are in correspondence box 9, Neergaard folders.

International Journal of Sex-Economy and Orgone Research, abbreviated IJSO.
Wilhelm Reich, *Beyond Psychology: Letters and Journals, 1934–1939*, abbreviated BP. (Letters published here are sometimes excerpted, but without ellipses. Therefore, I always cite from the archival original letter. I give reference to the published, edited versions merely for the reader's convenience. I refer to these as "partial translations" when editing has occurred.)
Wilhelm Reich, *The Bioelectric Investigation of Sexuality and Anxiety*, abbreviated BISA.
Wilhelm Reich, *Bion Experiments on the Cancer Problem*, abbreviated BECP.
Wilhelm Reich, *The Bion Experiments on the Origin of Life*, abbreviated BE after first usage.
Wilhelm Reich, *The Cancer Biopathy*, abbreviated CB.
Wilhelm Reich, *Character Analysis*, abbreviated CA.
Wilhelm Reich, *The Function of the Orgasm*, abbreviated FO.

ARCHIVES CONSULTED

Zeitschrift für Politische Psychologie und Sexual-ökonomie, abbreviated ZPPS.

N.B. Since the U.S. Food and Drug Administration destroyed so many of the first editions of these works, they are now difficult to find. When Ms. Higgins and Reich's Trust brought new editions back into print with Farrar, Straus and Giroux after 1960, new translations were often done for copyright purposes. Since those are much easier to obtain, I usually try for the reader's convenience to give page numbers from the new edition in citations, even when I also give page numbers from the original editions.

NOTES

Introduction

1. Among historians of psychoanalysis, one who does take Reich's research program seriously is Norwegian Håvard Nilsen, whose works are cited below in Chapter 6, notes 38, 39, and 146. Most sources on Reich's career are highly unreliable. The only overall chronology of Reich's career I have found that takes the trouble to (at least) get all the facts right is the one published on the website of the Wilhelm Reich Trust; see http://www.wilhelmreichtrust.org/biography.html.

2. In addition to 25 to 30 physicians who became practicing orgone therapists, well-known *New Yorker* cartoonist and children's book author (including *Shrek*) William Steig remained convinced of the validity of Reich's science and used an orgone energy accumulator until his death in 2003. See Steig, *The Agony in the Kindergarten* (New York: Duell, Sloan and Pearce, 1950); Steig's *New York Times* obituary, http://www.nytimes.com/2003/10/05/obituaries/05STEI.html; also http://www.boston.com/news/globe/obituaries/articles/2003/10/05/william_steig_new_yorker_cartoonist_wrote_shrek/; and Lee Lorenz, *The World of William Steig* (New York: Artisan, 1998), pp. 25, 74, 77–78, 80, 82; also Jonathan Cott, *Pipers at the Gates of Dawn* (New York: Random House, 1983), pp. 86–133. Educator A. S. Neill, founder of the Summerhill school in England, also credited his therapy with Reich for much of his personal insight into children. Another student who respected Reich throughout a long, productive career was Michael Rothenberg, MD, pediatric psychiatrist and professor at University of Washington Medical School in Seattle. Rothenberg coauthored two editions of the "Dr. Spock" book on childrearing, beginning in 1985. Strick oral history interview with Rothenberg, 25 Apr., 12 May, and 19 June 1999. Some sources—perhaps embarrassed that important figures supported Reich—simply choose to leave that fact out. Publisher Roger Straus, however, was dedicated to bringing an unprecedented twenty-two titles of Reich's work back into print over a period of forty plus years because he hated government censorship. Of those 22 titles, 10 were books that had been banned

and burned by the U.S. government. R. D. Laing has some insightful thoughts on this phenomenon in "Why Is Reich Never Mentioned?," *Pulse of the Planet* 4: 76–77 (1993), reprint of a 1968 article. Leo Raditsa expresses other insights on the topic in "A Lost Genius?," *Midstream: A Monthly Jewish Review* (Dec. 1999), pp. 37–40.

3. The best account of the legal case against Reich is Jerome Greenfield, *Wilhelm Reich vs. the USA* (New York: Norton, 1974). Some useful information can also be found in James Turner, *The Chemical Feast; The Ralph Nader Study Group Report on Food Protection and the Food and Drug Administration* (New York: Grossman, 1970), pp. 30–48. On the Stalinist ties of Brady and her husband, see Jim Martin, *Wilhelm Reich and the Cold War* (Ft. Bragg, CA: Flatland Books, 2000), esp. pp. 280–307.

4. A recent piece on Reich's SexPol work, for example, is rather sympathetic to that work, but nonetheless takes the two false claims as truth not needing fact-checking, simply because they have been repeated so many times. See http://www.theguardian.com/commentisfree/video/2013/jul/24/william-reich-sex-pol-video-radical-thinkers.

5. The derisive "orgone box" narrative is already firmly in place in Henry A. Grunwald, "The Second Sexual Revolution," *Time*, 24 Jan. 1964, pp. 54–59. This narrative is picked up repeatedly from secondary sources and repeated uncritically without any evidence of the writer ever reading Reich himself. For a recent example, see Miriam Reumann, *American Sexual Character: Sex, Gender, and National Identity in the Kinsey Reports* (Berkeley: U. of California Press, 2005). Reumann presents Fromm and Marcuse as level-headed, astute observers of society while taking Grunwald's account of Reich uncritically at face value and thus putting him in a "wild loner" category. This is regrettable given her otherwise meticulously scrupulous history, which urges us not to take such accounts at face value—particularly when, as with Grunwald, they so clearly have an axe to grind.

6. As an example, Myron Sharaf, author of a highly regarded biography of Reich, *Fury on Earth: A Biography of Wilhelm Reich* (New York: St. Martin's, 1983, p. 276), did not even correctly understand the reasons why Reich chose the name "orgone" for the energy he thought he had discovered: because it was capable of charging organic (insulating) substances such as rubber or cotton, and because he was led to it by research that originated in the study of the function of the orgasm. Sharaf here repeated the mistaken view of Elsworth Baker that "orgone" was derived from the word "organism" rather than because of the energy's ability to charge organic substances. Baker had notably tried to soften or water down Reich's claims about orgone energy, in order to make "orgonomy" a more respectable science to the scientific establishment, according to Chester Raphael in *Wilhelm Reich: Misconstrued, Misesteemed*

(Forest Hills, NY: privately published, 1970), so this is not a trivial semantic distinction. Leaving out the central role of the evidence of physical experiments with orgone (which I discuss in Chapter 7) makes the work seem more arbitrary and opinionated (and thus easier to dismiss as "pseudoscience") than it actually was.

7. For a thoughtful reflection on the recent revival of scientific biography and an appreciation of the merits of the genre, see Richard Holmes, "The Scientist Within," *Nature* 489: 498–499 (27 Sept. 2012). See also Neeraja Sankaran, "Scientific Biography: Still Fertile Soil for Cycling History," *Stud. Hist. Biol.* 6 (2): 104–110 (2014), which includes some consideration of the shortcomings of this literary form as well.

8. James Strick, "Swimming against the Tide: Adrianus Pijper and the Debate over Bacterial Flagella, 1946–1956," *Isis* 87: 274–305 (1996); Nicolas Rasmussen, *Picture Control: The Electron Microscope and the Transformation of American Biology, 1940–1960* (Stanford, CA: Stanford U. Press, 1997).

9. See Julie Edgeworth, Nathalie Gros, Jack Alden, Susan Joiner, Jonathan Wadsworth, Jackie Linehan, Sebastian Brandner, Graham S. Jackson, Charles Weissmann, and John Collinge, "Spontaneous Generation of Mammalian Prions," *Proc. Nat. Acad. Sci. USA*, www.pnas.org/cgi/doi/10.1073/pnas.1004036107; see also John D. Young and Jan Martel, "The Rise and Fall of Nannobacteria," *Scientific American*, Jan. 2010, pp. 52–59; idem, "Purported Nanobacteria in Human Blood as Calcium Carbonate Nanoparticles," *Proc. Nat. Acad. Sci. USA* 105 (14): 5549–5554 (8 Apr. 2008). In a more recent paper, these authors actually refer to the structures they describe as "bions," Reich's term, though apparently by independent coinage since they do not cite Reich. See C-Y. Wu, L. Young, D. Young, J. Martel, J. D. Young, "Bions: A Family of Biomimetic Mineralo-Organic Complexes Derived from Biological Fluids," *PLoS ONE* 8 (9): e75501. doi:10.1371/journal.pone.0075501 (25 Sept. 2013); http://www.plosone.org/article/info:doi/10.1371/journal.pone.0075501. Young and Martel's work is discussed further in the Epilogue of this book.

10. Reich's journal was titled *Zeitschrift fur Politische Psychologie und Sexual-Ökonomie* [Journal of political psychology and sex-economy] (hereafter, ZPPS).

11. John Prebble and Bruce Weber, *Wandering in the Gardens of the Mind: Peter Mitchell and the Making of Glynn* (New York: Oxford U. Press, 2003); James Lovelock, *Homage to Gaia: The Life of an Independent Scientist* (New York: Oxford U. Press, 2000).

12. Wilhelm Reich, *Die Bione: Zur Enstehung des vegetativen Lebens* (Oslo: Sexpol Verlag, 1938), quote from Eng. trans. Derek and Inge Jordan as *The Bion Experiments on the Origin of Life* (hereafter, *BE*) (New York: Farrar, Straus and Giroux, 1979), p. 152.

13. Evelyn Fox Keller, *A Feeling for the Organism: The Life and Work of Barbara McClintock* (San Francisco: W. H. Freeman, 1983).
14. Microbiologist Carl Woese, for example, liked to tell a Kuhnian story about Kuhn's initial difficulties convincing colleagues of the importance of the Archaea. See, e.g., Virginia Morell, "Microbiology's Scarred Revolutionary," *Science* 276: 699–702 (2 May 1997). The story of his marginalization before ca. 1978 was somewhat more complicated; see Steven Dick and James Strick, *The Living Universe: NASA and the Development of Astrobiology* (New Brunswick, NJ: Rutgers U. Press, 2004), p. 108. Historians and philosophers of science have widely critiqued Kuhn's categories as over-simplistic. Kuhn responded to many of these critiques in the Postscript to the 1970 second edition of his book. Nonetheless, these categories are widely used, if with some caution. It is in that qualified sense that I use them here.
15. Mary Boyd Higgins, "Introduction" to Wilhelm Reich, *Beyond Psychology: Letters and Journals, 1934–1939* (hereafter, BP) (New York: Farrar, Straus and Giroux, 1994), p. xi, my emphasis.
16. Yirmiyahu Yovel has pointed out that Freud's "libido" bears much resemblance to Spinoza's "conatus." "Human beings are fully integrated in nature and are moved by a dominant natural striving or source of energy (conatus in Spinoza, libido in Freud), which streams forth from a number of outlets, and of which knowledge and the products of high culture are but sublimated configurations." Yirmihahu Yovel, *Spinoza and Other Heretics: The Adventures of Immanence* (Princeton, NJ: Princeton U. Press, 1989), p. 142. One of Spinoza's core ideas, "the claim that nature is self-moving and creates itself," led many followers of Spinoza to conclude that spontaneous generation of life from nonliving matter and evolution of that living matter to more and more complex forms must be inherent in nature, not requiring any supernatural intervention. See Jonathan Israel, *The Radical Enlightenment: Philosophy and the Making of Modernity, 1650–1750* (New York: Oxford U. Press, 2001), pp. 160, 236–238, 708, quote on 160. For a modern philosopher's view of these issues, see David M. Brahinsky, *Reich and Gurdjieff: Sexuality and the Evolution of Consciousness* (Xlibris, 2011, see http://www.akhaldan.com/).
17. See, e.g., discussion of Schjelderup's views in the popular press in Norway, before Reich came there to carry out his bioelectric experiments: *Dagbladet* n.s. nr. 93, anon., "Hvordan voksne mennesker kan få et nytt jeg: Og stryken av vårt raseri måles i volt og ampère; Professor Schjelderup om moderne psykologis resultater" (23 Apr. 1934) [How adults can get a new ego. And the strength of our rage is measured in volts and amperes: Prof. Schjelderup about the results of modern psychology].
18. Freud used this analogy to describe his idea of Besetzung (cathexis), the process of extending our libido out into the world and investing it in an object

of our desire. See, e.g., 1914 "On Narcissism," in *The Complete Psychological Works of Sigmund Freud* (The Standard Edition) (New York: W.W. Norton, 1976), vol. 14: 75. "Thus we form the idea of there being an original libidinal cathexis of the ego, from which some is later given off to objects, but which fundamentally persists and is related to the object-cathexes much as the body of an amoeba is related to the pseudopodia which it puts out." Later Freud used the same analogy in lecture 26 of *Introductory Lectures in Psychoanalysis from 1916* (New York: W.W. Norton, 1965). My thanks for advice on this point from historian of psychoanalysis Håvard Nilsen.
19. Reich, *BP*, p. 7.
20. Wilhelm Reich, "Zur Triebenergetik" [On the energy of drives], *Zeitschr. f. Sexualwissenschaft* 10: 99–106 (1923).
21. On the importance in psychology at this time of conservation of energy ideas, see, e.g., David Millett, "Hans Berger: From Psychic Energy to the EEG," *Persp. Biol. Med.* 44: 522–542 (Fall 2001).
22. Chester M. Raphael, introduction to Wilhelm Reich, *Early Writings*, vol. 1 (New York: Farrar, Straus and Giroux, 1976), p. ix.
23. Ibid. A nuanced study of Reich's early career in Vienna is Karl Fallend, *Wilhelm Reich in Wien: Psychoanalyse und Politik* (Vienna: Geyer Edition, 1988).
24. Wilhelm Reich, "The Bioelectrical Function of Sexuality," unpublished ms., May 1938, Wilhelm Reich Archives, Countway Library of Medicine, Boston (hereafter, RA), Orgone Institute (hereafter, OI) box 9, manuscripts, emphasis in original. Freud's earlier use of this metaphor: Sigmund Freud, *Three Contributions to the Theory of Sex*, [1905] pt. 1, sec. 6, in A. A. Brill, ed., *The Basic Writings of Sigmund Freud* (New York: Random House, 1938), p. 577: Describing the origins of psychoneurosis via the inability to attain normal sexual gratification, Freud adds that "the libido behaves like a stream, the principal bed of which is dammed; it fills the collateral roads which until now perhaps have been empty."
25. Wilhelm Reich, "Dialectical Materialism and Psychoanalysis," in Lee Baxandall, ed., *Sexpol Essays* (New York: Vintage, 1972), pp. 15–17; originally pub. in *Unter dem Banner des Marxismus* 3: 736–771 (1929). In other places Freud had spoken of libido as "like an electrical field that spread itself out over the memory traces of an idea," saying, "It is a quantitative energy, directed to an object" (*Three Contributions, op. cit.*, p. 553). For the state of ideas about hormones at this time, see Nelly Oudshoorn, "Endocrinologists and the Conceptualization of Sex, 1920–1940," *J. Hist. Biol.* 23: 163–186 (1990); Diana Long Hall, "Biology, Sex Hormones, and Sexism in the 1920s," in Carol Gould and Marx Wartofsky, eds., *Women and Philosophy: Toward a Theory of Liberation* (New York: Putnam, 1976), pp. 81–96; also C. Sengoopta, "Glandular Politics: Experimental Biology, Clinical Medicine, and Homosexual

Emancipation in Fin-de-Siècle Central Europe," *Isis* 89: 445–473 (1998); and idem, *The Most Secret Quintessence of Life: Sex, Glands and Hormones, 1850–1950* (Chicago: U. of Chicago Press, 2006).

26. Walter and Käthe Misch, "Die vegetative Genese der neurotischen Angst und ihre medikamentöse Beseitigung," *Der Nervenarzt* 8: 415–418 (1932).

27. Friedrich Kraus and Samuel G. Zondek, "Über die Durchtränkungsspannung," *Klinische Wochenschrift* (Berlin) 1: 1778–1779 (1922); see also Kraus, *Allgemeine und Spezielle Pathologie der Person. I. Teil: Tiefenperson* (Leipzig: Thieme, 1926). Reich reviewed the latter book for a psychoanalytic journal in 1926. On Kraus (1858–1936), see Martin Lindner, *Die Pathologie der Person: Friedrich Kraus' Neubestimmung des Organismus am Beginn des 20. Jahrhunderts* (Berlin: Verlag f. Geschichte der Naturwissenschaften und der Technik, 1999); see also Manfred Stürzbecher, "Kraus, Friedrich" in *Neue Deutsche Biographie*, vol. 12 (Berlin: Duncker and Humboldt, 1980), p. 685.

28. Max Hartmann, "Relative Sexualität und ihre Bedeutung für eine allgemeine Sexualitäts- und Befruchtungstheorie," *Naturwiss.* 19: 8–16 (1931); pt. 2 in *Naturwiss.* 19: 31–37 (1931). See Alfred Henry Sturtevant, *A History of Genetics* (New York: Academic Press, 1965), pp. 118–119. Reich came across these papers in 1933 or late 1932.

29. Wilhelm Reich, "Der Urgegensatz des vegetativen Lebens," *Zeit. f. Polit. Psychol.u. Sex.-Ökon.* (hereafter, ZPPS) 1 (2): 125–142 (1934); Eng. trans. as "The Basic Antithesis of Vegetative Life Functions," in Wilhelm Reich, *The Impulsive Character and Other Writings*, ed. B. G. Koopman (New York: Random House, 1977), pp. 83–121. Reich prepared an updated 1945 version of the article, available in Wilhelm Reich, *The Bioelectrical Investigation of Sexuality and Anxiety* (New York: Farrar, Straus and Giroux, 1982) (hereafter, *BISA*).

30. Wilhelm Reich, *Reich Speaks of Freud* (New York: Farrar, Straus and Giroux, 1967), pp. 200–201. For more on relations between Reich and Fenichel, see Benjamin Harris and Adrian Brock, "Freudian Psychopolitics: The Rivalry of Wilhelm Reich and Otto Fenichel," *Bull. Hist. Med.* 66: 578–612 (1992).

31. In an article whose ms. was dated Nov. 1941, Reich called it "biophysical functionalism." See Wilhelm Reich, "Biophysical Functionalism and Mechanistic Natural Science," *International J. Sex-Economy and Orgone Research* (hereafter, *IJSO*) 1: 97–108 (1942).

32. Reich, *BE*, 1947 introduction to Eng. trans., p. v.

33. Wilhelm Reich, *Maase und Staat* (Oslo: SexPol Verlag, 1935), included in *The Mass Psychology of Fascism* (New York: Farrar, Straus and Giroux, 1969); idem, *Die Sexualität im Kulturkampf* (Oslo: SexPol Verlag, 1936).

34. Interestingly, even modern artists interested in artificial life have taken an interest in Reich's bion experiments. See http://www.cal.msu.edu/bion-adam-brown/, an installation from 2006.

1. Reich's Background, Origins of His Research Program, and Relevant Context

1. Wilhelm Reich, "Background and Scientific Development of Wilhelm Reich," *Orgone Energy Bulletin* 5: 5–9 (1953), quotes on p. 6.
2. Most of this biographical material is from Wilhelm Reich, *Passion of Youth: An Autobiography, 1897–1922* (New York: Farrar, Straus and Giroux, 1988), esp. pp. 3–67.
3. Reich, "Background and Development," *op. cit.*, p. 7. For more on the state of sexual science at this time, see Kevin Amidon, "Sex on the Brain: The Rise and Fall of German Sexual Science," *Endeavour* 32 (2): 64–69 (2008).
4. Wilhelm Reich, *The Function of the Orgasm* (hereafter, *FO*), Eng. trans. Theodore P. Wolfe (New York: Orgone Institute Press, 1942), p. 77; this discussion of orgastic potency is from pp. 72–78. This is on pp. 95–101 in the more recent edition of 1973, trans. by Vincent Carfagno (New York: Farrar, Straus and Giroux; hereafter, FSG ed.).
5. T. P. Wolfe, "Table of Events", app. to Reich, *FO* (1942 ed.), p. 355.
6. Reich, *FO* (1942 ed.), pp. 112–114, quote on pp. 113–114 (FSG ed., pp. 136–138). Freud's eventual rejection of Reich's orgasm theory must have been the same kind of personal blow, deeply shaking his self-confidence, as that received by Darwin when Herschel and Whewell severely criticized his theory of natural selection. My reading is that Reich—like Darwin—was able to keep moving forward with his theory despite such a personal shock not because of personal arrogance but because his theory was rooted in the bedrock of clinical observations he had made himself, and the theory resulted from his struggle to make sense of those observations.
7. S. Freud to Reich, 27 July 1927 (Reich Archives [hereafter, RA], correspondence box 2). Freud assured Reich that he valued the younger man's enthusiasm, energy, and work. He told Reich that his talents were recognized even by psychoanalysts who sharply opposed Reich—such as Paul Federn—especially Reich's talent for training new analysts. For this, Freud said the general opinion was that nobody in Vienna could do the training any better than Reich. Freud noted that some analysts thought Reich's focus on the orgasm amounted to riding a hobby (*Aufzucht von Steckenpferden*). But he also told Reich that he had no problem at all with the younger man being an innovator, encouraging Reich to go on innovating, because if his insights were valid, Freud said, eventually that would be recognized, and in any case the discussions would be very fruitful.
8. Wilhelm Reich, "Zur Triebenergetik," *Zeitschr. f Sexualwissenschaft* 10: 99–106 (1923); Eng. trans. in *Early Writings*, vol. 1, pp. 144–150. Reich had given a draft of this paper to Freud to critique by 5 Dec. 1920; see Reich, *Passion of Youth*, *op. cit.*, p. 144.

9. See Erna Lesky, *The Vienna Medical School of the Nineteenth Century* (Baltimore: Johns Hopkins U. Press, 1976), esp. pp. 355–360 on Freud and 360–363 on Wagner von Jauregg, under both of whom Reich trained extensively.
10. Stefan Zweig gives a frank and revealing glimpse into just how dramatic was the change in Viennese sexual mores from ca. 1900 to 1940 in *The World of Yesterday: An Autobiography* (New York: Viking Press, 1943), chap. 3, "Eros Matutinis."
11. Elizabeth Danto, *Freud's Free Clinics: Psychoanalysis and Social Justice, 1918–1938* (New York: Columbia U. Press, 2005).
12. Wilhelm Reich, *CB* on genetics of cancer, race biology generally (including the Kallikak study), p. 10 in 1973 edition. See also Helmut Gruber, *Red Vienna: Experiment in Working Class Culture, 1919–1934* (Oxford: Oxford U. Press, 1987); Loren Graham, "Science and Values: The Eugenics Movement in Germany and Russia in the 1920s," *Am. Hist. Rev.* 82: 1133–1164 (1977); Nils Roll-Hansen, "Eugenics in Norway," *Hist. Phil. Life Sci.* 2: 269–298 (1980); G. Broberg and N. Roll-Hansen, eds., *Eugenics and the Welfare State* (East Lansing: Michigan State U. Press, 2005); Atina Grossman, *Reforming Sex: The German Movement for Birth Control and Abortion Reform, 1920–1950* (Oxford: Oxford U. Press, 1995).
13. Richard Stites, *The Women's Liberation Movement in Russia* (Princeton, NJ: Princeton U. Press, 1978), pp. 383, 392; Wilhelm Reich, *Die Sexualität im Kulturkampf* (Oslo: SexPol verlag, 1936).
14. Benjamin Harris and Adrian Brock, "Freudian Psychopolitics: The Rivalry of Wilhelm Reich and Otto Fenichel, 1930–1935," *Bull. Hist. Med.* 66: 578–612 (1992). On use of the term "Red fascism," see L. K. Adler and T. G. Paterson, "Red Fascism: The Merger of Nazi Germany and Soviet Russia in the American Image of Totalitarianism, 1930s–1950s," *Am. Hist. Rev.* 75: 1046–1064 (1970).
15. Loren Graham, *Science in Russia and the Soviet Union* (Cambridge: Cambridge U. Press, 1994), p. 101; see also Garland Allen, *Life Science in the Twentieth Century* (Cambridge: Cambridge U. Press, 1975), pp. 103–106; and more recently idem, "Mechanism, Vitalism and Organicism in Late Nineteenth and Twentieth Century Biology: The Importance of Historical Context," *Stud. Hist. Phil. Biol. Biomed. Sci.* 36: 261–283 (2005) on "holistic materialism" and dialectical materialism; also Donna Haraway, *Crystals, Fabrics, and Fields: Metaphors of Organicism in Twentieth-Century Developmental Biology* (New Haven, CT: Yale U. Press, 1976) on "organicism." See Anne Harrington, *Reenchanted Science: Holism in German Culture from Wilhelm II to Hitler* (Princeton, NJ: Princeton U. Press, 1996) on "holism." For a recent take, by a developmental biologist, on "holism" and "organicism," see Scott Gilbert and Sahotra Sarkar, "Embracing Complexity: Organicism for the 21st Century," *Developmental Dynamics* 219: 1–9 (2000).

16. Wolfe, "Translator's Preface," to Reich, *The Function of the Orgasm* (New York: Orgone Institute Press, 1942), pp. vii–viii. When Wolfe (1902–1954) came across Reich's concept, he found it extremely useful and eventually came to study with Reich in Oslo during the bion experiments there.
17. Reich, "Zur Triebenergetik," *op. cit.* For more on Semon, see Daniel Schachter, *Forgotten Ideas, Neglected Pioneers* (Philadelphia: Psychology Press, 2001).
18. Paul Kammerer, *Allgemeine Biologie*, 1st ed. (Stuttgart: Deutsche Verlags Anstalt, 1915), pp. 6–9.
19. For more on reductionism, see William Bechtel, *Discovering Cell Mechanisms: The Creation of Modern Cell Biology* (Cambridge: Cambridge U. Press, 2005); also Marie Kaiser, "The Limits of Reductionism in the Life Sciences," *Hist. Phil. Life Sci.* 33: 453–476 (2011). Notwithstanding Kaiser's claim that many life scientists actively question the limits of reductionism, the idea of cells composed of "nothing but molecular machines" has gained strength over time with many others in life sciences. See, e.g., Bruce Alberts, "The Cell as a Collection of Protein Machines: Preparing the Next Generation of Molecular Biologists," *Cell* 92: 291–294 (6 Feb. 1998). For a thoughtful deconstruction of some of the modern "machine" imagery in cells, see Maria Trumpler, "Converging Images: Techniques of Intervention and Forms of Representation of Sodium-Channel Proteins in Nerve Cell Membranes," *J. Hist. Biol.* 30: 55–89 (1997).
20. Joseph Heckman, "History of Organic Farming: Transitions in Organic Farming from Sir Albert Howard's *War in the Soil* to USDA Certified Organic," *National Organic Program: Renewable Agriculture and Food Systems* 21: 143–150 (2006); see also J. von Liebig, *Organic Chemistry in Its Applications to Agriculture and Physiology* (London: Taylor and Walton, 1840) and idem, *Animal Chemistry* [1843] (New York: Johnson Reprint, 1964).
21. Richard Eyde, "The Foliar Theory of the Flower," *Am. Sci.* 63: 430–437 (July–Aug. 1975), p. 435.
22. L. Pearce Williams, *The Origins of Field Theory* (Ithaca, NY: Cornell U. Press, 1965), p. 78.
23. See Joan Steigerwald, "Rethinking Organic Vitality in Germany at the Turn of the Nineteenth Century," in Sebastian Normandin and Charles T. Wolfe, eds., *Vitalism and the Scientific Image in Post-Enlightenment Life Science, 1800–2010* (New York: Springer, 2013), pp. 51–75. Steigerwald shows that not all Romantic biology, nor even all *Naturphilosophen* invoked a "vital force," contrary to the straw man claims of Helmholtz and the 1847 "physicalists" who were reacting so strongly against the earlier vitalism (especially of their mentor, Johannes Müller).
24. See Philip Pauly, *Controlling Life: Jacques Loeb and the Engineering Ideal in Biology* (Oxford: Oxford U. Press, 1988).

25. Garland Allen, "Rebel with Two Causes: Hans Driesch," in Oren Harman and Michael Dietrich, eds., *Rebels, Mavericks and Heretics in Biology* (New Haven, CT: Yale U. Press, 2008), pp. 37–64. See Hans Driesch, *Der Vitalismus als Geschichte und als Lehre* (Leipzig: J. A. Barth, 1905; Eng. trans. 1914 by C. K. Ogden as *The History and Theory of Vitalism*).
26. H. Bergson, *L'Evolution Creatrice* (Paris: Felix Alcan, 1907; Eng. trans. 1911 as *Creative Evolution*); for more on Bergson, see F. Burwick, ed., *The Crisis of Modernity: Bergson and the Vitalism Controversy* (London: Cambridge U. Press, 1992).
27. Albrecht Hirschmüller, *The Life and Work of Josef Breuer: Physiology and Psychoanalysis* (New York: NYU Press, 1989), esp. pp. 40, 222–225. On Kammerer, see idem, "Paul Kammerer und die Vererbung erworbener Eigenschaften," *Medizinhist. J.* 26: 26–77 (1991); also Sander Gliboff, "The Case of Paul Kammerer: Evolution and Experimentation in the Early Twentieth Century," *J. Hist. Biol.* 39: 525–563 (2006); idem, "'Protoplasm Is Soft Wax in Our Hands': Paul Kammerer and the Art of Biological Transformation," *Endeavour* 162–167 (2005); and idem, *The Pebble and the Planet: Paul Kammerer, Ernst Haeckel, and the Meaning of Darwinism*, unpublished PhD diss., Johns Hopkins U., Jan. 2001; on Freud's Lamarckism, see Eliza Slavet, "Freud's Lamarckism and the Politics of Racial Science," *J. Hist. Biol.* 41: 37–80 (2008).
28. Hirschmuller, *op. cit.*, pp. 39–40. For more on Hering's less mechanistic physiology, see R. Steven Turner, "Vision Studies in Germany: Helmholtz vs Hering," *Osiris* n.s. 8: 80–103 (1993).
29. Breuer to Theodor Gomperz, 25 July 1895, in Hirschmüller, *op. cit.*, pp. 222–225.
30. Ibid., p. 223.
31. Ibid., p. 224.
32. Ibid., pp. 224–225.
33. A new study of vitalism—despite much interesting insight—unhelpfully confuses this distinction. *Vitalism and the Scientific Image in Post-Enlightenment Life Science, 1800–2010*, ed. Sebastian Normandin and Charles T. Wolfe (New York: Springer, 2013), argues (introduction, p. 10) that by 1900 or so one began to see "new vitalisms . . . vitalisms of a theoretical or even a material (physical) sort." Reich clarified this by recognizing that this meant a crucial break from the mystical or spiritualistic vitalisms of the past. That is why he insisted on calling his approach by a new name, at first "dialectical materialism," later "functionalism," precisely to emphasize that orgone energy could be investigated and understood by the methods of laboratory science, while Bergson's elán vital, Driesch's entelechy, or Blumenbach's Bildungstrieb could not be. Yet Normandin and Wolfe pointedly elide this distinction, insisting

that "Jung and Freud's essential fascination with the libido can be seen as the groundwork for vitalism in Reich's 'orgone' and 'life energy.'" This might be thought merely an academic or semantic distinction. But the widespread understanding in life sciences that "vitalism" means mystical forces, not accessible to laboratory investigation, has been widely used to caricature Reich's orgone energy. So for the vast majority of life scientists, continued insistence that Reich's work counts as vitalism can only lead—with the exception of a few academics who appreciate Normandin's historical attempt to redefine it—to the continued dismissal of Reich's experimental studies as worthless. This is regrettable, particularly since Normandin and Wolfe's project of recovering antireductionist ideas is otherwise quite valuable for biology in the age of evo-devo and epigenetics. (See also, by Wolfe, https://www.academia.edu/7275464/HOLISM_ORGANICISM_AND_THE_RISK_OF_BIOCHAUVINISM.)

34. Freud, *Three Contributions, op. cit.*, p. 553n2.
35. Ibid., p. 577.
36. Sigmund Freud, *Collected Papers*, vol. 1 (London: Hogarth Press, 1924), p. 75.
37. F. J. Allen, "What Is Life?" *Proc. Birmingham Nat. Hist. Phil. Soc.* 11: 44–67 (1899), p. 45.
38. John Cronin and Sandra Pizzarello, "Enantiomeric Excesses in Meteoritic Amino Acids," *Science* 275: 951–955 (14 Feb. 1997).
39. Allen, *op. cit.*, p. 45.
40. Ibid., p. 46.
41. Ibid., pp. 46–47.
42. Ibid., p. 52.
43. Ibid., p. 47. Some aspects of this idea about nitrogen as "forming very unstable compounds and consequently of releasing energy" are still in circulation and repeated by Reich in *BE*, p. 126.
44. Ibid., p. 55.
45. Eduard, Pflüger, "Über die physiologische Verbrennung in den lebendigen Organismen," *Arch. Gesam. Physiol. (Pflügers Archiv)* 10: 251–367 (1875). On Pflüger's (1829–1910) career, see W. D. H., "E. F. W. Pflüger," *Proc. Roy. Soc. (London)* 82B: i–iii (1910); also Charles Culotta, "Tissue Oxidation and Theoretical Physiology: Bernard, Ludwig and Pflüger," *Bull. Hist. Med.* 44: 109–140 (1970).
46. Allen, *op. cit.*, 1899, pp. 55–56.
47. Ibid.
48. Ibid.
49. Ibid.
50. Robert Kohler, "The Enzyme Theory and the Origin of Biochemistry," *Isis* 64: 181–196 (1973).

51. Graham had written, in a vein Allen echoes repeatedly, "another and eminently characteristic quality of colloids is their mutability. . . . The colloidal is, in fact, a dynamical state of matter; the crystalloidal being the statical condition. The colloid possesses ENERGIA. It may be looked upon as the probable primary source of the force appearing in the phenomena of vitality." *Phil. Trans. Roy. Soc. Lond.* 151: 183–224 (1861), emphasis in original.
52. James Strick, *Sparks of Life* (Cambridge, MA: Harvard U. Press, 2000), chap. 5, esp. pp. 105–119.
53. Reich, *BE*, pp. 135–139, 153–157. Much of this section seems to have been written out in draft form around 1934 or 1935; see RA, OI box 5: ntbk 3, "Zur Theorie der psychische-physische-funktional Identität, Sexökonomie." This notebook opens with the back-and-forth between mechanist and vitalist that eventually ended up in *Die Bione*, pt. 2. Another entry here is titled "Weitere Problematisch zur Enstehung des Lebens" [still about 1935—the next p. has 1935 written on it]. This passage suggests that Reich may have been thinking explicitly of the origin of life problem as early as 1935.

    "Hartmann, *Allgemeine Biologie* (3rd ed. [Jena: Gustav Fischer, 1933]) refs Sex, p. 411 ff," 18 pp. (long section working through, mostly from Hartmann, the phases of mitosis and ideas about chromosomes). Reich has detailed notes on anaphase and telophase (with diagrams of chromosomes and spindle fibers) that continue occasionally through the rest of this notebook. Reich was very much up to date on genetics and chromosomes as of Hartmann's biology textbook. He knows the details of mitosis well. In this section is an inserted Danish newspaper clipping (1933), "Det evige Køngsspørgsmaal: En russisk Professor har fundet en Methode til at forudbestemme Dyrenes Køn ved Hjælp af Electricitet." See Reich, *FO* (1942 ed.), p. 283 (FSG ed., p. 253 [The eternal question of gender/sex: A Russian professor has found a method to determine/ influence the gender of animals by the use of electricity]).
54. J. A. T., "Radiobes and Biogen," *Nature* 74 (3 May 1906): 1–3. See also Luis Campos, *Radium and the Secret of Life*, unpublished PhD diss., Harvard U., May 2006, in press with U. of Chicago Press. For more on the "protoplasm theory of life" and its previous implications—unraveling in light of post-1900 biochemistry, see G. Geison, "The Protoplasmic Theory of Life and the Vitalist-Mechanist Debate," *Isis* 60: 273–292 (1969).
55. See, e.g., Benjamin Moore, *The Origin and Nature of Life* (London: Williams and Norgate, 1912), esp. pp. 17–19, 44, 225–226.
56. Benjamin Moore, in "British Association for the Advancement of Science, Combined Meeting of the Sections of Zoology and Botany: The Origin of Life," *Brit. Med. J.* 2 (21 Sept. 1912): 722.
57. Benjamin Moore, *The Origin and Nature of Life* (New York: Henry Holt, 1912), pp. 225–226, italics in original.

58. Ibid., pp. 226–227, italics in original.
59. Strick, *Sparks, op. cit.*, chap. 5; Graham, *op. cit.* (1861); see also A. I. Kendall, "Bacteria as Colloids," in H. N. Holmes, ed., *Colloid Symposium Monograph*, vol. 2 (New York: Chemical Catalogue Co., 1925); and Neil Morgan, "The Strategy of Biological Research Programmes: Reassessing the 'Dark Age' of Biochemistry, 1910–1930," *Ann. Sci.* 47: 139–150 (1990); also Scott Podolsky, "The Role of the Virus in Origin of Life Theorizing," *J. Hist. Biol.* 29: 79–126 (1996).
60. Charles Darwin, *On the Origin of Species* (facsimile of 1st ed., Cambridge, MA: Harvard U. Press, 1964), pp. 134–139; see also Peter Bowler, *Evolution: The History of an Idea* (Chicago: U. of Chicago Press, 2002), pp. 160, 236. Darwin's dearly held pangenesis hypothesis had as one of its main goals to account for how inheritance of acquired characters could be physiologically possible.
61. Paul Kammerer, *Allgemeine Biologie*, 1st ed. (Stuttgart: Deutsche Verlags Anstalt, 1915), pp. 7–8. It is worth noting that the opening passage was retained unchanged in the 1920 2nd ed. of Kammerer's book; however, the passage ("Were I also to finally make known . . .") about a specific life energy had been removed. In the copy of the first edition (1915) in Reich's personal library, this entire (later deleted) passage is underlined in red pencil with marginal marks for further emphasis. My thanks to Mary Boyd Higgins, trustee of Reich's estate, for permission to photograph Reich's marginalia in this book; also to Dr. Philip Bennett for preparing a catalog of Reich's entire personal library (available through the Wilhelm Reich Museum bookstore). Reich quotes the passage he underlined in full in his 1948 book *The Cancer Biopathy*, Eng. trans. Theodore P. Wolfe (New York: Orgone Institute Press; a more widely available translation by Andrew White was published in 1973 by Farrar, Straus and Giroux).
62. Reich, *FO* (1942 ed.). The book was also titled *The Discovery of the Orgone*, vol. 1. The 1940 German ms. from which the translation was made is sold through the Wilhelm Reich Museum bookstore. The discussion of his early views on the mechanism-vitalism controversy, on Kammerer, and on other topics is on pp. 6–12 of the published (1942) version (pp. 22–26 in the more recent, more widely available ed. from Farrar, Straus and Giroux, 1973, trans. Vincent Carfagno from the same 1940 German ms.).
63. Ibid., p. 7 (FSG ed., p. 24).
64. Wilhelm Reich, *The Cancer Biopathy* [1948] (New York: Farrar, Straus and Giroux, 1973; hereafter, *CB*), pp. 30, 48–49. Üxküll's book was published by Julius Springer, Berlin, in 1921. A 1934 work of Üxküll, also from Springer, *Streifzüge durch die Umwelten von Tieren und Menschen*, is now available in Eng. trans. (Joseph O'Neil, trans.) as *A Foray into the Worlds of Animals and Humans* (Minneapolis: U. of Minnesota Press, 2010).

65. Ibid.
66. Ibid. Reich's comments on such a life energy remind the modern reader of "the Force" from George Lucas's *Star Wars* movies, i.e., a natural force which one can learn to handle and use, but not in a mechanistic way—only where an intact organism, by trusting and relying on its own feelings, can master and manipulate this energy. Given the popularity of Reich's writings during the 1960s and 1970s, one can only wonder whether Lucas was inspired by Reich's orgone energy.
67. Wilhelm Reich, "The Developmental History of Orgonomic Functionalism, pt. 1," *Orgonomic Functionalism* 1: 4–5 (1990).
68. Marie Bardy calls Reich's approach "materialistic vitalism." A close reading of Reich does not validate such a description, but she seems intent on portraying Reich as more influenced by du Teil's vitalism than Reich's writings can support. Du Teil had vitalist beliefs. See her article "Reich and du Teil," *op. cit.*, p. 40, and Roger du Teil, "Leben and Materie," pp. 117–125 in Wilhelm Reich, ed., *Die Bione* (Oslo: SexPol Verlag, 1938). Eng. trans. as "Life and Matter," published in 1977, by anonymous ed. (along with the 1938 and 1944–1945 versions of *Die Bione*), Freedom Press, Los Angeles. This lecture, delivered by du Teil to the Academie des Sciences Morales, Nice, on 18 Sept. 1937, was included in the original text of *Die Bione* but excluded from the 1944–1945 translation and the 1947 second edition at Dr. Reich's direction, though an unpublished English translation by D. and I. Jordan does exist. More recently, see a similar argument by Sebastian Normandin, *"Vitalism and Wilhelm Reich: From Hippocrates to Holistic Health,"* talk in Countway Library History of Medicine series, Nov. 2010. See note 33 above on more recent work by Normandin and Wolfe that perpetuates this confusion.
69. Wilhelm Reich, "The Developmental History of Orgonomic Functionalism, pt. 4," *Orgonomic Functionalism* 4: 1–18 (1992), quote on p. 8.
70. Ibid.
71. Lloyd Ackert, *Sergei Vinogradskii and the Cycle of Life: From the Thermodynamics of Life to Ecological Microbiology, 1850–1950* (New York: Springer, 2013), p. xiv.
72. Ibid.
73. Reich, *FO, op. cit.* (1942 ed.), p. 8 (FSG ed., p. 25). On Kammerer's skill at getting animals to breed in captivity, see Arthur Koestler, *The Case of the Midwife Toad* (New York: Random House, 1971), pp. 23–25.
74. See, e.g., Kammerer, "Steinachs Forschungen uber Entwicklung, Beherrschung, und Wandlung der Pubertät," *Ergeb. d. inneren Medizin und Heilkunde* 17: 295–398 (1919); also idem, *Rejuvenation and the Prolongation of Human Efficiency: Experiences with the Steinach Operation on Man and Animals*, Eng. trans. A. P. Märker-Branden (New York: Boni and Liveright, 1923), which

included discussion of Steinach and Kammerer's 1920 "Climate and Puberty" study, arguing that interstitial cells are the source of a sex drive physiologically separate from the reproductive function of gonads, and that this sex drive is heightened by temperature at the expense of the reproductive function. Once rats have grown at elevated temperatures for a time, they claimed, the resulting increased density of interstitial cells and increased sex drive are passed down heritably, without further heat exposure to offspring for several generations. See Cheryl Logan "Overheated Rats, Race and the Double Gland: Paul Kammerer, Endocrinology and the Problem of Somatic Induction," *J. Hist. Biol.* 40 (2007): 683–725.

In 1920 Steinach announced his bilateral vasectomy rejuvenation operation in a monograph on rejuvenation *(Verjungung)*; supposedly vasectomy causes atrophy of germ cells but hypertrophy of hormone-secreting interstitial cells. See Eugen Steinach and Josef Loebel, *Sex and Life: Forty Years of Biological and Medical Experiments* (London: Faber and Faber, 1940), p. 18: "Just as the sex glands are situated in the middle of the body, so too are they at the hub of life itself." Freud, feeling loss of his own vitality, underwent the Steinach operation in 1923, hoping it would help avoid a recurrence of his oral cancer. See Chandak Sengoopta, "Glands of Life: Eugen Steinach, Rejuvenation, and Experimental Endocrinology in the Early Twentieth Century," paper presented at AAHM Annual Meeting, New York, 30 Apr. 1994; idem, "'Dr. Steinach Coming to Make Old Young!': Sex Glands, Vasectomy, and the Quest for Rejuvenation in the Roaring Twenties," *Endeavour* 27: 122–126 (2003).

75. For good surveys of the eugenics movement and the excesses of "hard" Mendelian heredity theory, see Diane Paul, *Controlling Human Heredity* (Amherst, NY: Humanities Press, 1995); also Dan Kevles, *In the Name of Eugenics: Genetics and the Uses of Human Heredity* (Cambridge, MA: Harvard U. Press, 1998).
76. Reich, *FO, op. cit.* (1942 ed.), pp. 8–9 (FSG ed., p. 26).
77. Ibid., p. 8, emphasis in original (FSG ed., pp. 25–26). Lest the reader think this has all changed in our more enlightened day, see Emily Martin, "The Egg and the Sperm: How Science Has Constructed a Romance Based on Stereotypical Male-Female Roles," *Signs*, 16(3): 485–501 (1991).
78. Ibid., pp. 9–10, emphasis in original (FSG ed., pp. 27–28). On the hypocrisy around sexuality in Vienna (and in the European academic world in general) at this time, see Zweig, *op. cit.*, n43.
79. Ibid., p. 11, emphasis in original (FSG ed., p. 29).
80. See Koestler, *Midwife Toad, op. cit.* The emergence of evo-devo biology and epigenetics, while not vindicating 1920s neo-Lamarckism of course, has shown the genome to have remarkably more plasticity than would have been believed

even 30 years ago after Kammerer's discrediting and under the reign of the molecular biology "central dogma," until the recognition of the importance of McClintock's transposition by the late 1970s.
81. This topic is treated at length in Garland Allen, *Life Science in the Twentieth Century, op. cit.*, and in Bowler, *Evolution, op. cit.* See also John Maynard Smith, introduction to J. B. S. Haldane, *On Being the Right Size and Other Essays* (London: Oxford U. Press, 1985).
82. See, for example, Herbert S. Jennings, "Heredity and Environment," *Scientific Monthly* 19: 234–238 (Sept. 1924).
83. Slavet, *op. cit.*, esp. p. 38n2.
84. In modern practice, see, e.g., John Ruscio, "The Emptiness of Holism," *Skeptical Inquirer* (Mar.–Apr. 2002), pp. 46–50.
85. Scott Gilbert and Sahotra Sarkar, "Embracing Complexity: Organicism for the 21st Century," *Devel. Dynamics* 219: 1–9 (2000), esp. pp. 4–5.
86. Allen, *Life Science, op. cit.*, p. 103.
87. An exchange between H. S. Jennings and Alexander Gurwitsch, well prior to the latter's mitogenetic radiation work, might be counted along with Henderson and Cannon; Gurwitsch emphasized that some key insights from vitalism ought not to be overlooked, despite embracing modern experimental work in biology. See Gurwitsch, "On Practical Vitalism," *Am. Nat.* 49: 763–770 (Dec. 1915), written in answer to H. S. Jennings's article of 29 months previous, "Doctrines Held as Vitalism," *Amer. Nat.* 47: 385–417 (July 1913).
88. Allen, *Life Science, op. cit.*, p. 104.
89. Ibid., p. 105.
90. Donna Haraway describes Weiss's work in *Crystals, Fabrics and Fields* (New Haven, CT: Yale U. Press, 1976), chap. 5. See also Jane Overton, "Paul Alfred Weiss, March 21, 1898–Sept. 8, 1989," *Proc. Nat. Acad. Sci. Biograph. Mem.* 72: 373–386 (1997).
91. Alexander Gurwitsch, "Über Ursachen der Zellteilung" [On the causes of cell division] in Roux's *Archiv d. Entwicklungsmechanik* 30: 167–181 (1923). Bacteriologist Otto Rahn at Cornell had several people working in his lab on MR. See, e.g., Rahn, *Invisible Radiations of Organisms* (Berlin: Borntraeger, 1936).
92. John Tyler Bonner, "Evidence for the Formation of Cell Aggregates by Chemotaxis in the Development of the Slime Mold *Dictyostelium discoideum*," *J. Exp. Zool.* 106: 1–26 (Oct. 1947).
93. Strick oral history interview with Bonner, 27 Jan. 1992.
94. This information is drawn from Nick Hopwood, "Biology between University and Proletariat: The Making of a Red Professor," *Hist. Sci.* 35: 367–424 (1997). Hans Driesch, Hans Przibram, and Ludwig von Bertalanffy are among those

who contributed to his Theoretical Biology series, which reached 30 publications by 1931.
95. Hopwood, *op. cit.*, p. 30n151. Compare this with Reich's "children of the future" and "We need more *fist*, more rational fighting knowledge." Like Schaxel, Reich's mental language and imagery draw strongly upon socialist models.
96. Hopwood, *op. cit.*, p. 30. For more on Schaxel and his early evo-devo research agenda, see Christian Reiss, "No Evolution, No Heredity, Just Development: Julius Schaxel and the End of the Evo-Devo Agenda in Jena, 1906–1933," *Theory Biosci.* 126: 155–164 (2007); also C. Reiss et al., "Introduction to the Autobiography of Julius Schaxel," *Theory Biosci.* 126: 165–175 (2007).
97. Julius Schaxel, "Das biologische Individuum," *Erkenntnis* 1 (1931): 467–492.
98. Donna Haraway, *Crystals, Fabrics and Fields* (New Haven, CT: Yale U. Press, 1976), p. 130. See also Ludwig von Bertalanffy, *Modern Theories of Development: An Introduction to Theoretical Biology*, Eng. trans. J. H. Woodger (Oxford: Oxford U. Press, 1933), esp. chap. 7, "Organismic Theories," in which he lauds Schaxel's work and Gurwitsch's "field theory." For more on Woodger, see also Nils Roll-Hansen, "E. S. Russell and J. H. Woodger: The Failure of Two Twentieth-Century Opponents of Mechanistic Biology," *J. Hist. Biol.* 17: 399–428 (1984).
99. Haraway, *op. cit.*, chap. 4. For more on crystals as analogous to life, see Hans Przibram, *Die anorganischen Grenzgebiete der Biologie, insbesondere der Kristallvergleich* (Berlin: Borntraeger, 1926); also Thomas Brandstetter, "Life beyond the Limits of Knowledge: Crystalline Life in the Popular Science of Desiderius Papp (1895–1993)," *Astrobiology* 12 (10): 951–957 (Oct. 2012).
100. The Soviet papers published in one volume as *Science at the Cross-Roads* (London: Kniga, 1931). A pointed disagreement broke out on 30 June between five of the Soviets and historian G. N. Clark and physiologist A. V. Hill, which, as Gary Werskey observes, partly motivated Clark's later response to the Hessen thesis in his 1937 book *Science and Social Welfare in the Age of Newton* (Oxford: Clarendon Press). See Werskey, *The Visible College: The Collective Biography of British Scientific Socialists in the 1930s* (New York: Holt, Rinehart, 1978), p. 141.
101. See Thomas Greenwood, "The International Congress of the History of Science and Technology," *Nature* 128: 78–79 (11 July 1931). Greenwood interestingly uses "organicist" to describe the essentially vitalist position of J. S. Haldane, while calling Joseph Needham's view "mechanistic." In modern usage, at least since Haraway (1976, *op. cit.*), "organicism" has been contrasted with vitalism. Haraway labels Needham's ideas by 1931 as moving toward this organicism, away from mechanism but explicitly rejecting vitalism. Needham

recalls his own reaction in "Foreword" to reissue of *Science at the Cross Roads* (London: Frank Cass, 1971), pp. vii–x.

102. Julian Huxley, *A Scientist among the Soviets* (London: Chatto and Windus, 1932); see esp. pp. 52–54; see also J. G. Crowther, *Fifty Years with Science* (London: Barrie and Jenkins, 1970), p. 77.

103. Greenwood, *op. cit.*, p. 79.

104. Werskey, *op. cit.*, pp. 141–142. In more recent years, why the Soviets presented as they did has been set more in the historical context of political developments under Stalin and scientists' struggles to protect their work; see, e.g., Loren Graham, "The Social and Economic Roots of Boris Hessen," *Soc. Stud. Sci.* 15: 705–722 (1985). Hessen may have espoused a Marxist line so forcefully in London because at home he was on very unsteady ground amid Stalin's consolidation of power, because of having supported Einsteinian relativity theory (Einstein had said that Ernst Mach's ideas were crucial, but Mach had been severely criticized by Lenin in *Materialism and Empirio-Criticism*).

105. On the "Lysenko Affair" and its reception by Western biologists, see Valery Soyfer, *Lysenko and the Tragedy of Soviet Science* (New Brunswick, NJ: Rutgers U. Press, 1994); see also Nils Roll-Hansen, *The Lysenko Effect: The Politics of Science* (Amherst, NY: Prometheus, 2005); and Nikolai Krementsov, *Stalinist Science* (Princeton, NJ: Princeton U. Press, 1997). For its relation to origin of life studies, see John Farley, *The Spontaneous Generation Controversy from Descartes to Oparin* (Baltimore: Johns Hopkins U. Press, 1977), pp. 168, 172–184.

106. Loren Graham, *Science in Russia and the Soviet Union* (Cambridge: Cambridge U. Press, 1994).

107. Most of what follows in this section is based on Loren Graham, *Science in Russia and the Soviet Union* (Cambridge: Cambridge U. Press, 1994), pp. 100–103.

108. Wilhelm Reich, *People in Trouble* [1953] (New York: Farrar, Straus and Giroux, 1976), chap. 2.

109. On scientific use of dialectical materialist thinking, Reich cites mostly from Engels, *Ludwig Feuerbach und der Ausgang der klassischen Philosophie* (Vienna and Berlin: Verlag für Literatur und Politik, 1927); also from V. I. Lenin, *Materialism and Empirio-Criticism* (1905), in idem, *Sämtliche Werke* (Vienna: Verlag für Literatur und Politik, 1927–1929). Reich does not seem to have read Engels's *Dialectics of Nature*, so important to Haldane; the book is not among the five books by Engels in Reich's library. The first German edition was in 1927; Haldane edited the first English edition of 1940.

110. Wilhelm Reich, "Dialectical Materialism and Psychoanalysis," in Lee Baxandall, ed., Wilhelm Reich, *SexPol Essays, 1929–1934* (New York: Random House, 1972). The original 1929 essay was published in the Communist Party

journal, *Unter dem Banner des Marxismus* 3: 736–771. The first English translation of the 1929 essay is by Anna Bostock in *Studies on the Left* 6: 5–46 (July–Aug. 1966). The Baxandall volume uses the updated second edition; I will cite from this one since the changes are almost entirely footnotes (useful ones) added by Reich in 1934.

111. Friedrich Lange, *Geschichte des Materialismus*, 2 vols. (Leipzig: Verlag J. Baedeker, 1902) is the edition in Reich's library; Eng. trans. Bertrand Russell, *History of Materialism* (New York: Harcourt Brace, 1925), Pasteur-Pouchet debate 3: 14, 18; Reich leans heavily on Lange's discussion of Democritus's "soul atoms" in his book *Ether, God and Devil* (New York: Orgone Inst. Press, 1949) and of the Pasteur-Pouchet debate in *CB, op. cit.*, pp. 10–11. Other important sources for this larger problem complex include Ernst Haeckel, "Beiträge zur Plastidientheorie," *Jenaische Zeitschrift f. Med. u. Naturwiss.* 5: 499–519 (1870); also idem, *Die Perigenesis der Plastidule* (Berlin: Georg Reimer, 1876), both inspired by Ewald Hering's famous 1870 lecture "Das Gedächtnis als allgemeine Funktion der organisierten Materie," Eng. trans. by Samuel Butler as "Memory as a Universal Function of Organized Matter," in Butler, *Unconscious Memory* (London: A. C. Fifield, 1910). Hering was one of the professors with whom Breuer trained most closely. By Richard Semon, Reich's library contains *Die Mneme: Als erhaltendes Prinzip im Wechsel des organischen Geschehens*, 2nd ed. (Leipzig: Wilhelm Engelmann, 1911), and *Die mnemischen Empfindungen in ihren Beziehungen zu den Originalempfindungen* (Leipzig: Wilhelm Engelmann, 1909). The first of these Semon books was in Eng. trans. by Louis Simon as *The Mneme* (New York: Macmillan, 1921). In Reich's library is also Henri Bergson, *Materie und Gedächtnis: Eine Abhandlung über die Beziehung zwischen Körper und Geist*. (Jena: Verlegt bei Eugen Diederichs, 1919).

112. Daniel Schachter, *Searching for Memory: The Brain, the Mind, and the Past* (New York: Basic Books, 1996), p. 57.

113. Max Verworn, *Irritability: A Physiological Analysis of the General Effect of Stimuli in Living Substance* (New Haven, CT: Yale U. Press, 1913), pp. 16–17.

114. Sungook Hong, "Semon's Mneme and Schrödinger's Gene: An Intellectual Influence," paper presented 4 Nov. 2007, History of Science Society Annual Meeting, Washington, DC.

115. See, e.g., Gary Werskey, *The Visible College: The Collective Biography of British Scientific Socialists in the 1930s* (New York: Holt, Rinehart, 1978).

116. Reich, *SexPol, op. cit.*, quote on pp. 8–9n7.

117. The same attitude is clearly on display in a tape-recorded seminar with his physician students in late 1949 or so, which has since been published on CD (from the Wilhelm Reich Museum Bookstore) as "Method of Approach," the first of a set of six such seminars.

118. See, e.g., Wilhelm Reich, "The Living, Productive Power, 'Work Power,' of Karl Marx," in *People in Trouble* [1953] (New York: Farrar, Straus and Giroux, 1976), pp. 48–76.
119. Reich, *SexPol, op. cit.*, p. 11. Ludwig Büchner (with his book *Kraft und Stoff*), Carl Vogt, and Jakob Moleschott led the rising school of German materialists in the 1850s. This is what Marxists referred to as "crude mechanistic materialism."
120. Ibid., pp. 12–13, long quote p. 13n14. Wilhelm Reich, "Der Urgegensatz des vegetativen Lebens," *Zeit. f. Polit. Psychol.u. Sex.-Ökon.* 1 (2): 125–142 (1934); Eng. trans. as "The Basic Antithesis of Vegetative Life Functions" in W. Reich, *The Impulsive Character and other Writings*, ed. B. G. Koopman (New York: Meridian, 1974), pp. 83–121. Reich prepared an updated 1945 version of the article, available in Wilhelm Reich, *The Bioelectrical Investigation of Sexuality and Anxiety* (New York: Farrar, Straus and Giroux, 1982).
121. Reich, *SexPol, op. cit.*, pp. 14–15. Of course, chemical metaphors for sexual attraction had been in use for a century or more. See, e.g., Johann von Goethe, *Elective Affinities* (*Die Wahlverwandtschaften*, 1809; numerous Eng. eds., e.g., New York: Penguin, 1978).
122. Reich, *SexPol, op. cit.*, p. 18.
123. Ibid., p. 15n20.
124. Reich, *FO, op. cit.* (1942 ed.), chap. 7, sec. 3. Theodore P. Wolfe says the reason he and other psychosomatic medicine researchers found Reich's concept attractive was because it was based on clinical facts, not on metaphysical concepts. See Wolfe, "Translator's Preface," to Reich, *FO* (1942 ed.).
125. Reich, *SexPol, op. cit.*, pp. 16–17.
126. Ibid., pp. 17–19.
127. Ibid., p. 17n24.
128. Ibid., pp. 18–19.
129. Wilhelm Reich, "Der masochistische Charackter: Eine sexualökonomische Widerlegung des Todestriebes und des Wiederholungszwanges," *Int. Zeitschr. Psychoanal.* 18: 303–351 (1932). Published in Eng. trans. as "The Masochistic Character," *International Journal of Sex-Economy and Orgone Research* (hereafter, *IJSO*) 3: 38–61 (1944). Freud had only acceded to publishing the paper in the official journal of the International Psychoanalytic Association if it were followed by a lengthy critique claiming that Reich's scientific views had been distorted by his Marxist politics; this represented the final theoretical break between Reich and Freud. This reply, by Siegfried Bernfeld, was "Die kommunitische Diskussion um die Psychoanalyse und Reichs 'Widerlegung der Todestriebhypothese,'" *Int. Zeitschr. Psychoanal.* 18: 352–285 (1932). See also Michael Molnar, ed. and trans., *The Diary of Sigmund Freud, 1929–1939* (London: Freud Museum, 1992); p. 119 discusses Freud's thoughts on taking

"step against Reich," i.e., intervening in the publication of Reich's article on masochism, 1 Jan. 1932.
130. Reich, FO, op. cit. (1942 ed.), p. 234 (FSG ed., p. 263).
131. Lore Reich Rubin, "Wilhelm Reich and Anna Freud: His Expulsion from Psychoanalysis," *Inter. Forum Psychoanal.* 12: 109–117 (2003).
132. See also Harris and Brock, op. cit.
133. Mary Boyd Higgins, introduction to Reich, *BP*, pp. xxii–xxiii; the lecture was expanded into a monograph in 1935, which was included in the 3rd and subsequent eds. of Reich's book *Character Analysis* (New York: Orgone Institute Press, 1945).
134. Reich to Lotte Liebeck, 7 Jan. 1935, *Reich Speaks of Freud* (New York: Farrar, Straus and Giroux, 1967), pp. 202–205; also Reich rundbrief to Marxist analysts, 16 Dec. 1934, ibid., pp. 194–199.
135. Reich to Felix Deutsch, 21 Jan. 1935, ibid., pp. 207–209.
136. Anne Harrington, *Reenchanted Science: Holism in German Culture from Wilhelm II to Hitler* (Princeton, NJ: Princeton U. Press, 1996), p. 189.
137. On Sigurd Hoel (1890–1960), see http://www.britannica.com/bps/edit/topic?topicId=268654# and http://snl.no/.nbl_biografi/Sigurd_Hoel/utdypning; Håvard Nilsen has pointed out that the protagonist of Hoel's last novel, *The Troll Circle*, was modeled on Wilhelm Reich (Eng. trans. Sverre Lyngstad, Lincoln: U. of Nebraska Press, 1991). He published in Reich's journal, e.g., "Kulturkampf und Literatur," *ZPPS* 3: 85–100 (1936), Eng. trans. as "The Place of Literature in the Cultural Struggle," in *IJSO* 1 (1942). On Nic Hoel (1905–1960), see her article "Unklarheit der Sexualpolitik und Sozialhygiene in England," *ZPPS* 3: 31–38 (1936). See also http://translate.google.com/translate?hl=en&sl=no&u=http://www.storenorskeleksikon.no/.nbl_biografi/Nic_Waal/utdypning&ei=BasBSvT8DILEM5OeodUH&sa=X&oi=translate&resnum=4&ct=result&prev=/search%3Fq%3Dnorsk%2Bbiografisk%2Bnic%2Bwaal%26hl%3Den%26client%3Dsafari%26rls%3Den%26sa%3DG; on Øverland (1889–1968), see http://translate.google.com/translate?hl=en&sl=no&u=http://www.snl.no/.nbl_biografi/Arnulf_%25C3%2598verland/utdypning&ei=iA4AStHxFsrJtgfpyv2IBw&sa=X&oi=translate&resnum=1&ct=result&prev=/search%3Fq%3Dnorsk%2Bbiografisk%2Barnulf%2B%25C3%2598verland%26hl%3Den%26sa%3DG; his famous antifascist poem "You Must Not Sleep!" was first published in the Nov.–Dec. 1936 issue of Reich's *Journal for Political Psychology and Sex-Economy* 3: 81–83 (with Ger. trans. pp. 83–84). See also Reich to Øverland, 29 Apr. 1938 (National Library of Norway, brevs 835). Øverland was captured by the Gestapo after the Nazi occupation of Norway and spent the war in Sachsenhausen camp.
138. See, e.g., Reich lecture notes for Character Analytic seminar, "Disposition des Kurses" (Sigurd Hoel papers, Norwegian National Library, ms. fol. 2351, "W. Reich Kurs papier" fld).

139. Reich, *BP*, pp. 89–90.
140. Ibid., p. 41.
141. Reich, *FO, op. cit.*, chaps. 7–9.
142. Mary B. Higgins and Chester Raphael, eds., *Reich Speaks of Freud* (New York: Farrar, Straus and Giroux, 1967), pp. 41–42. See also Prof. Arthur Kronfeld, Berlin, 1934, in a letter to the editors of the journal *Psychotherapeutische Praxis* (U.S. State Dept. Reich files, Jerome Greenfield papers, Taminent Library and Robert F. Wagner Labor Archives, NYU).
143. T. P. Wolfe, translator's footnote in Reich, *FO, op. cit.* (1942 ed.), p. 232n; 1948 p. 203n.
144. Wilhelm Reich, "The Orgasm as an Electrophysiological Discharge," ZPPS 1: 29–43 (1934). Eng. trans. with 1945 notes and edits added by Reich in *BISA*, pp. 3–19. A direct Eng. trans. of the original 1934 article is in Barbara Koopman, trans. and ed., *The Impulsive Character and Other Writings* (New York: Meridian, 1974), pp. 123–138.
145. Reich, *BISA*, pp. 13–14.
146. On Hartmann's theory, Reich wrote "Fortpflanzung eine Funktion der Sexualität" [Reproduction as a Function of Sexuality], ZPPS 3: 24–31 (Mar. 1936). Reich had read Hartmann, Möwus, and Jollos by 1933 already. Reich completed the ms "Fortplflanzung eine Funktion der Sexualität" in 1934 (Mss Box 3, RA). In August 1944 he added a two-page postscript at the end of the article.

  Hartmann's relative sexuality theory later fell into disrepute because of its association with the work on the topic by Franz Moewus, considered by many to be discredited. See Jan Sapp, *Where the Truth Lies: Franz Moewus and the Origins of Molecular Biology* (Cambridge: Cambridge U. Press, 1990). Nonetheless, at least some geneticists remained intrigued by it as a possible "baby thrown out with the bath water" that deserved another look. See, e.g., Alfred Henry Sturtevant, *A History of Genetics* (New York: Academic Press, 1965), p. 119: "There is no doubt about the reality of the phenomenon of relative sexuality, and it is clearly something that needs study and analysis; but the papers of Moewus have unfortunately given it unpleasant associations." When I asked protozoologist David Nanney (oral history interview, June 1998) if the relative sexuality theory was abandoned because of disreputable associations with Moewus rather than based on factual evidence, Nanney replied, "I think so. You certainly don't see that theory even *mentioned* today in any texts. It's *forgotten*, not just crowded out by the genetics of *E. coli*, bacteriophage, etc. that took over microbial genetics in the'50s and'60s. I think there may have been some other factors involved. For one, the overall anti-German sentiment, because Hartmann, too, was seen as a collaborator with the Nazis. His 1943 book *Sexuality* did not even mention [Tracy] Sonneborn once, or cite any of his extensive work on *Paramecium*, though Hartmann definitely knew of it." [Sonneborn was Jewish.] For Hartmann's ability to work at the Kaiser Wilhelm Gesellschaft under the Nazis, see Kristie Macrackis, *Surviving the Swastika: Scientific Research in Nazi Germany* (New York: Oxford U. Press, 1993).

147. Reich, *FO, op. cit.* (FSG ed.), pp. 102. Hereafter, FSG ed. page numbers are cited unless otherwise indicated.
148. Reich, *BE*, p. 20; *BP*, pp. 63–64.
149. Wilhelm Reich, "The Basic Antithesis of Vegetative Life," *ZPPS* 1: 125–142 (1934). Eng. trans. with 1945 notes and edits added by Reich in *BISA*, pp. 3–19. A direct Eng. trans. of the original 1934 ZPPS article is in B. Koopman, trans. and ed., *The Impulsive Character and Other Writings, op. cit.*, pp. 83–121.
150. Their concern was well-founded. See, e.g., Lee Baxandall's 1966 mistranslation of the title of Reich's 1934 article as "The Fundamental Contradiction of Vegetable Life" in Reich, *SexPol, op. cit.*, p. 13n14.
151. Walter and Käthe Misch, "Die vegetative Genese der neurotischen Angst und ihre medikamentöse Beseitigung" *Der Nervenarzt* 8: 415–418 (1932).
152. See, e.g., F. Kraus and S. G. Zondek, "Über die Durchtränkungsspannung," *Klinische Wochenschrift* (Berlin) 1: 1778–1779 (2 Sept. 1922), which also refers to the colloid work of Wolfgang Ostwald; also see S. G. Zondek, "Die Bedeutung der Antagonismus von Kalium und Calcium für die Physiologie und Pathologie," *Klin. Woch.* 2: 382–385 (23 Feb. 1923). In the Reich Archive (OI box 5, file "Versuche Bioelektrik") is also a reprint of B. Wichmann, "Das vegetative Syndrom und seine Behandlung," *Deutsche Med. Woch.*, nr. 40: 1500–1504 (5 Oct. 1934).
153. The collected articles were published as S. Bernfeld and S. Feitelberg, *Energie und Trieb: Psychoanalytische Studien zur Psychophysiologie* (Vienna: Int. Psychoanal. Verlag, 1930) (repr. from *Imago: Zeitschrift für die Anwendung der Psychoanalyse auf die Natur- und Geisteswissenschaften*, vol. 15 and vol. 16, 1929 and 1930). This bound collection is in Reich's personal library.
154. Reich, *BISA*, p. 29.
155. Ibid., pp. 35–41. Note Reich's familiarity with animal life such as snails and hedgehogs and how it behaves in nature. It is worth recalling that Reich was raised on a farm, where he was tutored extensively on zoology and botany and encouraged to observe nature directly. See Wilhelm Reich, *Passion of Youth: An Autobiography, 1897–1922* (New York: Farrar, Straus and Giroux, 1988).
156. Sept. 1922, at the Seventh International Psychoanalytic Congress in Berlin, Reich says (*FO*, 1942 ed., pp. 233–234), "As a result of studying Semon and Bergson, I had engaged in a scientific phantasy. One should, I said to some of my friends, take Freud's picture of 'sending out of libido' literally and seriously. Freud had likened the sending out and retracting of psychic interest to the putting out and retracting of pseudopodia in the amoeba. The putting forth of

sexual energy is plainly visible in the erection of the penis. I thought that erection was functionally identical with the putting out of pseudopodia in the amoeba, whereas conversely, erective impotence, due to anxiety and accompanied by shrinking of the penis, was [p. 234] functionally identical with the retraction of the pseudopodia. My friends were horrified at such muddled thinking. They laughed at me, and I was offended" (pp. 262–263 in 1973 FSG ed. of *Function*). (In *Beyond Psychology*, he says this was 1923 at Seefeld, rather than 1922).

157. Reich, *BISA*, pp. 123–125.
158. Ibid., p. 38.
159. Ibid., p. 45.
160. Ibid., pp. 45–46.
161. Otto Bütschli, *Untersuchungen über mikroskopische Schäume and das Protoplasma* (Leipzig: W. Engelmann, 1892). Referred to by Reich in *BISA*, p. 46, and in *BE*, p. 129. A compelling summary of the remarkably lifelike "pseudoamoeboid" behavior simulated in such experiments through the 1930s, including experiments by H. S. Jennings, can be found in Reinhard Beutner, *Life's Beginnings on the Earth* (Baltimore, MD: Williams and Wilkins, 1938), pp. 157–174.
162. Reich, *BISA*, p. 47.
163. Reich emphasized that this problem was not resolved even by the "collation of the many brilliant insights in modern physiological literature that . . . were condensed and compiled in Müller's *Die Lebensnerven*" (*FO, op. cit.*, FSG ed., p. 269). Reich refers here to L. R. Müller, *Lebensnerven und Lebenstriebe* (Berlin: Springer, 1931), the edition in Reich's library.
164. Reich summarizes the separate effects of parasympathetic and sympathetic, as well as his insight about how these effects relate to the functioning of the total organism in *FO, op. cit.* (FSG ed.), pp. 290–295.
165. Reich discusses this view regarding a wide range of diseases—including many questions for which his theory had as yet no answer—in a seminar with his physician students in 1950, which is available from the Wilhelm Reich Museum under the title "Problems of a New Medicine," 6th in a set of such recorded seminars with his students.

2. Reich's Move toward Laboratory Science

1. Wilhelm Reich, *Experimentelle Ergbnisse über die elektrische Funktion von Sexualität und Angst* (Copenhagen: SexPol Verlag, 1937), pp. 6–7; Eng. trans. as *BISA*, pp. 73–74 (henceforth, this English translation will be cited).
2. Reich, *BISA*, p. 75.
3. See, e.g., Otto Veraguth, "Measuring Joy and Sorrow: The Psychogalvanic Reflex as a Detector of Emotion," *Sci. Am. Suppl.* 1727: 87 (6 Feb. 1909). For

detailed background on physiologists' study of emotions from 1850 until the time of Reich's experiments, see Otniel Dror, *Blush, Flush, Adrenalin: Science, Modernity and Paradigms of Emotions, 1850–1930* (under revision with U. of Chicago Press). Dror's earlier PhD dissertation is "Modernity and the Scientific Study of Emotions, 1880–1950" (Princeton U., 1999).
4. Reich, *BISA*, p. 76.
5. On Brandt's (1913–1992) time in Norway and his connections there, see Willy Brandt, *In Exile: Essays, Reflections and Letters, 1933–1947* (Philadelphia: U. of Pennsylvania Press, 1971), chap. 5; also see the interview with Brandt by Digne Meller-Markovicz in her film *Wilhelm Reich: Viva Little Man* (1987). The experiments moved Reich's "vegetotherapy" further in the direction of "body therapy"; other schools of therapy in Norway were substantially influenced by Reich's work in this direction, e.g., Aadel Bülow-Hansen's. See Michael C. Heller, *Body Psychotherapy: History, Concepts and Methods* (New York: Norton, 2012), pp. 84–85.
6. By the autumn of 1935, disagreements with Hoffmann and Löwenbach over many issues led Reich to break off their involvement in the work. See, e.g., Reich to Hoffmann and Löwenbach, 6 Sept. 1935, *BP*, pp. 48–50; also see Reich to Schjelderup, 26 Sept. 1935, in which he is particularly critical of (as he sees it) Löwenbach's lack of objectivity about the experiments. *BP*, pp. 51–55. This will be discussed further in Chapter 6.
7. Reich, *BISA*, pp. 126–128; Reich, *FO* (1942 ed.)], *op. cit.*, pp. 328–329.
8. Reich, "Versuche am Radio," 17 pp. ms. laboratory notes, 28 Feb.–10 Mar. 1936 (RA, OI box 6).
9. Reich, *BISA*, pp. 85–86, 96–97, figs. 5, 6, 15.
10. Ibid., pp. 111–112, figs. 26–29.
11. Ibid., pp. 112–113, fig. 30.
12. Reich, *FO* (1942 ed.), *op. cit.*, p. 329.
13. Ibid.; Reich, *BISA*, pp. 92–93.
14. Reich, *BISA*, p. 104.
15. Ibid.
16. Ibid., p. 6.
17. Ibid., pp. 107–108.
18. See, e.g., William Broad, "I'll Have What She's Thinking," *New York Times*, 29 Sept. 2013, p. 8 (Sunday Review), or http://www.nytimes.com/2013/09/29/sunday-review/ill-have-what-shes-thinking.html?pagewanted=all&gwh=397E39BFC231E9B78DE3909BD631771C.
19. Reich, *BISA*, pp. 113–118.
20. Ibid., pp. 116–117 (1937 original pp. 34–35).
21. Ibid., p. 96, fig. 15.
22. Ibid., pp. 113–114.

23. Ibid., p. 119 (1937 original, p. 36).
24. Ibid., pp. 120–121.
25. Ibid., p. 119.
26. Ibid., p. 125.
27. Ibid., emphasis in original.
28. William H. Masters and Virginia E. Johnson, *Human Sexual Response* (Boston: Little, Brown, 1966).
29. Marvin Zuckerman, "Physiological Measures of Sexual Arousal in the Human," *Psycholog. Bull.* 75: 297–329 (1971).
30. Sharaf, *Fury on Earth, op. cit.*, p. 215, emphasis in original. More recently, see Byron Braid and R. A. Dew, "Reich's Bioelectric Experiments: A Review with Recent Data," *Annals of the Institute for Orgonomic Science* 5: 1–18 (September 1988).
31. W. Hoffmann, "Dr. Reich og hans elektro-fysiologi" [Dr. Reich and his Electrophysiology], *Arbeiderbladet*, 8 June 1938.
32. W. Hoffmann to [Rolf] Gjessing, 28 May 1936 (RA, OI box 6, "Versuche" fld).
33. Hoffman, "Dr. Reich og," *op. cit.*; also idem, "Dr. Reich og vi andre" [Dr. Reich and we others], *Dagbladet*, 21 June 1938.
34. Wilhelm Reich, "Orgonotic Pulsation," *IJSO* 3: 97–150 (1944), pp. 139–140, emphasis in original. Reich also responds to their critiques in detail in "Der Besuch Professor Roger du Teil's von der Universität Nizza in Oslo vom 26.7 bis 7.8. 1937" (RA, correspondence box 10, du Teil flds). A detailed response is also in Reich to N. Hoel, S. Hoel, O. Havrevold, 13–14 June 1938 (RA, correspondence box 9, "1936–39" flds).
35. Reich to Schjelderup, 26 Sept. 1935 (RA, correspondence box 9), published in *BP*, pp. 51–55, quote on p. 52.
36. The 1935 and 1937 papers were published as separate pamphlets by SexPol Verlag, Copenhagen. The masochism and "Psychic Contact" papers were later included (in edited form) as chaps. 11 and 13 in *CA*. This sequence of developments was described by Reich in chap. 7 of *FO*, including most of the findings of the "*Orgasmusreflex*" paper.
37. Reich, *BISA*, pp. 128–129.
38. Ibid., p. 121.
39. Reich, *BE*, p. 20; *BP*, pp. 63–64.
40. Ibid., p. 152.
41. See, e.g., Rhumbler, "Aus dem Lückengebiet zwischen organismischer und anorganismischer Materie," *Ergebnisse der Anatomie und Entwicklungs.* 15: 1–38 (1906).
42. Reich, *BE*, p. 26.
43. Reich, *FO* (1942 ed.), p. 338, emphasis in original.
44. Ibid., pp. 338–339, emphasis in original.

45. Ibid., p. 339, emphasis in original.
46. See, e.g., Ludwig Rhumbler, "Aus dem Lückengebiet," *op. cit.*; idem, "Anorganisch-organismische Grenzfragen des Lebens," in Hans Driesch and Heinz Woltereck, eds., *Das Lebensproblem im Lichte der modernen Forschung* (Leipzig: Quelle & Meyer, 1931).
47. Judy Johns Schloegel and Henning Schmidgen, "General Physiology, Experimental Psychology, and Evolutionism: Unicellular Organisms as Objects of Psychophysiological Research, 1877–1918," *Isis* 93: 614–645 (2002); see also H. S. Jennings, *Behavior of the Lower Organisms* (New York: Columbia U. Press, 1906).
48. An excellent description of these is given in Reinhard Beutner, *Life's Beginnings, op. cit.*, pp. 103–180. Sydney Payne uses the term "simulacra" for the structures created in Bastian's silica experiments; *Ann. of Botany* 30: 383 (July 1916). Evelyn Fox Keller has recently made the case for historical reexamination of this work, long thrown into shadow after the development of molecular biology. She evaluates one strand, the experiments by Stefan LeDuc on "osmotic life forms." See Keller, *Making Sense of Life: Explaining Biological Development with Models, Metaphors, and Machines* (Cambridge, MA: Harvard U. Press, 2002). Du Teil told Reich about LeDuc's work in a letter of 3 Feb. 1937 (RA, Reich–du Teil correspondence).
49. Reich, *BISA*, p. 46, and Reich, *BE*, p. 129.
50. See Beutner, *op. cit.*, pp. 126–147.
51. Alfonso Herrera's experiments on plasmogeny were described in "Los Protobios 1917," published in *La Plasmogenia* by I. Castellanos (Havana, 1921); see also idem, "Plasmogeny," in Jerome Alexander, ed., *Colloid Chemistry*, vol. 2 (New York: Chemical Catalog Co., 1928), pp. 81–92; and idem, "On the Origin of Life," *Science* 96: 14 (1942). See also Beutner, *op. cit.*, pp. 142, 161–163.
52. See H. James Cleaves, Antonio Lazcano, Ismael Ledesma, Alicia Negrón-Mendoza, and Juli Pereto, eds., *Herrera's "Plasmogenia" and Other Writings* (New York: Springer, 2014); see also Ismael Ledesma-Mateos and Ana Barahona, "The Institutionalization of Biology in Mexico in the Early 20th Century: The Conflict between Alfonso Luis Herrera and Isaac Ochoterena," *J. Hist. Biol.* 36 (2003): 285–307. On Herrera's early support of Darwinism, analogous to Ernst Haeckel's role in Germany, see Roberto Moreno, "Mexico," in Thomas Glick, ed., *The Comparative Reception of Darwinism* (Chicago: U. of Chicago Press, 1987), 346–374, esp. 346–350, 368–370. Historian Adolfo Olea-Franco informs me that "Herrera underwent a kind of revival during the government of Lazaro Cardenas, the most revolutionary of Mexican presidents (1934–1940). During his regime, materialist and socialist thought was supported more than ever. [This is when Mexico welcomed Trotsky.] In contrast, Herrera's ideas were not consonant with the conservative atmosphere that

prevailed in Mexico in the late 1920s and early 1930s—a time during which the Mexican Communist Party, founded in 1919, was forbidden and many of its militants (more than was usual before) were jailed."

53. See G. W. Crile, M. Telkes, and A. F. Rowland, "Autosynthetic Cells," *Protoplasma* 15: 337–360 (1932); Beutner, *op. cit.*, pp. 174–186. See also Peter English, *Shock, Physiological Surgery, and George Washington Crile* (Westport, CT: Greenwood Press, 1980).

54. See Steven Dick and James Strick, *The Living Universe: NASA and the Development of Astrobiology* (New Brunswick, NJ: Rutgers U. Press, 2004), chap. 8; see also Iris Fry, *The Emergence of Life on Earth: A Historical and Scientific Overview* (New Brunswick, NJ: Rutgers U. Press, 2000), chap. 14.

55. Reich, *BE, op. cit.*, p. 15 and fig. 2.

56. Reich to du Teil, 22 Mar. 1937; also Reich diary, 15 Oct. 1938 (Personal box 2, ms p. 95). Reich had begun living on Drammensvejen in Mar. 1935, planning to buy it: "I am becoming quite a bourgeois in my old age. But Bert Brecht has also got a small house." Reich, *BP*, pp. 33–34.

57. See 15 Sept. 1936, entry by Reich and Havrevold, Bion lab ntbk 1 ("Entwicklung der Bionenforschung vom April 1936," OI box 6, fld of same name), ms. pp. 100–103, including Protocol on sending 1mA current through bion preparations; the same notebook on 24 Sept. 1936, ms. p. 115 records his first observations on org-tierchen *(Vorticella)* and their movements.

58. Reich, *BE, op. cit.*, p. 25.

59. Ibid., pp. 28, 31.

60. Ibid., p. 31. See also Reich, *BP*, pp. 64–65, 6 May 1936 diary entry.

61. Reich, *BE*, pp. 31–37, "Org-Tierchen" in the original German. Bion lab ntbk 1 ("Entwicklung der Bionenforschung vom April 1936," OI box 6, fld of same name, 24 Sept. 1936, ms. p. 115) records his first observations on org-tierchen *(Vorticella)* and their movements.

62. Reich, *BE*, p. 39.

63. Ibid. Replication of these experiments, with similar results, was reported by biologists Sol Kramer and Bernard Grad in July–Aug. 1950 (Notes from cancer course taught by Reich, Kramer papers, U. of Florida); also by Dr. Charles Oller in 1970–1978, in a series of captioned microphotos later donated to and available through the Wilhelm Reich Museum. Some other bion experiments were the inspiration to physicist Adolph Smith at Sir George Williams (Concordia) U. in Montreal, in origin-of-life experiments he conducted from 1966 to 1981 (Strick oral history interview with Smith, 30 Jan. 1997, 27 Sept. 1998; oral history interview with Grad, 5 Jan. 2000). See, e.g., A. Smith and F. T. Bellware, "Dehydration and Rehydration in a Prebiological System," *Science* 152: 362–363 (15 Apr. 1966); see also Smith and Dean Kenyon, "Is Life Originating *De Novo?*," *Persp. Biol. Med.* 15: 529–542 (Aug. 1972); and Smith,

Clair Folsome, and Krishna Bahadur, "Nitrogenase Activity of Organo-Molybdenum Microstructures," *Experientia* 37: 357–359 (1981). See also Sept. 1990 talk by Smith in Princeton, NJ, "The Bioenergetics of Cancer: Reich's Ideas from the Perspective of 1990," including a discussion of Otto Warburg's "energy flow model." My thanks to Dr. Smith for a copy of this ms.

64. Reich, *BE*, p. 26. That literature review was published as A. Hahn, "Die Geschichte der Auffasungen seit dem 17 Jahrhundert über den Ursprung des organischen Lebens," pp. 137–205 in *Die Bione* (1938, *op. cit.*); Eng. trans. D. and I. Jordan as "The History of Concepts of the Origin of Organic Life since the 17th Century" was not included in the 1979 Eng. trans. of *Die Bione (BE)* but is available as a separate reprint through the Wilhelm Reich Museum.

65. See, e.g., Bion lab ntbk 1, ms. p. 13, 4 May 1936, on tulip preps; Reich, *BE*, p. 40.

66. Bion lab ntbk 1, ms. pp. 14–15, 18, 20, dated 7–10 May 1936, on earth preps; Reich, *BE*, pp. 40–45, quote on p. 40.

67. Reich, *BE*, p. 46.

68. A 0.1 normal KCl solution is the same concentration as 0.1 molar, since one mole of potassium ion is released into solution for every one mole of potassium chloride dissolved.

69. Reich, *BE*, p. 38.

70. Ibid.

71. For a sense of the cutting edge of what was possible at this time, see Hannah Landecker, "The Lewis Films: Tissue Culture and 'Living Anatomy,' 1919–1940," in Jane Maienschein and Marie Glitz, eds., *History of the Carnegie Institute Laboratory of Embryology* (Cambridge: Cambridge U. Press, 2004), pp. 117–144; James Strick, "Swimming against the Tide: Adrianus Pijper and the Debate over Bacterial Flagella, 1946–1956," *Isis* 87: 274–305 (1996); also Osborne O. Heard, "Some Practical Considerations on Time-Lapse Motion Photomicrographic Devices," *J. Biol. Phot. Assoc.* 1 (4): 4–18 (1932); and S. Zimmer, "The Photomicrography of Blood," *J. Biol. Phot. Assoc.* 5: 113–119 (1937). The Heard and Zimmer articles explain some points of care that must be taken in photomicrography in order to obtain accurate results, such as the use of apochromatic lenses insisted on by Reich and claimed by some of his contemporaries to be excessively meticulous.

72. Reich, *BE*, p. 38. For a description of the equipment, see pp. 8, 15 and figs. 3–9. The equipment in its original configuration can be examined at the Wilhelm Reich Museum. See also Kari Berggrav, "Personal Recollections of Reich and His Work," *J. Orgonomy* 8 (1): 19–26 (May 1974); and John Bell, "Working with Wilhelm Reich: An Interview with Kari Berggrav," *J. Org.* 31: 37–51 (1997), esp. pp. 40–41.

73. Reich, *BE*, p. 38.

74. Berggrav, *op. cit.* See Strick, "Swimming against the Tide," *op. cit.*
75. Bion lab ntbk 2, ms. p. 190, RA.
76. Interestingly, Reich found that grass or moss collected in spring and summer did not produce protozoa; only moss collected in autumn gave unequivocal results (bion lab ntbk 2, ms. pp. 144–146, published in *BP*, p. 105). See John Tyndall, *Essays on the Floating Matter of the Air in Relation to Putrefaction and Infection* (London: Macmillan, 1881), p. 150, for a different explanation of the fresh versus old, dried hay or moss phenomenon.
77. Berggrav, *op. cit.* See Reich, *CB*, *op. cit.*, pp. 239–240.
78. Reich, *BE*, p. 46.
79. RA, OI box 5, small black notebook, under "L" divider, 15 June 1936 entry; also published in *BP*, p. 66. Also described in Reich, *BE*, pp. 46–49.
80. The boiled preparations were first created in May 1936 according to Reich in Reich to du Teil, 22 Mar. 1937.
81. Reich, *BE*, p. 50.
82. RA, Bion lab ntbk 1, ms. p. 16, 16 May 1936; Reich, *BE*, p. 50, emphasis in original.
83. Ibid., p. 51, emphasis in original.
84. Ibid., emphasis in original.
85. Ibid.
86. RA, bion lab ntbk 1, ms. pp. 11–12, 29 Apr. 1936, "*Beobachtungen an Farnblättern:* Farnblätter vor zwei Tagen in Wasser gelegt, zeigen im Mikroskop die typische pflanzliche Zellenstruktur."
87. Reich, *BE*, p. 51, emphasis in original.
88. Ibid., p. 52, emphasis in original.
89. Ibid., emphasis in original.
90. Ibid., p. 53.
91. Ibid., pp. 52–53, emphasis in original.
92. Ibid., p. 53.
93. RA, OI box 7, bion lab ntbk 2, ms. pp. 159–165, 18–24 May 1937, "Kohleglühversuch am 18.5.1937"; this work is described by Reich in *BE*, pp. 110–114, almost verbatim from the entries in the laboratory notebook. The same, almost verbatim details were reported by Reich in letters to Roger du Teil and Paul Neergaard, both on 24 May 1937.
94. Reich, *BE*, pp. 54–55; see entries in Bion lab ntbk 2, 16 Jan. 1937 (ms. pp. 7–15).
95. Reich, *BE*, pp. 54–57, quote on p. 57.
96. Ibid., pp. 58–59.
97. Ibid., p. 59.
98. Ibid., p. 60.
99. Ibid.
100. Ibid., pp. 60–61.

101. Bion lab ntbk 2, 19 Jan. 1937, ms. p. 18; also Reich to du Teil, 22 Mar. 1937.
102. Reich diary entry, 7 Dec. 1936, Personal box 2, fld "May 34–Feb. 37, ms p. 27; an edited version is in *BP*, p. 77.
103. Reich, *Die Bione op. cit.*, pp. 87–88; Reich, *BE*, pp. 141–142. For similar arguments in the nineteenth-century debates, see John Farley, *Spontaneous Generation Controversy, op. cit.*, esp. pp. 105–107 and William Bulloch, *The History of Bacteriology* (Oxford: Oxford U. Press, 1938), pp. 101–105 on the glacier episodes, which Tyndall repeated at Bel Alp in Aug. 1876; also Strick, *Sparks of Life, op. cit.* Pouchet and his collaborators, of course, did find microbial growth in the boiled infusions they opened atop the Pyrenees.
104. Reich to du Teil, 16 Feb. 1937, in a PS added "19 Jan." [*sic*; actually 19 Feb.]. On the early history of microbial culture media, see William Bulloch, *op. cit.*, chapt. 9, esp. pp. 218–227.
105. See, e.g., Reich to du Teil, 19 Jan. [*sic*; actually Feb.] 1937 (published in Eng. in *BE*, pp. 91–92); also see du Teil to Reich, 13 Sept. 1937 and 6 Apr. 1938; Olga Amsterdamska has made a very illuminating study of the reasons why the "pleomorphist" position could be revived in the 1920s–1940s in "Stabilizing Instability: The Controversy over Cyclogenic Theories of Bacterial Variation during the Interwar Period," *J. Hist. Biol.* 24: 191–222 (1991).
106. See, e.g., Harris Coulter, *Divided Legacy IV: Medicine and Science in the Bacteriological Era* (Berkeley, CA: North Atlantic, 1994), esp. pp. 181–211 on the Kendall-vs.-Zinsser debate; also Thomas Rivers and Saul Benison, *Tom Rivers : Reflections on a Life in Medicine and Science* (Cambridge, MA: MIT Press, 1967). For Rivers linking these pleomorphist ideas to a revival of spontaneous generation claims, see Thomas Rivers, "Spontaneous Generation and Filterable Viruses," *Northwest Med.* 29: 555–561 (1930). In a letter of 7 May 1938, du Teil repeatedly reports pleomorphist observations; du Teil to Reich (RA, correspondence box 10, du Teil flds) "Retrospectives des Operations ayant et faites aux mois de Mars et Avril 1938" [Summary of the experiments done in March and April 1938], 4 pp.
107. Olga Amsterdamska gives a good overview of this pleomorphist revival; see "Stabilizing Instability," *op. cit.*; idem, "Medical and Biological Constraints: Early Research on Variation in Bacteriology," *Social Studies of Science* 17: 657–687 (1987). See also Milton Wainwright, "Extreme Pleomorphism and the Bacterial Life Cycle: A Forgotten Controversy," *Perspect. in Biol. and Med.* 40: 407–414 (1997). For primary sources, see F. Löhnis, "Studies upon the Life Cycles of the Bacteria: Pt. 1; Review of the Literature, 1838–1918," *Memoirs of the National Academy of Sciences*, vol. 16, 2nd memoir (1921); a detailed review of the evidence in favor of this view is Philip Hadley, "Microbic Dissociation: Instability of Bacterial Species with Special Reference to Active Dissociation and Transmissible Autolysis," *J. Infect. Dis.* 40: 1–312 (1927); another is Joseph

A. Arkwright, "Variation," in *A System of Bacteriology in Relation to Medicine*, vol. 1 (London: HMSO, 1930), 311–374. The story is also thoroughly covered in Pauline Mazumdar, *Species and Specificity* (Cambridge: Cambridge U. Press, 1995), pp. 15–97; and in William Summers, "From Culture as Organism to Organism as Cell: Historical Origins of Bacterial Genetics," *J. Hist. Biol.* 24: 171–190 (1991).

108. James G. Hanley, "Models and Microbiology: Pasteur and the Body," *Canadian Bull. Med. Hist.* 20 (2): 419–435 (2003), see esp. pp. 425–426. Quote from Reich to Emil Walter, 1 June 1938 (Correspondence box 9, RA), *BP* pp. 153–154. For more on thinking about the role of viruses in origin of life debates at this time, see Scott Podolsky, "The Role of the Virus in Origin of Life Theorizing," *J. Hist. Biol.* 29:79–126 (1996); also Angela Creager, *The Life of a Virus* (Chicago: U. of Chicago Press, 2001).

109. Theobald Smith, "Koch's Views on the Stability of Species among Bacteria," *Ann. Med. Hist.* 4: 524–530 (1932); A. I. Kendall, "Bacteria As Colloids," in H. N. Holmes, ed., *Colloid Symposium Monograph* (New York: Chemical Catalog Co., 1925), vol. 2, p. 195; Ludwik Fleck [1934], *Entstehung und Entwicklung einer wissenschaftliche Tatsache*, Eng. trans. *The Genesis and Development of a Scientific Fact* (Chicago: U. of Chicago Press, 1979), pp. 27–31, 92–94. On the DNA story, see Maclyn McCarty, *The Transforming Principle* (New York: W. W. Norton, 1985). In many bacteriology texts, such as E. O. Jordan's (by the 1920s), and Park and Williams (by the early to mid-1930s), variation was treated as an important topic. See Strick, "Evolution of Microbiology as Seen in the Textbooks of Edwin O. Jordan and William H. Park," *Yale J. Biol. and Med.* 72: 321–328 (1999); see also Edwin O. Jordan, "Presidential Address to Soc. of Amer. Bacteriologists: Variation in Bacteria" [1905], 13 pp., Amer. Soc. for Microbiol. Archives.

110. Thomas Rivers, "Spontaneous Generation," *Northwest Med.* 29: 555–561 (1930); Saul Benison, *Tom Rivers: Reflections on a Life in Medicine and Science* (Cambridge, MA: MIT Press, 1967), pp. 597–598; also Harris Coulter, *Divided Legacy: Medicine and Science in the Bacteriological Era* (Berkeley, CA: North Atlantic, 1994).

111. The death-knell of these revived doctrines was sounded in Werner Braun, "Bacterial Dissociation: A Critical Review of the Phenomenon of Bacterial Variation," *Bacteriol. Rev.* 11 (2): 75–114 (June 1947).

112. Jan Sapp, *New Foundations of Evolution: On the Tree of Life* (New York: Oxford U. Press, 2009), p. 116.

113. Ibid., pp. 119–120, quote on 119.

114. *S. ventriculi* was first described by John Goodsir; see John Lowe, "On *Sarcina ventriculi*, Goodsir," *Edin. New Phil. J.* 12 (n.s.): 58–64 (1860). For more recent photographs, see *J. Gen. Microbiol.* 10: 452–456, esp. figs. 4, 6. Many changes

in bacterial nomenclature have occurred since the 1930s; in the text, whenever possible, I give the modern name in parentheses. Henry Charlton Bastian has interesting remarks on *Sarcina*; he believed it was not necessarily a fully living microbe since it had "only been seen to undergo a process of growth and development [like a crystal], never one of spontaneous fission." See Bastian, *Beginnings of Life* (London: Macmillan, 1872), vol. 2: app. A, pp. ii–vii, quote on pp. iii–iv.

115. Du Teil to Reich, 31 Aug. 1937.
116. Reich, *BE*, pp. 72–73 and fig. 45 (taken from bion films).
117. Ibid., p. 112. The point is also made clearly in the bion films.
118. Ibid., p. 73.
119. Reich to President of Commission on Biological Studies, CEP, Paris, 5 Aug. 1937 (RA, correspondence box 10, du Teil flds).
120. Reich to du Teil, 24 Aug. 1937; 13 Sept. 1937.
121. Reich to du Teil, 7 Sept. 1937.
122. Report on bion types, 1 Oct. 1938 (RA, OI box 8, "Präparatherstellung und Protokoll" fld, ms. pp. 1–12).
123. Kreyberg to Reich, 23 Sept. 1937; Thjøtta to Reich, 26 Apr. 1938 (RA, correspondence box 9, "1936–39" flds).
124. Reich to du Teil, 15 Oct. 1937.
125. Paulo and Alexandra Correa, "The PA and SAPA Bion Experiments and Proto-Prokaryotic Biopoiesis," *J. Biophys. Hematol. Oncol.* 1 (2): 1–49 (2010), pp. 4–5.
126. Du Teil to Reich, 11 Sept. 1937.
127. Bon to Reich, 23 Mar. 1939 and 12 May 1939; Reich reply, 17 May 1939 (RA, correspondence box 9, Bon flds).

3. Reich and du Teil

1. Roger du Teil, *Specialisation et Evolution: Essai Pragmatique sur la Transcendance de la Loi Morale* (Nice: Librairie Félix Alcan, 1935).
2. The German text of the monograph had only just been completed on 20 Jan. 1936. See Reich, *BP*, p. 61, Reich to du Teil, 29 July 1936.
3. Marie-Chantal Benoit Bardy, "Wilhelm Reich and Roger du Teil," *Energy and Character* 6: 39–42 (1975), quote from p. 40. Bardy was a student of du Teil much later in his life.
4. Reich to Albert Fischer, 25 Sept. 1936 (RA, correspondence box 9, "1936–39" fld). Dr. J. H. Leunbach in Copenhagen had written Reich that Fischer would be willing to make some sort of film apparatus of his institute available (Reich thanked him and arranged for his photographer Kari Berggrav to go to Copenhagen for a few days to learn the process). Fischer was also a researcher at the Biological Institute of the Carlsberg Foundation in Copenhagen. Fritz

Lipmann worked in his lab from 1932 to 1939. Jonathan Hoegh "Joyce" Leunbach (1884–1955) was a Danish physician, sex-educator, and advocate of liberalizing abortion laws and a staunch supporter of Reich. He was prominent in the World League for Sexual Reform.

5. Leunbach to Reich, 8 January 1937 (RA, correspondence box 9, "1936–39" fld). " 'The structures that we saw were just simple cholesterol-or-lecithin-formations.' 'And he claims to have observed spindle formations during cell division—this is absolutely not visible even with the largest magnification.' 'All movements were movements of liquid in the preparation.' 'Those are fairy-tales from the time before Pasteur, when rats and mice were created by leaving old rags with flour and moisture lying around in the dark.' 'You know, we haven't been so flabbergasted in a long time!' 'So, please, spare us from such things!' "

6. Reich to Albert Fischer, 9 Jan. 1937 (RA, correspondence box 9, "1936–39" fld), most of this letter published in *BP*, pp. 91–93, quote from p. 92, italics in original.

7. Ibid. For more on the relationship between Brownian movement and diffusion, see Denys Wheatley and Paul Agutter, "Historical Aspects of the Origin of Diffudion Theory in 19th-Century Mechanistic Materialism," *Perspec. Biol. Med.* 40: 139–156 (1996), esp. p. 148–149; also D. N. Wheatley, "On the Vital Role of Fluid Movement in Organisms and Cells," *Medical Hypotheses* 52: 275–284 (1999). A related article is P. C. Malone and D. N. Wheatley, "A Bigger Can of Worms?" *Nature* 349: 373 (31 Jan. 1991).

8. See Reich, *BE*, fig. 40, compared with figs. 41, 42.

9. Reich, *BP*, p. 93.

10. See, e.g., Stanley Klosevych, "Photomicrography—Resolution and Magnification," *J. Biol. Phot. Assoc.* 32: 133–147 (Nov. 1964).

11. Reich began Bion lab ntbk 2 at this time, noting on the title page, "Begun, 9 Jan. 1937, *Decision to maintain a compulsion-neurotic protocol, resulting from Leunbach's report about the unscientific, narrow-minded remarks made by Albert Fischer concerning the successful demonstration* [emphasis in original: "*Beschluss, ein zwangsneurotisch genaues Protokoll zu führen, erfolgt auf Bericht Leunbachs über die unwissenschaftlichen, borniertern Ausserungen Albert Fischers zur erfolgten Demonstration*"]. Elsewhere around 1950, Reich recounted, "To prevent false rumors, I refuted all Fischer's claims [reported to me] in a letter to Leunbach . . . although I continued to believe (inexcusably) in the objectivity of natural scientists. It would have been better to act defensively and curtly. Fischer had simply attempted to explain away obvious facts. He denied the Giemsa stain, paid no attention to the appearance of rods and cocci a few minutes after the preparation had been made, and resorted to portraying the entire experiment in a ridiculous light. Unfortunately, I did not

dream of how new, how revolutionary and comprehensive my experiment was." Wilhelm Reich, *People in Trouble* (New York: Farrar, Straus and Giroux, 1976), p. 269; originally published as "Wilhelm Reich on the Road to Biogenesis (1935–1939)," *Orgone Energy Bull.* 3: 146–162 (1950), quote from p. 156.

12. Reich to du Teil, 12 Dec. 1936 and 22 Mar. 1937; Reich diary, 20 Nov. 1936 (Personal box 2, fld "May 34–Feb. 37, ms p. 25–26). The Reichert microscopes had not yet arrived as of 9 Jan. 1937, when Reich wrote that he was still using a binocular Leitz model; see *BP*, p. 92. They must have come within the next month or so. For the capabilities of the Reichert "Z" model Reich used, see the user's manual for that model, *Large Universal Microscope "Z": 1937 Pattern* (Vienna: Reichert Optical Works, 1937), 24 pp.

13. In one microphotograph, for example (Wilhelm Reich, *Bion Experiments on the Cancer Problem* [Oslo: SexPol Verlag, 1939; hereafter, *BECP*], fig. 28, p. 16), Reich uses the 150× objective, an 8× ocular and the 50 percent magnification factor built into the Reichert binocular tube, to obtain approximately 1800× at very high resolution, so that the photograph can be enlarged 3× for a total final image size 5400× greater than the actual object. Using the 150× objective with 20× compensating oculars and a 50 percent factor in the inclined tube, he could view objects under the microscope at 4500×. Alternatively, a 100× objective with 25× oculars and a 50 percent inclined tube factor yields 3750×.

14. The detailed description of the preparation was reprinted in *Die Bione*; see Eng. trans., *BE*, pp. 61–63.

15. Ibid., pp. 87–88.

16. Ibid., pp. 88–89.

17. Ibid., p. 89.

18. Ibid.

19. Du Teil to Reich, 3 Feb. 1937 (RA, partial Eng. trans. in box 9, "English translations" fld). LeDuc's experiments are discussed in Evelyn Fox Keller, *Making Sense of Life: Explaining Biological Development with Models, Metaphors, and Machines* (Cambridge, MA: Harvard U. Press, 2002).

20. Reich, Bion lab ntbk 2, ms. p. 37 (RA, OI box 7, "Entwicklung der Bionenforschung," 2nd vol. of this title, this one starting from 1 Jan. 1937 also titled "Geschichte der Erforschung die Bione"; English trans. of this passage published in *BP*, p. 96). Reich describes these culturing difficulties in *BE*, bottom of p. 72 to the top of p. 73.

21. Reich to du Teil, 8 Feb. 1937; *BE*, p. 90.

22. Reich to du Teil, 16 Feb. 1937; *BE*, pp. 90–91; most of the details of these experiments are in Bion lab ntbk 2, ms. pp. 57–61.

23. On Havrevold, see Norsk Biografisk Leksikon: http://translate.google.com/translate?hl=en&sl=no&u=http://www.snl.no/.nbl_biografi/Johan_Vogt/utdypning_%25E2%2580%2593_2&ei=5yL_SZ_cM5agMt2L0LAE&sa=X&oi

=translate&resnum=7&ct=result&prev=/search%3Fq%3Dnorsk%2Bbiografisk%2Bjohan%2Bvogt%26hl%3Den%26sa%3DG. Havrevold's scientific interests later in his career lay in the fields of neurology, endocrinology, and pharmacology.

24. On 11 Apr. 1938, K. Schreiner to Havrevold (RA, correspondence box 9, "1936–39" flds) recounted the conversation Schreiner had with Havrevold a year earlier.

25. Reich, *BP*, p. 102.

26. Quotes from Reich, *People in Trouble* (New York: Farrar, Straus and Giroux, 1976), p. 271. According to Thjøtta, in his account in the 1952 Kerrigan interview (U.S. State Department investigation of Reich, Jerome Greenfield papers, Taminent Library and Robert F. Wagner Labor Archives, New York University), p. 8, Thjøtta says he telephoned Reich. According to Thjøtta, "Reich reacted by becoming irritated and saying that he, Reich, or one of his assistants must have sent the wrong cultures to Professor Thjøtta. Professor Thjøtta stated that he then invited Dr. Reich to come to his office and examine his, Professor Thjøtta's, findings. Dr. Reich, however, according to Prof. Thjøtta, wanted Thjøtta to go to Reich's office and apparently the conversation ended with both men standing on their dignity and refusing to visit the other." *Bacillus proteus* is now called *Proteus vulgaris*, with other species also included in the genus, among a gram-negative group, Proteobacteria.

27. Bion lab ntbk 2, ms. p. 96.

28. Ibid., ms. p. 99, published in *BP*, p. 103.

29. Ibid., ms. p. 155, published in *BP*, p. 106.

30. Reich, "Die soziale Macht des Gerüchts in der Bürokratie" [The social power of gossip in bureaucracy], RA, Personal box 4, p. 3.

31. Ibid.

32. H. B. from Socialdepartment, Medizinalabteilung, Oslo to Reich, Lindenberg and Motesiczky, 25 [Apr.] 1938 (*sic*; original Norwegian version in next fld is dated Oct. 1938 and placed in that order in the collection), (RA, correspondence box 9, "1936–39" flds).

33. Du Teil to Reich, 14 Mar. 1937, "Regarding the gelatin, a member of our Society, a physician whom I invited to be present on March 7, reported that during his time as an assistant doctor in Paris he had seen several cases of infections caused by gelatin, for example when treating hemorrhages, although the gelatin had been sterilized [prior to usage]. He asked me to inquire if you know something about this, in other words, if the sterilization techniques used on the gelatin are now so advanced that you can guarantee, in other words, if you are certain, that germs could not get into the mixture via the gelatin. He did, however, recognize that, should this be the case, you had

the prodigious credit of discovering and cultivating germs that were, until now, unrecognizable and invisible to us."
34. Reich to du Teil, 22 Mar. 1937. See also bion lab ntbk 2, ms. pp. 99–107, 16 Mar. 1937.
35. Bion lab ntbk 2, ms. p. 120 (previous Eng. trans. published in *BP*, p. 104; the idiomatic expression *"Ei des Columbus"* is left out of the published translation).
36. Ibid., pp. 121–122.
37. Strick, *Sparks, op. cit.*, chap. 4.
38. Reich, *BP*, pp. 93–94.
39. Reich, *BE*, p. 130. Sidney Fox makes an argument in some ways similar, distinguishing between random Brownian movements and other nonrandom movements in his proteinoid microspheres. See Fox, *The Emergence of Life: Darwinian Evolution from the Inside* (New York: Basic Books, 1988), p. 98.
40. Ibid., pp. 149–150.
41. Reich to Paul Neergaard, 24 Feb. 1937, Eng. trans. in *BP*, pp. 97–98. See also Roger du Teil in Reich, *Die Bione* (1938), p. 51 (*BE*, p. 96).
42. Ludwik Fleck [1934], *The Genesis and Development of a Scientific Fact* (Chicago: U. of Chicago Press, 1979), pp. 27–31, 92–94. Physiological chemist Reinhard Beutner, *Life's Beginnings, op. cit.*, pp. 157–174, discusses contemporary (1938) research on genuine living motility and the imitations of it produced in cell model experiments.
43. Reich later discussed this phenomenon of dismissing something by saying, "Oh, that's just x, that's nothing new." In a letter to Ola Raknes (1887–1975), he pointed out that an opponent saying a discovery of his (an X-ray photograph suggesting an orgone energy field) "was 'nothing new' does not correspond to the truth. . . . Did he prove it? And even if he proved the existence of such a photograph, there would still be no explanation [of its meaning]. . . . In addition, we have to consider the fact that phenomena seen but not understood are not really seen in the strict scientific sense of the word. Thousands of biologists have seen the bions, yet nobody has seen them." Reich to Raknes, 3 Mar. 1950 (Raknes papers, ms. 4° 2956: 8, National Library, Oslo). See also Lynn Margulis, "Big Trouble in Biology" and "Swimming against the Current," in L. Margulis and Dorion Sagan, *Slanted Truths: Essays on Gaia, Symbiosis, and Evolution* (New York: Springer, 1997), esp. pp. 273–280. In cell biology and microbiology Margulis took a great interest in observations on the living state, and in preserving an archive of microcinematography documenting motility. She felt that similar prejudices were at work in rejecting her views on spirochete symbionts as the origin of eukaryotic flagella and kinetosomes.

44. Reich, *BE*, pp. 110–111. Reich's claim about Robert Brown is thoroughly advocated for by Peter Jones in his book *Artificers of Fraud: The Origin of Life and Scientific Deception* (Preston, UK: Orgonomy UK, 2013). For a contemporary (1930s) summary of mainstream views on Brownian movement, see Paul Heyl, "Atoms," *Sci. Monthly* 38: 493–500 (June 1934), pp. 495–498, including the statement, "Brown appears to have believed at first that in his active molecules he had found these [previously supposed] atoms of life" (p. 495).
45. Ian Simmons, "Going, Going, (Or)gone," *Fortean Times* (London), 29 Aug. 2013, p. 56.
46. Robert Brown, "A Brief Account of Microscopical Observations Made on the Particles Contained in the Pollen of Plants, and on the General Existence of Active Molecules in Organic and Inorganic Bodies," *Phil. Mag.* 4: 161–173 (Sept. 1828); also distributed by Brown as a separate pamphlet beginning on 30 July 1828.
47. Strick, *Sparks, op. cit.*, p. 55, quote from Brown's laboratory notes, Robert Brown Papers, British Museum of Natural History, Solander box 24, file 70 Onagrariæ, pp. 24/224v.
48. Robert Brown Diary, "Maton's Trip through France with Mr. Brown, 1828, Part 1"; also Brown's diary (1829) (BMNH, Brown Papers); Brown showed the molecules to Prof. Robert Graham of Edinburgh, botanists George Benson and John Lindley, Baron Cuvier in Paris, Lorenz Oken in Munich, Prof. Sömmering in Frankfurt, and many others at the Heidelberg meeting of the Deutscher Naturforscher Versammlung from 18 to 24 Sept. 1829.
49. Philip Sloan, "Organic Molecules Revisited," in *Buffon 88: Actes du Colloque international pour le bicentenaire de la mort de Buffon* (Paris: Vrin, 1992), pp. 415–437; John Turberville Needham, "A Summary of Some Late Observations upon the Generation, Composition, and Decomposition of Animal and Vegetable Substances," *Phil. Trans. Roy. Soc. (Lond.)* 45: 615–666 (1748); G.-L. de Buffon, *Histoire Naturelle, Générale et Particulière*, 1st ed. (Paris, 1749), vol. 2, *Histoire générale des Animaux*, pp. 24–32. Needham tried to avoid the materialist "self-active matter" interpretation by asserting that a vital "plastic force" was already in the matter since it was previously living—essentially the position of heterogenesis.
50. For responses to Brown, showing that his claim was interpreted as spontaneous generation, see C. A. S. Schultze, *Mikroskopische untersuchungen über des herrn Robert Brown entdeckung lebender, selbst im feuer unzerstörbarer Teilchen in allen Körpern und über Erzeugung der Monaden* (Karlsruhe: Herder'schen Kunst u. Buch, 1828); H. Muncke, "Über Robert Brown's mikroskopische Beobachtungen, über den Gefrierpunkt des absoluten Alkohols, und über eine sonderbare Erscheinung an der Coulomb'schen Drehwaage," *Ann. d. Physik u. Chemie* 17: 159–165 (1829), Eng. abstract of which appears in *Edin. New Phil.*

J. (July 1830) and *Phil. Mag.* (Sept. 1830); David Brewster, "Observations Relative to the Motions of the Molecules of Bodies," *Edin. J. Sci.* 10: 215–220 (1829); R. Bakewell, "An Account of Mr. Needham's Original Discovery of the Action of the Pollen of Plants with Observations on the Supposed Existence of Active Molecules in Mineral Substances," *Mag. Nat. Hist.* 2: 1–9 (1829); and idem, "Active Molecules," *Mag. Nat. Hist.* 2: 213–214 (1829). David C. Goodman, "The Discovery of Brownian Motion," *Episteme* 6: 12–29 (1972) summarizes these responses well; he shows that the belief that Brown had implied self-active molecules capable of spontaneous generation lasted for several years.

51. Adrian Desmond and James Moore, *Darwin* (London: Michael Joseph, 1991), p. 223. It was to Brown that Henslow sent the young Charles Darwin before the *Beagle* voyage for advice on what kind of microscope would be useful for the marine invertebrate studies Darwin wished to pursue while away. Darwin obtained the same type of microscope suggested by Brown, since studying the "atoms of life" (in an attempt to develop some of Grant's hypotheses about zoophytes) seems to have been specifically what he had in mind. Particularly in what he saw as the connection to transmutationism, Darwin certainly realized the explosive potential of Brown's "molecules," if only because of Brown's secretiveness about them, and thus kept all his notes and ideas on this (and the early "monad theory of evolution" to which it led him) completely private for almost another forty years. See *The Autobiography of Charles Darwin*, ed. Nora Barlow (New York: Norton, 1958), pp. 103–104. See also Philip Sloan, "Darwin, Vital Matter, and the Transformism of Species," *J. Hist. Biol.* 19: 369–445 (1986); and idem, "Darwin's Invertebrate Program, 1826–1836: Preconditions for Transformism," in David Kohn, ed., *The Darwinian Heritage* (Princeton, NJ: Princeton U. Press, 1985), pp. 71–120. I have argued elsewhere that Brown's Active Molecules are the intellectual progenitors of Darwin's "gemmules" from pangenesis; see *Sparks of Life, op. cit.*, chap. 2. On the highly charged valence of "self-active matter," see also Leland Rather, *The Genesis of Cancer* (Baltimore: Johns Hopkins U. Press, 1978), p. 212n23.

52. Robert Brown, "Additional Remarks on Active Molecules," *Phil. Mag.* 6: 161–166 (1829); also Christian G. Ehrenberg, "Über das Entstehen des Organischen aus einfacher sichtbarer Materie," *Ann. d. Physik u. Chemie* 24 (n.s.): 1–48 (1832). Eng. trans. appeared in *Scientific Memoirs*, vol. 1, ed. Richard Taylor, pp. 555–583 (London: R. and J. Taylor, 1837); and idem, "On the Magnitude of the Ultimate Particles of Bodies," *Edinb. New Phil. J.* 13: 319–328 (1832). Most of these sources from the 1828–1829 controversy are republished in J. Strick, ed., *The Origin of Life: Cells, Molecules and Generation*, 6 vols. (Bristol, UK: Thoemmes Press, 2004).

53. J. B. Dancer, "Remarks on Molecular Activity as Shown under the Microscope," *Proceedings of the Literary and Philosophical Society of Manchester* 7: 162–164 (1868); Emil DuBois-Reymond, "Leibnizische Gedanken in der neueren Naturwissenschaft," *Monatsberichte der Königlich Preussischen Akademie der Wissenschaft (Berlin)* July 1870: 835–854. The narrative changes very little in Steven Brush, "Brownian Movement from Brown to Perrin," *Archives of History of the Exact Sciences* 5: 1–36 (1968), a century later. D'Arcy Wentworth Thompson tells a very similar story in *On Growth and Form*, in the original 1917 edition and with little change in 1961 (Cambridge: Cambridge U. Press, 1961), pp. 44–48, saying, e.g., that "while the Brownian movement may thus simulate in a deceptive way the active movements of an organism, the reverse statement also to a certain extent holds good." Thompson is pointing out that a lot of living movement is what recent biology calls "biased random movement." See, e.g., Howard Berg, *Random Walks in Biology* (Princeton, NJ: Princeton U. Press, 1993).
54. This episode is discussed in detail in Strick, *Sparks, op. cit.*, chap. 4.
55. See Strick, *Sparks, op. cit.*, esp. chap. 2; and idem, introduction to *The Origin of Life: Cells, Molecules, op. cit.* Also L. J. Rather, *Addison and the White Corpuscles: An Aspect of Nineteenth-Century Biology* (London: Wellcome Inst., 1972) and *The Genesis of Cancer: A Study in the History of Ideas* (Baltimore: Johns Hopkins U. Press, 1978); Arthur Hughes, *A History of Cytology* (London: Abelard-Schuman, 1959), pp. 118–119; L. C. Dunn, "Ideas about Living Units, 1864–1909: A Chapter in the History of Genetics," *Perspec. Biol. Med.* 8: 335–346 (1965). By Béchamp, see, e.g., *The Blood and Its Third Anatomical Element*, Eng. trans. M. R. Leverson (Philadelphia: Boericke and Tafel, 1911).
56. Reich, *BE*, pp. 110–114. (The reader may judge by viewing these films on the website that accompanies this book.)
57. Strick, *Sparks, op. cit.*, pp. 54–55.
58. See, e.g., Reich, *BE*, p. 96. Ronchese appears to be Dr. Angel Denis Ronchese (1882–1967).
59. Steven Vogel, *Life's Devices: The Physical World of Animals and Plants* (Princeton, NJ: Princeton U. Press, 1988), p. 165.
60. Du Teil to Reich, 31 Aug. 1937. The surgeon was Raoul-Charles Monod (b. 1887), author of several texts including *Tactique opératoire des annexes de l'utérus* (Paris: G. Doin, 1931). René Allendy (1889–1942) was a homeopathic physician and psychoanalyst and a previous acquaintance of Reich. One of the founding members of the Société psychoanalytique de Paris, Allendy and his wife, Yvonne, wrote *Capitalisme et Sexualité* in 1932. The physicist Bourdet was probably Claude Bourdet (1909–1996), whom du Teil says was also a member of the Front Populaire government.

61. For more background on Marcel Martiny, Allendy, and some of the others at this meeting, see George Weisz, "A Moment of Synthesis: Medical Holism in France between the Wars," in Christopher Lawrence and George Weisz, eds., *Greater Than the Parts: Holism in Biomedicine, 1920–1950* (New York: Oxford U. Press, 1998), Martiny on pp. 78, 82–83.
62. Ibid., p. 73. Allendy's 1922 book was *Les Tempéraments: Essai sur une théorie des temperaments et de leurs diathèses* (Paris: Vigot, 1922).
63. James Strick, "Of Microbes, Medicine and Ecology," *Science* 311: 1243 (3 Mar. 2006).
64. 19 Aug. 1937, "Auszüge aus den Briefen von Prof. du Teil" of 10–19 Aug. 1937, reporting on the reaction to his presentation in Paris (RA, correspondence box 10, du Teil flds). See also du Teil report concerning a lecture on "The Bions" delivered in Paris on 9 Aug. 1937 (RA, correspondence box 9, "English translations" fld, taken mostly from long du Teil letter to Reich of 19 Aug., box 10, du Teil flds.)
65. Ibid., (Auszüge aus den Briefen").
66. Du Teil to Reich, 8 Oct. 1937; also "Auszüge aus den Briefen von Prof. du Teil" of 10–19 Aug. 1937, reporting on the reaction to his presentation in Paris (du Teil flds); du Teil to Reich, 21 June 1938; and Reich to Annaeus Schjødt, 16 June 1938 (*BP*, pp. 156–158, original in RA, correspondence box 9, "1938–1939" flds). Robert Debré (1882–1978) was professor of bacteriology since 15 Mar. 1933 at Paris U. Medical School and author of *Travaux pratiques de Bacterologie* (Paris: Masson, 1936).
67. Du Teil to Reich, 1 Oct. 1937. See also du Teil to Odd Havrevold, 7 Jan. 1938 (RA, correspondence box 10, du Teil flds).
68. Du Teil to Reich, 31 Aug. 1937.
69. Reich to du Teil, 7 Sept. 1937.
70. Ibid., 9 Sept. 1937.
71. Ibid.
72. Du Teil to Reich, 11 Sept. 1937.
73. Ibid.
74. Ibid.
75. Reich to du Teil, 16 Sept. 1937.
76. "Louis Lapicque, 1866–1952" (obituary), *J. Neurophysiol.* 16: 97–100 (1 Mar. 1953), quote on p. 97.
77. Lapicque to Reich, 25 Jan. 1938 (RA, correspondence box 9, "English translations" fld); Reich reply, 28 Jan. 1938 (copy on back of 5 Feb. Reich letter to du Teil).
78. Reich telegram to Lapicque, 2 Feb. 1938; Lapicque to Reich, 3 Feb.; Reich to Lapicque, 5 Feb. 1938 (all in RA, correspondence box 9, "Lapicque" fld).

79. Reich diary, 15 Feb. 1938 (Personal box 2, fld "May 34–Feb. 37," ms. p. 81), partial Eng. trans. in *BP*, p. 136.
80. Reich to du Teil, 5 Feb. and 9 Feb. 1938; Du Teil reply, 12 Feb. and 17 Feb. 1938.
81. Lapicque to Reich, 7 Feb. 1938 (RA, correspondence box 9, "Lapicque" fld).
82. Reich to Lapicque, 15 Feb. 1938 (RA, correspondence box 9, unlabeled Lapicque fld). Reich published this letter in English in full in his book *The Cancer Biopathy* (New York: Farrar, Straus and Giroux, 1973), p. 20.
83. Reich to du Teil, 16 Feb. 1938.
84. Du Teil to Reich, 10 Mar. 1938.
85. G. Van Den Areud to du Teil, 8 Feb. 1938 (box 10, du Teil flds).
86. Van Den Areud to du Teil, 18 Feb. 1938 (box 10, du Teil flds).
87. Du Teil to Reich, 10 Mar. 1938 (box 10, du Teil flds).
88. Du Teil to Reich, 30 Apr. 1938.
89. See, e.g., Ola Raknes to Allendy, 14 June 1938 (RA, correspondence box 10, du Teil flds).
90. Reich lab ntbk entry, 14 Sept. 1938, bion lab ntbk 3, ms. pp. 152–153.
91. Reich to French Academy, 20 Feb. 1939 (RA, correspondence box 9, unlabeled fld); Eng. trans. of this letter published in full in *BP*, p. 190.
92. Bion lab ntbk 2, 8 Feb. 1937, ms. p. 41.
93. Reich, *Die Bione, op. cit.*, p. xii (*BE*, pp. 7–8).
94. Reich to du Teil, 26 Apr. 1937. Reich emphasized the need for the inclined binocular tubes in *BE*, p. 115; see the user's manual, *Large Universal Microscope "Z": 1937 Pattern* (Vienna: Reichert Optical Works, 1937), 24 pp., posted on the website accompanying this book, http://wilhelmreichbiologist.org/. Note that even in the 1860s and 1870s Lionel Beale used light microscope magnification up to 3000× for studying particles much like Reich's bions. Dallinger and Drysdale saw crucial structures others had not seen in spontaneous generation debates, using magnifications up to 2500×. See Strick, *Sparks, op. cit.*, pp. 47, 121.
95. Reich, *Die Bione*, p. xi (*BE*, p. 7).
96. Ibid. Also idem, *Bion Experiments on the Cancer Problem* (Oslo: SexPol Verlag, 1939), p. 13. A most useful reference for issues of light microscopy is Douglas Murphy, *Fundamentals of Light Microscopy and Electronic Imaging* (New York: Wiley-Liss, 2001).
97. Neergaard to Reich, 21 May 1937 (RA, correspondence box 9, "Neergaard" flds).
98. See, e.g., Stanley Klosevych, "Photomicrography—Resolution and Magnification," *J. Biol. Phot. Assoc.* 32: 133–147 (Nov. 1964).
99. L. C. Dunn to O. L. Mohr, 25 Sept. 1935 (Mohr papers, Inst. of Medical Genetics Archives, Oslo U.); on the Rife microscope, see R. E. Seidel and M. E. Winter, "The New Microscopes," *J. Franklin Inst. (Philadelphia)* 237:

103–130 (1944); the Rife microscope and the pleomorphism controversy are discussed on pp. 119–123.
100. Bion lab ntbk 2, 17 Feb. 1937 (ms. pp. 62–63), entry about assistant Berle (communicated to P. Neergaard in letter of 24 Feb. 1937, from which the quote comes).
101. Neergaard to Reich, 14 Jan. 1938 (RA, correspondence box 9, "Neergaard" flds).
102. Strick, *Sparks, op. cit.*, chap. 5.
103. Thomas Kuhn, *The Structure of Scientific Revolutions*, 2nd ed. (Chicago: U. of Chicago Press, 1970), pp. 62–64. The original publication (from which Kuhn appears to have cribbed the term "paradigm") is J. S. Bruner and Leo Postman, "On the Perception of Incongruity: A Paradigm," *J. Pers.* 18: 206–223 (1949).
104. See Carl Zimmer, "It's Science, but Not Necessarily Right," *New York Times*, 26 June 2011, http://www.nytimes.com/2011/06/26/opinion/sunday/26ideas.html?_r=0.
105. Ilse Ollendorff [Reich's wife and lab assistant in the United States from 1940 onward] to C. Grier Sellers, 21 July 1979; my thanks to Sellers for sharing this letter.
106. "From the Archives of the Orgone Institute: A Laboratory Manual for Bion Experiments" (Rangeley, ME: Wilhelm Reich Infant Trust, 2009). See also the recipes for meat broth and for phosphate bouillon (RA, OI box 7, bion lab ntbk 2, ms. p. 108, 22 Mar. 1937, in Norwegian), "*Kjöttvann* [meat broth] Skjærer kjöttet fri for fett og sener. Maler det og tilsetter 2 liter vann pr. kgr. [2 liters of distilled water per kg of meat, cleaned and trimmed of fat] Kokes 1–2 timer uten trykk. Presses gjennem et osteklede og det eventuelle vannerstattes. Filtreres gjennem fint filtrerpapir til det er helt klart.

Fosfatbuljong efter Kaufmann
1 liter kjöttvann
1% pepton (Park Davis)
0.2% sekundært Fosfat (Na_2HPO_4) nach Sörense
0.3% NaCl
25 cm N/1 NaOH
****

Kokes ca. 20 min.
pH korrigeres til 7.8
Autoklaveres en halv time 120 grader
Kontrollerer pH som nu skal være gått ned til 7.4
Filtreres.
Fordeles i flasker.
Autoklaveres en halv time 120 grader
Kokes ved strömmende vanndamp en halv time den fölgende dag.

107. Reich to du Teil, 17 Mar. 1937, for example. Thus, the entire Reich-du Teil correspondence is posted on the website accompanying this book.
108. The original file is in RA, OI box 7, "Notes by a Laboratory Worker, 1938" (Bion cookbook), p. 21. *BECP*, p. 9.
109. Ibid. Similarly, on p. 30 we have the specific formula (numerous variants exist) used by Reich's lab for Gram-staining microorganisms and bions:

> "*Grams Farbemethode*
> 1) Farben mit Karbolmethylviolet 2 Minuten lang [Stain for 2 mins. with solution of carbolmethylviolet]
> 2) Jod- Jodkalium [Flood slide with iodine-potassium iodide solution for 1 minute]
> 3) Absprulen mit Alkohol bis keine Farbe mehr abgeht [Flood slide with 95 percent ethanol until no color remains visible] *Gegenfarbung* [Counterstain]
> Abspruling mit verdrinutem Karbolfüsein." [Flood slide with carbol fuchsin. The next page has the recipe to mix all the stains from scratch, according to the procedure of Folkehelsen.]

110. Du Teil to Reich, 8 Oct. 1937; in the 1870s see William Roberts, "Studies on Abiogenesis," *Phil. Trans. Roy. Soc. (London)* 164: 457–477 (1874).
111. Reich to du Teil, 15 Oct. 1937.
112. Wilhelm Reich, "The Basic Antithesis of Vegetative Life," inWilhelm Reich, *The Impulsive Character and Other Writings*, ed. B. G. Koopman, ed. (New York: Meridian, 1974), p. 196.
113. Du Teil to Reich, 12 Aug. and 19 Aug. 1937.
114. "Does Life Form under As-Yet Unknown Conditions? In Dr. Reich's Llaboratory in Oslo, Three Experiments Raise a Furor—French Professor Visits Oslo for the Purpose of Collaboration and Control," *Dagbladet*, 20 Aug. 1937. The article states, "The renowned analyst and biologist, Dr. Wilhelm Reich, who has lived in Oslo for many years where he treats patients in character analysis, has been occupied for some time with certain biological experiments that are surrounded by great secrecy. They concern the development of life under as yet unknown conditions, so-called bion experiments.

The purpose of Prof. Du Teil's visit was to personally participate in Dr. Reichs' experiments performed here in this laboratory. For several months, Prof. Du Teil has performed control experiments on the results of Dr. Reichs' biological tests. During his visit in Oslo, he has completed several experimental sequences and could compare his control methods with those of Dr. Reich.

In particular, he had orders to submit a report to the Biologic [sic] Commission, which was appointed by the 'Center for Studies of Human Problems' at a meeting on Aug. 9, relating the results of the experiments that he and Dr. Reich had completed. The commission consists of bacteriologists, biologists, and surgeons from the University of Paris. According to reports we received, the commission showed great interest concerning the experiments that were made in Oslo. In France, laboratories and means have been put at the professor's disposal, so that further experiments can be completed. Likewise, he should submit a report on Sept. 18 addressing further natural philosophic consequences of the Oslo experimentations."

115. "Protokolle der ausgeführten Versuche," RA, OI box 7, and "Partial Eng. trans." fld in same box, which translates all except 2 pp. from 28 July 1937 and 4 pp. (2 in German, 2 of the same experiments, in French) from 6 Aug. 1937. I have posted this document on the website accompanying this book. See also the schedule of activities with du Teil from Mon., 26 July through Sun., 1 Aug. 1937, in bion lab ntbk 2, ms. p. 191.

116. Mohr to Dunn, 9 Sept. 1937 (Dunn papers, Mohr 37–9 file, APS), describes fantastic summer weather in Oslo: "Great sun every day, almost too much of it." Also observes, "Faussen who just returned from a month's stay in Berlin was perfectly horrified at the wild and barbarous spirit down there. It was much worse than anything he had expected. . . . The end result has been much worse than I could dream of. You should hear the Nürnberger [Nazi Party Congress] addresses on the radio, it is simply unbelievable!" Mohr had spoken to a British neurologist interested in genetics who just returned from Moscow. The neurologist told Mohr that it had been impossible to talk to Gershenson, Levit, and others without an "interpreter" being present all the time. "Suspicion everywhere, and all conversations started with cliché declarations about the Marxist science being the only science, etc. etc."

117. Reich to du Teil, 15 May 1937.

118. Ibid.

119. Reich to du Teil, 24 May (RA, correspondence box 10, du Teil flds, some of this letter published in BP, pp. 106–107); *Die Bione*, pp. 59–63 (BE pp. 107–114); Bion lab ntbk 2, ms. pp. 154–165. Perhaps the most detailed, week-by-week account of the bion experiments is, indeed, Reich and du Teils's letters from Jan. 1937 through Sept. 1938. Hence, I have posted the entire correspondence on the website accompanying this book. Some excerpts from this lab notebook material: [ms p. 159] "Incandescence tests with coal on 05.18.1937" [ms p. 164] "4. Coal-incandescence experiment in bouillon and in polyvalent medium, microscopic finding. Cloudy turbidity in all preparations. . . . When the glowing red coal was added to the liquid, it first stayed suspended. The liquid is blackish thick. During about 10–30 minutes the solution clearly

lightened. Individual particles sank, but in place of the thick, blackish turbidity was a cloudy gray-whitish haziness. This phenomenon can only be explained as the release of electrical energy when the connections between the individual coal particles dissolved when heated to incandescence. The freed particles voraciously absorb the liquid and swell as a result. The color of the liquid changes due to this." See also 19 May 1937 (OI box 7, bion lab ntbk 2), ms. p. 154.
120. Reich to du Teil, 24 May 1937.
121. Ibid. Details for the polyvalent culture medium can be found in his lab notebook for 21 May 1937 (bion lab ntbk 2, ms. p. 157): "Polyvalentes Nährboden: 50% Fleischbouillon, 25% KCl 0.1 N, 25% Ringer, 2% glukose." One of Reich's lab assistants, chemist Odd Wenesland, did much of the polyvalent culture media work; see "Research Notebook 1937–38" in RA, OI box 8. See also OI box 7, "Journal, Odd Wenesland" fld, entry of 27 Apr. 1937, reporting on Wenesland conducting Preparation 6 variations, results in his lab notebook.
122. Reich to du Teil, 24 May 1937. Compare with *BE*, p. 108.
123. OI box 7, bion lab ntbk 2, ms. p. 165, "*Versuch mit Kieselsteinen vom 22.5.37*; see also *BE*, p. 114 [Bion lab ntbk 2, ms. p. 165], "Experiments with Pebbles from May 22, 1937.

A pebble (silica or flint [*Kieselstein*]) is grated into fine dust. One sample in potassium chloride, a second heated to incandescence over a flame, then in potassium chloride.

The pebble dust that was not heated to incandescence displays vesicular structures on the rim of the crystals when dry. There is no motion in potassium chloride, not even twitchy Brownian movement. On the other hand, the pebble dust heated to incandescence shows the complete image of a bacteria culture when viewed at a magnification of 250× during dark-field observations. Movement from one position to the next, dancing, twitching movements, and amoeboid vesicle clusters. After a few minutes, it was apparent that the moving individual vesicles are drawn to the larger crystals, and that then the movements gradually stop, obviously discharge. A sample was added to bouillon-potassium chloride."
124. RA, du Teil flds, app. (pp. 11–16 of Eng. trans.) is dated 1 Dec. 1937. From the same original German notebook of du Teil's visit: "Zur Mitnahme für Herrn Professor Roger du Teil, Nizza," Oslo, 6 Aug. 1937 (posted on book website). This report lists, among many others, the following bion preparations they made:
Copy designated for Professor Roger du Teil, Nizza: Protocols of completed experiments from notebook

Vials			
1.	Soil db	in KCl N/10	db: Dry sterilized at 180°C for two hour, boiled in KCl for ½ hour
2.	Coke db	in KCl N/10	db: Dry sterilized at 180°C for two hours, boiled in KCl for ½ hour
3.	Soil c	in KCl N/10	c: Autoclaved non-sterilized soil in KCL at 120°C for ½ hour
4.	Soot e	in bouillon + KCL N/10	e: Annealed (heated to incandescence)
5.	Soot...	in bouillon	e: Annealed (heated to incandescence)
6.	Bions:	6ab XXXIII	ab: Boiled mixed, non-sterilized materials for ½ hour
7.	Bions:	6ac XXXIII	ac: Autoclaved mixed, non-sterilized materials for ½ hour at 120°C
8.	Bions:	6ae XXXV August 2	ae: Mixed non-sterilized materials, autoclaved ½ hour at 120°C
		August 4	Autoclaved again two days later at 120°C for ½ hour

125. Just after 6 Apr. 1938 lab ntbk entry, Krebsmodell 14 (OI box 8, "Präparatherstellung und Protokoll" fld, ms. pp. 62–63, n.d.), "*Präparat 17 Eisenbione* [iron bions]. On 3–16 Aug. 1938, Reich writes about creating iron bions (bion lab ntbk 3, ms. pp. 139–144); more on iron bions in same ntbk, ms. pp. 145–146, 15 Aug. 1938.
126. Reich to du Teil, 24 May 1937 (RA, correspondence box 10, du Teil flds; partial trans. published in *BP*, pp. 106–107).
127. Reich entry in bion lab ntbk 3, 14 May 1938 ("Basic Research III, ms. p. 88, OI box 7): "Beobachtung!!!! *Lava* aus Island zeigts durch gehend bläsige Strukturen. Bei KCl zeigt starke *elektrisch Anziehung*." [Lava from Iceland shows vesicular structures continuously throughout. A strong electrical attraction when KCl is added.] The lava had been brought to him by Norbert Ernst, a former German Trotskyite contact, who had escaped to Iceland in 1933 (Håvard Nilsen interview with Ernst, 2001; my thanks to Nilsen for the information).
128. Reich to du Teil, 24 Aug. 1937 (RA, correspondence box 10, du Teil flds).
129. One is reminded of Robert Brown personally preparing pollen grains for scientists all over Europe to view active molecules in 1828 and 1829, taking along his own high-quality single-lens microscope for the purpose since he knew few of them would possess one of comparable quality. Du Teil planned to return the microscope to Reich (it was only loaned to him); however, the threat of losing his job in June 1938 and the resulting personal financial crisis for him, followed by the Munich crisis and preparations for war in France, threw du Teil's life into sufficient disarray that he never did get the microscope back to Reich. Presumably it was lost in the war, during which Reich permanently lost contact with du Teil.

130. See, e.g., Stephen Nagy, "Rediscovering Dr. Reich's 150× Microscope Objective," unpublished ms., available on the website accompanying this book. In this article, Nagy, an MD with extensive training in optical microscopy, reports in detail confirming Reich's use of a very high-magnification fluorite objective lens manufactured by W. and H. Seibert of Wetzlar, Germany. When tested (I was present and observed this personally), it formed an excellent image with the 160 mm body tube length possessed by Reich's Reichert microscopes. Reich had mistakenly referred to it as an apochromat (fluorites are considered "semiapochromatic"; they transmit near-UV wavelengths in addition to visible light) and had mistakenly stated it was manufactured by Leitz (*BE*, p. 7). On fluorite lenses, see D. B. Murphy, *Fundamentals of Light Microscopy and Electronic Imaging* (New York: Wiley-Liss, 2001), pp. 53–54.
131. Sloan, "Organic Molecules," *op. cit.*; further discussed in James Strick, "Spontaneous Generation," in Moselio Schaechter, ed., *Encyclopedia of Microbiology*, 3rd ed. (New York: Academic Press, 2009), vol. 5, pp. 101–112.
132. See, e.g., E. M. Purcell, "Life at Low Reynolds Number," *Am. J. Phys.* 45: 3–11 (1977). Also Howard Berg, "How Bacteria Swim," *Sci. Amer.* 233 (Aug.): 36–44 (1975).
133. Reich, *BE*, p. 182. Du Teil's full description of the Sy-Clos system is on pp. 181–185.
134. Ibid., pp. 186–187.
135. Ibid., pp. 189–192.
136. Ibid., p. 192.
137. Reich to du Teil, 4 Oct. 1937.
138. Ibid. Reilly appears to be James Reilly (1887–1974). See D. A. Buxton and Robert Laplane, "James Reilly and the Autonomic Nervous System: A Prophet Unheeded?," *Ann. Roy. Coll. Surg. England* 60: 108–116 (1978), http://europepmc.org/articles/PMC2491967/pdf/annrcse01487-0031.pdf.
139. Du Teil to Reich, 8 Oct. 1937.
140. Reich to du Teil, 14 Oct. 1937.
141. Reich to du Teil, 9 Sept. 1937.
142. Reich, *BP*, p. 156.
143. Reich to du Teil, 15 Oct. 1937.
144. Ibid.
145. Reich, OI box 8, "Präparatherstellung und Protokoll" fld, 1 Oct. 1938, ms. pp. 1–6, "Die am 1.10.38 vorhandenen Bion-Kulturen."
146. Reich, *BECP*, p. 10. Photo 18 shows BOPA bions.
147. Ibid., p. 11.
148. End of Jan. 1938, bion präparat 12, Präparat SEKO (OI box 8, "Präparatherstellung und Protokoll" fld, ms. pp. 50–52; see also Bion lab ntbk 3, ms. p. 71. Reich discusses them in Reich to du Teil, 22 Mar. 1938; also in OI box 8,

"Research Preparations 1937–39" fld, ms. p. 12) *"Das Seko-Präparat in Bouillon + KCl."* See photos 19, 20, and 27, Reich, *BECP*, p. 13.
149. OI box 8, "Präparatherstellung und Protokoll" fld, ms. p. 8, 1 Oct. 1938.
150. Reich, *BECP*, photos 26, 27. The mouse injection experiments and the autopsy results on each mouse are documented in OI Box 11, small black notebook "Geschichte der Krebsforschung Nov.1936–Nov. 1938 and in the subsequent fld, "29.X.38–15.XII.39," RA). See also fld "Mice Protocols 1937–1939" in OI box 11.
151. Ibid., p. 13; see photos 21, 22, and 25.

4. An Independent Scientist

1. Reich, *BP*, pp. 65–66, undated but sequence in ntbk indicates May or June 1936.
2. The first indication of Reich thinking that PA and T (or S-bacilli as he called them until 1938) were opposites is in "Eigenschaften des S-bacillus" [Properties of the S-bacillus], n.d. [but ca. late 1937 or early 1938 since he is still calling them "S-bacilli"], ts. p. 10 (RA, mss. box 9).
3. Reich, *BP*, p. 66; original in RA, OI box 5, small black ntbk, under "L" divider.
4. Reich to du Teil, 22 Mar. 1937 (RA, Reich–du Teil correspondence), ms. pp. 3–4, emphasis in original.
5. Wilhelm Reich, "Der Dialectical Materialismus in der Lebensforschung," *ZPPS* 3: 137–148 (Sept. 1937), quotes on p. 138, emphasis in original. This paper is the text of a talk given 1 May 1937 to mark the opening of Reich's laboratory for biological research in Oslo.
6. By 1940 or so, Reich was explicitly avoiding the language of dialectical materialism and using the new term "energetic functionalism" to emphasize the break that he now recognized from his earlier intellectual roots. He left instructions for future editors of a second edition of *Die Bione*, insisting that "dialectical materialism" should be changed throughout to "energetic functionalism." See Reich, *BE*, p. v. On Schaxel's version of dialectical materialist biology, see Nick Hopwood, "Biology between University and Proletariat: The Making of a Red Professor," *Hist. Sci.* 35: 367–424 (1997).
7. Reich, *BE*, pp. 125–126, quote on 125.
8. Much of this background on colloid chemistry is based on Pauline Mazumdar, "The Template Theory of Antibody Formation and the Chemical Synthesis of the Twenties," in P. Mazumdar, ed., *Immunology, 1930–1980* (Toronto: Wall and Thompson, 1989), pp. 12–32; Pauli's work discussed on pp. 17–18. Also, for colloid chemistry and origin-of-life ideas, see John Farley, *The Spontaneous Generation Controversy from Descartes to Oparin*, chap. 8. See also W. Pauli, *Physical Chemistry in the Service of Medicine*, Eng. trans. M. H. Fischer (New York: Wiley, 1907). Work by Pauli more contemporary with the bion experiments includes Pauli, "Proteins as Colloids," in Jerome Alexander, ed., *Colloid Chemistry*, vol. 2 (New York: Chemical Catalog Co., 1928), pp. 223–234; and W.

Pauli and R. Weiss, "Über beziehungen zwischen Kolloiden and konstitutiven anderungen einer Proteine I," *Biochemische Zeitschr.* 233: 381–443 (1931).

9. Mazumdar, *op. cit.*, pp. 17–18. On the Whiggish tradition, see, e.g., Marcel Florkin, "The Dark Age of Biocolloidology," in M. Florkin and E. Stoltz, eds., *Comprehensive Biochemistry*, vol. 30 (Amsterdam: Elsevier, 1972), pp. 279–284; see also Joseph Fruton, "Early Theories of Protein Structure," in P. K. Srinivasan, J. S. Fruton, and J. T. Edsall, eds., *Origins of Modern Biochemistry*, Ann. NY Acad. Med. 325 (8): 53–73 and (9): 1–15 (1979). But for a recent reevaluation, see Neil Morgan, "The Strategy of Biological Research Programmes: Reassessing the 'Dark Age' of Biochemistry," *Ann. Sci.* 47: 139–150 (1990). On the origins and gradual progress of the macromolecule concept, see Robert Olby, *The Path to the Double Helix* (New York: Dover, 1994) and idem, "Structural and Dynamic Explanations in the World of Neglected Dimensions," in T. J. Horder, J. A. Witkowsky, and C. C. Wylie, eds., *A History of Embryology* (Cambridge: Cambridge U. Press, 1986), pp. 275–308.

10. Reich, *BE*, p. 127. For more on lyophilic colloids, see Martin Fischer, "Lyophilic Colloids and Protoplasmic Behavior," in *Colloid Chemistry*, vol. 2, *op. cit.*, pp. 235–254.

11. This was most dramatically clear with carbon or soot bions, where the colloidally turbid solution was densely black at first. These bions are described in *BE*, pp. 107–108, illustrated in figs. 50, 52–53, 56; the turbidity and their striking motility can be seen in the film Reich made of this process, which can be viewed on the website that accompanies this book, http://wilhelmreich biologist.org/. My thanks to Mary Boyd Higgins and the Wilhelm Reich Infant Trust for permission to post films and manuscripts by Reich.

12. *BE*, p. 129.

13. Ibid., p. 132.

14. Pflüger's cyanogen theory was discussed in Chapter 1; Reich also refers to it in *The Cancer Biopathy* (hereafter, *CB*; New York: Orgone Institute Press, 1948), more recent trans. Andrew White (New York: Farrar, Straus and Giroux, 1973), pp. 9, 255. The version of Arrhenius in Reich's library is *Das Werden der Welten*, 7th ed. (Leipzig: Akademische Verlagsgesellschaft, 1921). See also Reich, *BP*, p. 165. An early entry indicating Reich's thinking about the cyanogen theory is from early June 1938 (n.d., bion lab ntbk 3, ms. pp. 120–121) "Verbrennung..."

15. *BE*, p. 133.

16. Ibid., pp. 132–133.

17. Reich, *FO*, *op. cit.*, pp. 257–263, 275–277 (pp. 225–234, 1942 ed.).

18. Reich, *BE*, pp. 134–135.

19. Reich, *FO*, *op. cit.*, pp. 282–286 (1942 ed., pp. 252–255); Reich, "Die Fortpflanzung als Funktion der Sexualität," *Zeitschr. f. Pol. Psych. Sex-Ökon.* 3: 25–31 (1936). An English translation of this paper as "Reproduction as a Function of

Sexuality" is on the website accompanying this book. On Hartmann, see Heng-an Chen, *Die Sexualitätstheorie und "Theoretische Biologie" von Max Hartmann in der ersten Hälfte des zwanzigsten Jahrhunderts* (Stuttgart: Franz Steiner Verlag, 2000), esp. chaps. 7 and 8.

20. Reich, *BE*, p. 157.
21. Ibid., p. 137; emphasis in original.
22. Ibid., pp. 153–154; emphasis in original.
23. Reich, *BISA*, pp. 24–25; Note that in the original 1934 version of the paper, Reich said this was a conclusion taught us by "dialectical observation." In the 1945 revision, published in 1982 in *BISA*, Reich changed this to "functional observation." This same substitution of wording was used repeatedly in this article by Reich in his 1945 revision.
24. Reich, *BE*, p. 154; emphasis in original.
25. Ibid., pp. 157–158.
26. Ibid., p. 158.
27. Ibid., pp. 158–159, quote on p. 159, emphasis in original.
28. Ibid., pp. 160–161, quote on p. 160.
29. Ibid., pp. 150–151.
30. Ibid., p. 151.
31. Ibid., p. 5; the article is in *ZPPS* 4: 137–148 (Sept. 1937).
32. Reich, "Dialektische Materialismus in der Lebensforschung," *op. cit.*, p. 139, emphasis in original.
33. Ibid., pp. 142–143, emphasis in original.
34. Ibid., p. 147.
35. See, e.g., J. B. S. Haldane, "Genetics in Madrid," *Nature* 13: 331 (20 Feb. 1937).
36. Reich, "Dialektische Materialismus in der Lebensforschung," *op. cit.*, p. 148.
37. Reich, *BE*, p. 147, emphasis in original.
38. Bion lab ntbk 2, ms. p. 69, emphasis in original.
39. LeComte DuNoüy, "Tissue Culture *In Vitro*" [1931], in *Between Knowing and Believing* (Paris: Hermann, 1966), pp. 21–38, quote on p. 22. See also idem, *Biological Time* (New York: Macmillan, 1937); DuNoüy was interested in Bergson. For more on DuNoüy, see Hannah Landecker, "Creeping, Drinking, Dying: The Cinematic Portal and the Microscopic World of the Twentieth Century Cell," *Sci. in Context* 24: 381–416 (2011). Reich later expressed the implication of this view for his theory of cancer: "The biopathic function of cancer cannot be understood by a description of the form and stain reaction of the cancer cells or of their position in relation to the cells of healthy tissue. Nor can the chemical composition of living protein, however sophisticated and complex, ever reveal anything about *living pulsation*." Reich, *CB*, *op. cit.* (1973 ed.)], pp. 244–245, italics in original.

40. Again, the reader is invited to view the films of, e.g., soot bions on the website accompanying this book, to judge this claim for herself. Note that embryologist Paul Weiss, who championed the "field theory" in embryology, was also impressed with the importance of the Lewises' films. See Hannah Landecker, "The Lewis Films: Tissue Culture and 'Living Anatomy,' 1919–1940," pp. 117–144 in Jane Maienschein and Marie Glitz, eds., *History of the Carnegie Institute Laboratory of Embryology* (Cambridge, UK: Cambridge U. Press, 2004); also James Strick, "Swimming against the Tide: Adrianus Pijper and the Debate over Bacterial Flagella, 1946–1956," *Isis* 87: 274–305 (1996).

41. Leland Rather, introduction to Virchow, *Cellular Pathology* (New York: Dover, 1971), pp. xxi–xxii. Physiologist Stephen Rothman has also, from the 1960s to the present, critiqued the artifacts in cell biology possible when experiments are not compared with controls run on cells in the living state. See his *Lessons from the Living Cell* (New York: McGraw-Hill, 2002); see also Strick oral history interview with Rothman, 28 July 2006. By about 1950, Shinya Inoué's development of the polarizing light microscope began to make it possible to see fine details of motion in living cells, such as spindle fibers moving during mitosis, that previously could not be seen without killing and staining the cells; see Shinya Inoué, "Microtubule Dynamics in Cell Division: Exploring Living Cells with Polarized Light Microscopy," *Ann. Rev. Cell Dev. Biol.* 24: 1–28 (2008). By 1980 or so, using video instead of the naked eye or film, Inoué, Robert Allen, and others found it was possible to dramatically improve the clarity of images made. Many would now refer to this technology, using electronic processing of images, as "imaging" rather than as simply "light microscopy."

42. Donald Fleming, "Emigré Physicists and the Biological Revolution," in D. Fleming and Bernard Bailyn, eds., *The Intellectual Migration* (Cambridge, MA: Harvard U. Press, 1969), pp. 152–189.

43. See E. O. Wilson, "The Molecular Wars," in his autobiography *Naturalist* (New York: Island Press, 1994).

44. Beverly Placzek, ed., *Record of a Friendship: The Correspondence of Wilhelm Reich and A. S. Neill* (New York: Farrar, Straus and Giroux, 1982), pp. 13–23; Jonathan Croall, *Neill of Summerhill: the Permanent Rebel* (New York: Pantheon, 1983), pp. 169–170ff.

45. Placzek, *Record of a Friendship, op. cit.*, pp. 92–93; Wells was dismissive. In his textbook (with Julian Huxley and G. P. Wells) *The Science of Life* (New York: Literary Guild, 1931), Wells shared Reich's criticism of Bergson's élan vital as mere metaphor, verbiage with which one could not work experimentally (pp. 638–639). However, he uncritically copied T. H. Huxley's Pastorian account of spontaneous generation (complete with swan-necked flasks) as a dead letter, plain and simple (pp. 436–438). Wells supported macromolecular (Svedberg) ideas about life's origin rather than colloid views; he also argues

there for an Oparin-Haldane "hot, thin primordial soup"–type origin of life over many millions of years (pp. 649–653), as described in Haldane's 1929 "Origin of Life" article.
46. Ibid., (Placzek, *Record of a Friendship*).
47. Bernal, *The Social Function of Science* (London: Macmillan, 1939); the preface is dated Sept. 1938, and the book appeared in print January 1939. On dialectical materialism, see esp. pp. 230–231, 237n26, 413–414. J. B. S. Haldane, *The Marxist Philosophy and the Sciences* (New York: Random House, 1939). See also Marcel Prenant, *Biology and Marxism*, with foreword by Joseph Needham (New York: International Pub., 1938).
48. Bernal, *Social Function, op. cit.*, pp. 2–3; Reich, *FO, op. cit.*, p. 23; idem, *CB, op. cit.*, p. 10. On fascist co-opting of "holist" ideas, see Harrington, *Reenchanted Science, op. cit.*; see also Gilbert and Sarkar, "Embracing Complexity," *op. cit.*, pp. 4–5. On H. J. Muller's combination of Communism and interest in eugenics, see his 1935 novel *Out of the Night* (New York: Vanguard Press, 1935). See also Elof A. Carlson, *Genes, Radiation and Society: The Life and Work of H. J. Muller* (Ithaca, NY: Cornell U. Press, 1981), esp. chap. 28; on pp. 262–263, Muller offers this Jan. 1939 opinion of Bernal: "One of the best, if not the best scientific mind in the world. . . . He is primarily a crystallographer, but he is actively interested in protein structure . . . by the Braggs' method of Xray diffraction and other methods."
49. Placzek, *Record of a Friendship, op. cit.*, pp. 21, 23. Otto Lous Mohr was a regular correspondent with Haldane, and both men were linked by the informal Rockefeller Foundation grantee network; it is possible that Haldane, at least, had heard from Mohr the story of Reich's difficulties with the scientific community in Norway. I found no mention of this, however, in the letters with Haldane available in Mohr's correspondence, e.g., J. B. S. Haldane to Mohr, 13 Oct. 1937 (Mohr papers, Inst. of Medical Genetics Archives, Oslo), or in Muller's archive.
50. Neill to Reich, 8 Feb. 1939 (RA, correspondence box 9, "1936–39" flds), published in Placzek, *Record of a Friendship, op. cit.*, pp. 26–27. Haldane's widely read article was "The Origin of Life," *Rationalist Annual* 148: 3–10 (1929), repr. in J. D. Bernal, *The Origin of Life* (New York: World Pub., 1967), pp. 242–249.
51. Original letter: Bernal to Neill, 28 Aug. 1938, (RA, correspondence box 9, "1936–39" flds). According to Croall (*op. cit.*, p. 262), Bernal's wife recalls him looking at the bion preparations under a microscope, then declaring that they were worthless and showed only Brownian movement. Croall did not have access to Bernal's detailed letter in the Reich Archives, since these opened to scholars only in Nov. 2007. Thus, Croall's suggestion that Bernal never took the experiments at all seriously seems entirely based on Eileen Bernal's

recollection. Bernal's original letter from the period is a more reliable assessment of his initial reaction than his wife's recollections many years later.
52. Second Bernal to Neill letter (n.d., ca. late Sept. 1938) (RA, correspondence box 9, "1936–39" flds, filed with Mar.–Apr. 1939 letters by mistake). In this letter, Bernal says he has left a detailed write-up of his opinions about the book with his secretary Paxton Chadwick, but I have been unable to locate that ms.
53. For more discussion of the "synthetic" versus the analytic approach in origin-of-life studies, see Steven Dick and James Strick, *The Living Universe: NASA and the Development of Astrobiology* (New Brunswick, NJ: Rutgers U. Press, 2004), chap. 3. Herrera remained active in the synthetic approach until his death in 1942, publishing "On the Origin of Life" *(op. cit.)* in *Science*. Later, in 1946, Bernal "had visited Princeton and talked to Einstein about the underlying unity, in terms of its biochemical processes, of life on Earth. As a result of their discussion, Sage came to the view that 'life involved another element, logically different from those occurring in physics at that time, by no means a mystical one, but an element of *history*. The phenomena of biology must be . . . contingent on events. In consequence, the unity of life is part of the history of life and, consequently, is involved in its origin'" (Andrew Brown, *J. D. Bernal: The Sage of Science* [New York: Oxford U. Press, 2005], p. 366). Bernal already shows he grasps this historical element and how it makes biology qualitatively different in *The Social Function of Science, op. cit.*, p. 413. In 1947 he suggested clay minerals might have acted as a template for the synthesis of proto-life; see note 61 below.
54. John Keosian, *The Origin of Life* (New York: Reinhold, 1964), pp. 2–3.
55. CB, p. 10 (1973 ed.). Biologist Sol Kramer, taking a course taught by Reich on bions and cancer in July–Aug. 1950, wrote in his notes for 11 Aug., "Find out whether genes are identical with bions (?) If a whole generation of [T. H.] Morgan followers have developed a gene theory, there must be something there which is real; but for the most part our geneticists do not know what they are talking about" (i.e., they are ignorant of bions, bionous origin of amoebae and of cancer cells). It is unclear whether this is a remark Reich made in the course or Kramer's personal reflection on something Reich said. It could represent a guess by Reich, attempting to figure out how to connect his work on bions with whatever was valuable (i.e., not moralism or "race biology") in the modern genetics of 1950 (Kramer papers, U. of Florida, Gainesville). Most of the content of this 1950 course was later published as Chester Raphael and Helen MacDonald, eds., "Orgonomic Diagnosis of the Cancer Biopathy," *Orgone Energy Bulletin* 4 (2): 65–128 (1952).
56. A. I. Oparin, *The Origin of Life* (New York: Dover, 1953). Louis Pasteur's highly selective 1861 account of the history of spontaneous generation ideas is a major contributor to creating the "taint" that made those ideas seem so outdated to

biologists forever after. This account was uncritically repeated by Huxley in his 1870 "Biogenesis and Abiogenesis" Address and copied from one textbook author to another almost verbatim thereafter. (This account completely overlooks, for example, the long and central debate over parasitic worms and dismisses Bastian's experiments out of hand despite strong evidence that his experimental technique was not sloppy, as Huxley and Tyndall imputed. See Strick, *Sparks, op. cit.*; Farley, *Spontaneous Generation Controversy*). Oparin's first chapter is yet another direct copying of Pasteur's account. On pp. 31–32 Oparin cites the 1873 line from Engels later published in his *Dialectics of Nature*: "It would be foolish to try and force nature to accomplish in twenty four hours, with the aid of a bit of stinky water, that which it took her many thousands of years to do."

57. See, e.g., A. I. Oparin, *The Origin of Life on the Earth* (New York: Academic Press, 1957), pp. 88–90 on LeDuc and Bütschli, pp. 90–92 on Herrera.
58. Andrew Brown, *J. D. Bernal: The Sage of Science, op. cit.*, esp. chap. 5. See also Brenda Swann and Francis Aprahamian, eds., *J. D. Bernal: A Life in Science and Politics* (London: Verso, 2000).
59. Robert Olby, *The Path to the Double Helix* (New York: Dover, 1994).
60. Kenneth Holmes, "The Life of a Sage," *Nature* 440: 149–150 (9 Mar. 2006), p. 149. Max Perutz and Rosalind Franklin also studied under Bernal.
61. For Bernal's mature views on the origin-of-life question, see, e.g., his books *The Physical Basis of Life* (London: Routledge and Kegan Paul, 1951) and *The Origin of Life* (New York: World Pub., 1967). The former was first published in *Proc. Phys. Soc. Lond.* 62: 537–558 (Sept. 1949), a Guthrie Lecture to the Physical Society of London, 21 Nov. 1947.
62. Reich, *BE*, p. v.
63. For example, Reich opened his highly regarded *Massenpsychologie des Faschismus* in 1933 by declaring that in the Nazi seizure of power, the German working class had suffered a major defeat. The German Communist Party (KPD) did not acknowledge Hitler's victory as more than a temporary setback and was deeply offended that Reich characterized them as oblivious of how serious a loss they had been dealt.
64. Brown, *Sage of Science, op. cit.*, esp. chap. 15.
65. On Haldane's ambivalence and delayed rejection of Lysenko, see Diane Paul, "Haldane and Lysenkoism in Britain," *J. Hist. Biol.* 16: 1–37 (1983). Not long before Haldane expressed skepticism to Neill about Reich's bions, H. J. Muller asked a correspondent not to share Muller's private opinion with Haldane, saying, "Haldane, especially, must *not* be informed—not now anyway—for I judge from the tone and content of his letters to me that he is at present having his political opinions impressed upon him with a rubber stamp [i.e., by the Moscow Central Committee's "party line" pronouncements] (greatly as I

admire his intellect and person), and could be influenced in the reverse direction from that which I intended. He would think I had gone over to the conservative or Fascist camp, which is the very impression I am trying to disprove." H. J. Muller to Julian Huxley, 9 Mar. 1937 (Muller papers, Lilly Library, Indiana U.). More recently, see Solomon O. Harman, "C. D. Darlington and the British and American Reaction to Lysenko and the Soviet Conception of Science," *J. Hist. Biol.* 36: 309–352 (2003).

66. Holmes, "Life of a Sage," *op. cit.*, p. 149. Holmes can only conclude that Bernal's embrace of Marxist philosophy "seems to have been an act of faith, a substitute for Catholicism." How he would explain the simultaneous embrace of these ideas by so many other great minds of such varying backgrounds (Haldane, Needham, etc.), one can only wonder. After the fall of the Berlin Wall, the Czech "velvet revolution," and other events in 1989 and the fall of the Soviet Union in 1991, a newly energized ahistorical discourse, fanned by neoconservatives, began describing Marxism as so thoroughly discredited a doctrine that it's hard to understand why anybody ever believed in it.

67. See L. J. Rather, *Addison and the White Corpuscles: An Aspect of Nineteenth-Century Biology* (London: Wellcome Trust, 1972), pp. 222–226; also Gustav Wetter, *Dialectical Materialism*, Eng. trans. Peter Heath (New York: Frederick Praeger, 1958), pp. 451–455, and Graham, 1987, *op. cit.*, pp. 83–84, 450n34; Krementsov, *op. cit.*, pp. 211–221, 281. For the full story including Lepeschinskaia's eventual downfall, see Soyfer, *op. cit.*, pp. 213–225; also J. N. Zhinkin and V. P. Mikhaylov, "On 'the New Cell Theory,'" *Science* 128: 182–186 (25 July 1958); and *Q. Rev. Biol.* 35: 51–53 (1960).

68. Graham, 1987, *op. cit.*, pp. 82–84.

69. Despite this extraordinary irony, on many occasions officials of the FBI, INS, and other U.S. government agencies displayed their incomprehension of how one who had a Communist past like Reich could in reality now be a very forceful anti-Communist. Philip Bennett, for example, shows this clearly in "The Persecution of Dr. Wilhelm Reich by the Government of the United States," *Int. Forum Psychoanal.* 19: 51–65 (2010), also http://dx.doi.org/10.1080/08037060903095366. The amateurish, "Keystone Cops" flavor of the FBI's simplistic understanding of Communism at this time is also captured well in Martin Kamen, *Radiant Science, Dark Politics: A Memoir of the Nuclear Age* (Berkeley: U. of California Press, 1985). For a far more nuanced account of distinctions among European left-wing emigrés to the United States, see Alan M. Wald, *The New York Intellectuals: The Rise and Decline of the Anti-Stalinist Left from the 1930s to the 1980s* (Chapel Hill: U. of North Carolina Press, 1987). Many of Reich's supporters and hangers-on after his arrival in America came from this group of Trotskyites. The fortunes of Reich's ideas in New York changed in parallel with the changes in intellectual climate within this group.

70. Werskey, *op. cit.*, p. 105.
71. Ibid., pp. 161–162.
72. Ibid., p. 127.
73. Wilhelm Reich, *The Sexual Struggle of Youth* [Jan. 1932], Eng. trans. (London: Socialist Reproduction, 1972); also idem, *The Sexual Revolution* [1945] (New York; Farrar, Straus and Giroux, 1971).
74. Wilhelm Reich, "Character and Society," *ZPPS* 3: 136–150 (Nov. 1936), Eng. trans. T. P. Wolfe in *IJSO* 1 (Nov. 1942): 247–256, quote on p. 256.
75. Compare this with the disillusionment of Reich's critic Otto Mohr in Dec. 1936, over the beginnings of the arrests and show trials; Mohr to L. C. Dunn, 6 Dec. 1936 (Dunn papers, APS, Mohr 1929–36 file, box 22). Fellow geneticist H. J. Muller had written Mohr, expressing a lack of enthusiasm about conditions in the Soviet Union. Mohr was quite a bit more aware than Muller of reactionary tendencies in the Soviet Union, but not as much so as Reich at this time. Mohr wrote, "On the whole, my own fears were confirmed. I think we have now had illustrations on a huge scale of what dictatorship leads to. Irrespective of the starting point, all dictatorships have outstanding features in common: without freedom of thought and utterance real humanity may not exist. The main achievement of science is that it has removed the fear of the unknown forces of nature. But when freedom is suppressed, this fear is replaced by fear among individuals and among nations. That's where we are standing now. The killing of Zinoviev, Kamenev and the others was a terrible blow to us all who looked to Russia for hopes in the future. Not the least here in Norway where we have Trotski as our guest."
76. Reich's foray into anthropology, much appreciated by Bronislaw Malinowski, who became his friend and colleague, was his book *Der Einbruch der Sexualmoral* [1932], Eng. trans. as *The Invasion of Compulsory Sex-Morality* (New York: Farrar, Straus and Giroux, 1971).
77. Sharaf, *Fury, op. cit.*, p. 66, for example, though this diagnosing of Reich as a zealot on a mission is a theme running throughout the book, as was noted in a review by David Brahinsky, "The Castration of Wilhelm Reich: A Critique of Myron Sharaf's *Fury on Earth*," unpublished ms.; I thank Dr. Brahinsky for sharing this review.
78. Sharaf, *op. cit.*; Harris and Brock, *op. cit.*
79. John Prebble and Bruce Webber, *Wandering in the Gardens of the Mind: Peter Mitchell and the Making of Glynn* (New York: Oxford, 2003); James Lovelock, "The Independent Practice of Science," *New Scientist* (6 Sept. 1979): 714–717. Mitchell came from a wealthy family; Lovelock earned patent royalties for his invention, the electron capture detector in gas chromatography, as well as from consulting work for industry and for NASA (Steve Dick and James Strick, *Living Universe, op. cit.*, chaps. 2–5).

80. Harris and Brock, *op. cit.*; see Anna Freud to Ernest Jones, 27 Apr. 1933 (CPA/F01/30, British Psychoanalytic Society Archives, London); their discussion about Reich continued in Jones to A. Freud, 2 May 1933 (CPA/F01/34, A34), Jones to Max Eitingon, 29 May 1933 (CEC/F01/69); A. Freud to Jones, 8 June 1933 (CPA/F01/35); Reich to Jones, 13 Aug. 1933 (CRA/F09/01); Jones to Freud, 2 Oct. 1933 (CPA/F01/43) re: German IPA branch: "They have absolutely no choice about the Gleichschaltung—it is no longer a debatable question"; [Such comments show that Jones and A. Freud had begun engineering Reich's expulsion from the IPA.] in Jones to Freud, 25 Oct. 1933 (CPA/F01/48); Jones to Freud, 7 Nov. 1933 (CPA/F01/52); Jones to Paul Federn, 29 Nov. 1933 (CFD/F03/04, FED 4); Jones to Freud, 2 Dec. 1933 (CPA/F01/56); Jones to Freud, 9 Dec. 1933 (CPA/F01/06); plans for upcoming Luzerne Congress discussed in Jones to Freud, 13 Dec. 1933 (CPA/F01/57); Freud to Jones, 1 Jan. 1934 (CPA/F01/59); Jones to Freud, 10 Jan. 1934 (CPA/F01/61); Freud to Jones, 28 May 1934 (CPA/F01/91); Freud to Jones, 7 Aug. 1934 (CPA/F02/08); and Jones to Freud, 15 Nov. 1934 (CPA/F02/16).
81. Ernest Jones, *The Life and Work of Sigmund Freud* (New York: Basic Books, 1953); in vol. 3, p. 166, for example, Jones claims Reich's 1932 masochism paper was "a nonsensical amalgamation of Communism and psychoanalysis." Interestingly, there is not a single word about Marxism in Reich's paper. Recall that Jones also actively tried to emasculate Freud's support (even as late as 1939) for Lamarckism. On this, see Eliza Slavet, *op. cit.*
82. Reich, BE, p. 150.
83. Ibid., pp. 147–148, emphasis in original.
84. Ibid., p. 148.
85. Reich, BP, p. 52.
86. Reich, BE, pp. 148–149.
87. Reich diary, 6 May 1936 (RA, Personal box 2, fld "May 34–Feb. 37, ms p. 17), partial Eng. trans. in BP, pp. 64–65.
88. Reich diary, 20 Nov. 1936 (RA, Personal box 2, fld "May 34–Feb. 37, ms pp. 25–26), partial Eng. trans. in BP, p. 76. Reich saw cancer cells in a film by French tissue culture specialist Alexis Carrel and in another film he describes only as "the English film" or the film "by the English delegation." See ms. R-3152 re "Bions and Cancer" (RA, mss. box 9), p. 2: "Cancer cells revealed a strong emergence of vegetative activity, as the Carrell film showed during fast-forward. Its form could be reproduced on film by accelerating the flow of *amoeba* plasma. It is reminiscent in some mystifying way of the vegetative movements and contractions of many a fermentation organism. These contractions grow more intense shortly before the cell divides. The English film showed this with respect to cancer cells as well." Ferdinand Blumenthal (1870–1941) was a prominent German cancer researcher; after the Nazis seized power, he

emigrated and taught at Belgrade University. He wrote *Ergebnisse der experimentellen Krebsforschung* (Leiden: A. W. Sijthoff's Uitgeversmaatschappij, 1934), a book much discussed by Reich in *CB, op. cit.* (see pp. 266 ff).
89. Reich, *BE*, p. 152.
90. See Wilhelm Reich, *Passion of Youth* (New York: Farrar, Straus and Giroux, 1988). Though not brought up on a farm, Erwin Chargaff was also raised in Czernowitz at this time. See his autobiography, *Heraclitean Fire: Sketches from a Life before Nature* (New York: Rockefeller U. Press, 1978).
91. Reich, "The Expressive Language of the Living," first published in *Character Analysis*, 3rd ed. (New York: Orgone Institute Press, 1949) and included in all subsequent editions of that book; also in Reich (M.B. Higgins, ed.), *Selected Writings* (New York: Farrar, Straus and Giroux, 1973), pp. 136–182. One of his physician-students asked him in a training seminar in 1949 why Reich felt compelled to focus on the physical rigidity of his patients, even though his psychoanalytic training called for sitting with his back to them, unable to even see them. Reich replied that the patients' physical immobility and rigidity "produced motor unrest in him," so sharply did it contrast with his own involuntary reactions. CD no. 1, "Method of Approach." Set of six CDs of Reich's training seminars, available from Wilhelm Reich Museum bookstore.
92. Reich, *BISA*, pp. 111–112, 127–128. Also in *FO* (1942 ed.), pp. 330–333; for example (*BISA* p. 129), "In several experiments, the subject was able, on the basis of his or her sensations, to tell what the apparatus in the adjoining room was showing."; also (*FO* p. 330): "We know from vegetotherapeutic clinical experience that sensations of sexual pleasure cannot always be brought about consciously. Similarly, an electro-biological charge cannot be provoked at an erogenous zone merely by pleasurable stimuli. Whether or not an organ responds with excitation to a stimulus, depends entirely on the attitude of the organ. This phenomenon has to be carefully taken into consideration in the course of the experiments" (p. 330). Or, one might add, taken into account in personal relationships.
93. Reich, *BISA*, pp. 121–122. See also *FO* (1942 ed.), *op. cit.*, p. 331; e.g., "*Emotionally blocked and vegetatively rigid individuals*, as for example catatonics, *show no or only very slight reactions*. With them, the biological excitation of the sexual zones lies within the range of that of the rest of the surface of the body. For this reason, the investigation of these electrical oscillation phenomena requires the choice of suitable experimental subjects. Reactions to anxiety in the form of rapid decreases of the surface charge were found at the mucous membranes of vagina and tongue, and at the palm. . . . Just as anxiety and pressure decrease the bio-electrical charge at the sexual zones, so does *annoyance*. In a state of *anxious anticipation* all electrical reactions are decreased; increases in potential cannot be brought about" (emphasis in

original). In another context, Reich said, "Microscopic observation of amoebae subjected to slight electrical stimulation renders the meaning of the term 'emotion' in an unmistakable manner. *Basically, emotion is an expression of plasmatic motion.* Pleasurable stimuli cause an 'emotion' of the protoplasm from the center towards the periphery. Conversely, unpleasurable stimuli cause an 'emotion'—or rather, 'remotion'—from the periphery to the center of the organism." Reich, "The Expressive Language of the Living," in *Character Analysis*, 2nd ed. (New York: Orgone Institute Press, 1945), p. 358, italics in original; pp. 137–138 in more recent version in *Selected Writings, op. cit.*

94. Reich, "Biophysical Functionalism," *op. cit.*, pp. 99–100. In some ways, this approach resembles investigations by Alexander von Humboldt in the immediate wake of the 1791 discovery of galvanism. See Joan Steigerwald, "The Subject as Instrument: Galvanic Experiments, Organic Apparatus and Problems of Calibration," forthcoming in L. Stewart and E. Dyck, eds., *Human Experimentation*. My thanks to Dr. Steigerwald for this ms. and for discussions on this topic at the Summer School on History of Life Sciences in Ischia, July 2013.

95. Leo Raditsa, "Wilhelm Reich's Social Thought," unpublished ms. in Raditsa papers, Harvard U. Archives, pp. 1–10. My thanks to Philip Bennett for bringing this to my attention. See also R. D. Laing, "Why Is Reich Never Mentioned?," *op. cit.*

96. Chester Raphael, foreword to Reich, *Early Writings*, vol. 1, *op. cit.*, p. xii.

97. Reich to du Teil, 4 Oct. 1937.

98. Ibid.

99. Du Teil to Reich, 8 Oct. 1937.

100. See, e.g., 6 Jan. 1935, Friedrich Kraus to Reich (U.S. State Dept. files 1952, J. Greenfield papers): "Many thanks for sending me your work, which has interested me tremendously. If ever you come through Berlin, I would like to discuss it with you." Unfortunately, Kraus died in Berlin 1 Mar. 1936, while Reich was still unable to return to Berlin because of Nazi rule.

101. Du Teil to Reich, 8 Oct. 1937, emphasis in original.

102. Ibid., emphasis in original.

103. Ibid.

104. Reich to du Teil, 15 Oct. 1937, not sent. An edited version of this letter published in Reich, *BP*, pp. 116–120.

105. Ibid.

106. Ibid.

107. Ibid.

108. Ibid.

109. Ibid.

110. Ibid.

111. Ibid.
112. Ibid.
113. Du Teil to Havrevold, 7 Jan. 1938 (RA, correspondence box 10, du Teil flds).
114. Ibid.
115. Ibid., emphasis in original.
116. Ibid.
117. Ibid.
118. Ibid.
119. Ibid.
120. Even Adrianus Pijper, for example, who shared Reich's medical training, and his deep commitment to studying motility and to giving priority to observations made on cells in the living state (in opposition to the emerging hegemony of electron microscopy), was at heart also a mechanist, not a dialectical materialist of Reich's unique variety.

5. Reich's Theory of Cancer

1. See, e.g., 6 Sept. 1934 article in Oslo's newspaper *Aftenposten* (a conservative, even Nazi-friendly newspaper) no. 448, p. 2, "Kreften må bekjempes med kirurgi og radiologi basert på en tidlig diagnose: 'Det er all grunn til å ta avstand fra disse løfter om sera og vaksiner.' Dr. med Leiv Kreyberg uttaler sig til Aftenposten" (Cancer is best treated with surgery and radiation based on early diagnosis), based on an interview with Leiv Kreyberg, Norway's only cancer researcher.
2. R. Kohler, "Rudolf Schoenheimer, Isotopic Tracers, and Biochemistry in the 1930s," *Historical Studies in the Physical Sciences* 8: 257–298 (1977), p. 269. See also E. Newton Harvey, "Biological Effects of Heavy Water," *Biol. Bull.* 66: 91–96 (Apr. 1934). The excitement about "life rays" was probably elevated by Niels Bohr's "Light and Life" speculations; see *Nature* (25 Mar. 1933), pp. 421–423, (1 Apr. 1933) 457–459.
3. Spencer Weart, *Nuclear Fear* (Cambridge, MA: Harvard U. Press, 1988), esp. pp. 36–50. See also Luis Campos, *Radium and the Secret of Life, op. cit.*, on Butler Burke and "radiobes."
4. Wilhelm Reich, "Charakter und Gesellschaft," first published in ZPPS 3: 136–150 (Nov. 1936), Eng. trans. Theodore P. Wolfe, IJSO 1: 247–256 (Nov. 1942), quote on p. 256. [Lebenstrahlen, italics in original].
5. Reich diary entry, 20 Jan. 1936 (Personal box 2, fld "May 34–Feb. 37, ms p. 14; BP, p. 61).
6. See, e.g., Maria Rentetzi, "Gender, Politics, and Radioactive Research in Interwar Vienna," *Isis* 95: 359–393 (2004). Radiochemistry, including an unusually prominent role for women, was also prominent at Oslo University when Reich arrived in Oslo; see Annette Lykknes, Lise Kvittingen, and Anne

Kristine Børresen, "Appreciated Abroad, Depreciated at Home: The Career of a Radiochemist in Norway; Ellen Gleditsch (1879–1968)," *Isis* 95: 576–609 (2004).

7. Lykknes et al., "Appreciated Abroad," *op. cit.*, p. 599.

8. Including the lab of Otto Rahn at Cornell University, discussed above; see, e.g., Rahn and J. B. Tuthill, "The Significance of the 'Induction Effect' in Mitogenetic Radiation," *J. Bact.* 25 (1933): 27; Rahn and Jean Ferguson, "Mitogenetic Radiation as the Cause of the Influence of the Number of Organisms upon the Lag Period," *J. Bact.* 25 (1933): 28; Rahn and Margaret N. Barnes, "On the Lethal Radiation from the Human Body," *J. Bact.* 25 (1933): 28–29. This last work, on mitogenetic radiation (MR) from menstrual blood, caused Rahn considerable controversy, mocking, and sensationalism in the press. He was forced to submit further work on the subject to approval by a supervisor. Nonetheless, Rahn was invited in 1934 to speak on MR at Cold Spring Harbor by Charles Davenport and at the International Congress of Radiobiology at Venice. He discussed MR at length in his 1936 book *Invisible Radiations of Organisms, op. cit.* and on pp. 53–54 of Rahn's unpublished 1952 autobiography (ts., American Society for Microbiology Archives).

See review article by Alexander Hollaender and E. Schoefl, "Mitogenetic Rays," *Q. Rev. Biol.*, 6: 215–222 (1931). (Note remark on p. 217: "The yeast technique has been tried with apparent success by the authors of the present article. Experiments were made using onion sole, muscle tissue, and cancer tissue as senders.") Gurwitsch stated, "Simultaneously and independently of us, the [MR] of malignant tumors was discovered by Siebert, and Reiter and Gabor." See A. Gurwitsch, *Die Mitogenetische Strahlung* (Berlin: Julius Springer, 1932). The 1928 discovery by Reiter and Gabor that tumor tissue emanates MR was also discussed in Hollaender and Schoefl's 1931 paper in *Q. Rev. Biol.*: "In the main their work substantiates that of Gurwitsch. . . . Like [Gurwitsch] Reiter and Gabor come to the conclusion that a radiation occurs, but they believe its effect is to retard mitosis with the result that more cells are to be found in the mitotic stage at any one time than would be found in this state under ordinary conditions. . . . [Also states:] Malignant cancerous tissue . . . is rich both in mitotic cells and in mitogenetic rays" (p. 218). J. G. Crowther's *The Progress of Science* (London: Kegan Paul,1934) carries an extensive section on MR and how new, more sensitive instruments and techniques are verifying what some with traditional techniques could not at first verify about Gurwitsch's claims (pp. 74–80). Crowther cites Hopkins's 1932 Royal Society address in support of MR. Hollaender had begun to have doubts about MR by Sept. 1935; by 1960 he considered it a completely dead issue, saying, however, that "one would like very much to see its reality proved, even though it fits into a pattern of thinking [field theories, biological rays] that was very prominent twenty-five years ago." See "Review of Gurwitsch's *Die*

Mitogenetische Strahlung," Q. Rev. Biol. 35: 246–247 (1960), p. 246. My thanks to Dr. Daniel Friedman for help with background on MR.

9. Reich, undated lab ntbk entry, ca. May or early June 1936, *BP*, pp. 65–66.
10. Reich, 15 June 1936 entry, OI box 5, small black ntbk under letter divider "L," published in *BP*, p. 66–67.
11. Bion lab ntbk 2, ms. p. 98a, May 13, 1937, "Mauser Krebs vom 13.V. zeigt ein deutige Pakettamöben in der Krebszellflüssigkeit!" (Mouse cancer from 13 May [*sic*; Mar.? This date seems more likely based on the surrounding dates in ntbk] reveals a clear packet amoeba in the cancer cell fluid!).
12. Leland J. Rather, *The Genesis of Cancer: A Study in the History of Ideas* (Baltimore: Johns Hopkins U. Press, 1978), pp. 67–68, 223–224n165. As Rather (pp. 82–83) makes clear, these "histological molecules" were not the same as earlier (and much larger) "globules" reported by, for example, Milne-Edwards as a result of the poor quality of his lenses. Bennett, like Schwann, saw the histological molecules with high-quality achromatic lenses. Reich observed his bions with both achromatic and even higher-quality apochromatic lenses. See also Rather, *Addison and the White Corpuscles, op. cit.*
13. Strick, *Sparks, op. cit.*, chap. 2.
14. Ibid., pp. 47, 133, 197.
15. On Warburg's theory, see O. Warburg, K. Posener, and E. Negelein, "Über den Stoffwechsel der Tumoren," *Biochemische Zeitschrift* 152: 319–344 (1924), repr. in Eng. in Warburg, *On the Metabolism of Tumors* (Constable: London, 1930). Warburg was still publicly promoting his theory in the 1950s and '60s. See, e.g., O. Warburg, "On the Origin of Cancer Cells," *Science* 123: 309–314 (1956), doi:10.1126/science.123.3191.309; also Hans Krebs, *Otto Warburg: Cell Physiologist, Biochemist and Eccentric* (Oxford: Clarendon Press, 1981), pp. 18–26; and http://en.wikipedia.org/wiki/Warburg_hypothesis. See also R. E. Kohler, "The Background to Otto Warburg's Concept of the *Atmungsferment*," *J. Hist. Biol.* 6: 171–192 (1972). For more recent evaluations of Warburg's theory, see S. Weinhouse, "The Warburg Hypothesis Fifty Years Later," *J. Cancer Res. and Clin. Oncol.* 87 (2): 115–126 (1976), doi:10.1007/BF00284370; and K. Garber, "Energy Boost: The Warburg Effect Returns in a New Theory of Cancer," *J. Natl Cancer Inst.* 96 (24): 1805–1806 (2004), doi:10.1093/jnci/96.24.1805. On the fascinating recent work on the methods behind Warburg's research, see Mathias Grote, "Vintage Physiology: Otto Warburg's 'Laborkochbücher' und Apparaturen," *NTM Zeitschrift für Geschichte der Wissenschaften, Technik, und Medizin* 21 (2): 171–185 (2013).
16. Reich, *CB, op. cit.*, pp. 239–240.
17. See Reich, "obituary of J. Neergaard" in *ZPPS* 4 (2): 65 (1937); see also Ellen Siersted, *Wilhelm Reich in Denmark* (Eng. trans. Karina Nilsson) in *Pulse of the Planet* 4: 44–69 (1993), pp. 45, 61.

18. Reich, *BECP, op. cit.*, p. 6; also idem, "Experimental Orgone Therapy of the Cancer Biopathy (1937–1943)," *IJSO* 2: 1–92 (1943), p. 24; Reich, *CB*, pp. 239–240.
19. Jørgen Neergaard (1910–2/2/1937) staph and strep experiments summarized as of 4 Jan. 1937 (or shortly after, RA, bion lab ntbk 1, ms. pp. 179–180). These notes appear to be in Danish, perhaps by Paul Neergaard (1907–1987), Jørgen's brother and a plant pathologist (letterhead says "Neergaard & Co., Oslo"). The notes document the results of bion injection experiments on mice from 22 Sept. 1936 through 4 Jan. 1937. "Injections tried: preparations 1–6, also Staphylococci from Jørgen." On the cancer conference, see Reich diary, 20 Nov. 1936 (Personal box 2, fld "May 34–Feb. 37," ms p. 25–26; *BP*, p. 76).
20. Reich, *CB*, pp. 156–158. Originally published in Reich, "The Carcinomatous Shrinking Biopathy," *IJSO* 1: 131–155 (1942), pp. 133–135.
21. Ibid.
22. Ibid., pp. 159–160. Original *IJSO*, p. 136.
23. Ibid., p. 160. Reich's understanding of the cancer cell, like his work with bions in general, bears some striking similarities to nineteenth-century pathologist John Hughes Bennett's observations and concepts of cancer cells. Bennett's description of the formation of cancer cells from histological "molecules" and their disintegration into such molecules is particularly reminiscent of Reich's descriptions involving bions—though Bennett did not have Reich's concept of "biopathy." Like Reich, Bennett also emphasized that cancer cells often displayed a characteristic caudate or spindle shape. Pathologist Leland Rather remarked about Bennett's technical skill that, notwithstanding the unavailability in the late 1840s of "aniline dyes and the rotary microtome (which were introduced in the third quarter of the nineteenth century), Bennett's punctilious procedure could well pass muster today." (Rather, *Cancer, op. cit.*, pp. 111–112, quote on p. 111; also p. 207n132). Given his expertise then, it is interesting that Bennett, like Reich 90 years later, came to the conclusion that cancer cells formed, as Schwann said all cells did, from an originally formless blastema. And he "doubted that the 'true cancer cell' was ever 'formed by the transformation of a previously existing one,'" as Virchow maintained (Ibid, p.114). This controversy was not just about experiment. In Bennett's time, as in Reich's, it was highly ideologically charged. Virchow insisted that such equivocal or spontaneous generation of cells "particularly when it is conceived of as self-arousal [i.e., self-active matter] is downright heresy or devil's work, and when we, of all people, defend not only the inheritance of generations on the whole but the legitimate succession of cell forms as well, this is truly trustworthy testimony." Rudolph Virchow, *Cellular Pathologie* [1858, p. 23], quoted in Rather, *Cancer, op. cit.*, p. 212n23. Feminist scholars or anthropolo-

gists interested in patrilineal descent might find Virchow's remarks thought provoking.
24. *BE*, p. 119.
25. Reich, *CB, op. cit.*, pp. 215–218.
26. Reich, *BECP*, p. 6.
27. Ibid., p. 8 (see, e.g., photo 11).
28. Ibid., p. 6.
29. First mention is Mon., 16 Aug. 1937 ("Research Notebook 1937–38," OI box 8), "Vorbereitungen für Ca-Versuche"; also Reich to du Teil, 24 Aug. 1937; also see 9 Sept. 1937, Reich entry in bion lab ntbk 3 ("Basic Research III," OI box 7, ms. pp. 34–42), which describes S-bacilli.
30. Reich to du Teil, 19 Apr. 1938 (*BP*, pp. 144–146; original in RA, correspondence box 10, du Teil flds).
31. Ibid.
32. Ibid. Reich gives more detail on observation methods with T-bacilli, including specifics of optics he uses, in Reich to Gonzalo Codina, "Farmacia Popular," Mexico City, 17 Jan. 1940 (RA, correspondence box 9, Codina flds). "I am sending you a t-bacilli gram preparation in the enclosure. You can observe it nicely at a magnification of 1800×, the magnification at which it was photographed. I request that you respect the following characteristics of t-bacilli:

> 1) They are not visible in a light field at a magnification of 3–400×. Whatever you observe at this magnification, no matter how small, are not t-bacilli.
>
> 2) One can just barely see the t-bacilli as the smallest visible points of light in a dark-field at this same magnification (3–400×), and that with great difficulty. *Clearly* visible circles of light or streaks are not t-bacilli. They are easier to differentiate in a strong, when possible pure culture application rather than when they lie scattered around amongst larger structures such as staphylococci or streptococci. They are much smaller than streptococci.
>
> 3) They are clearly visible in gram preparations and can be photographed at different sizes, some crooked, some straight, lancet-like structures, at a magnification of 1800×. Larger rods such as those about the width of a staphylococcus are usually two closely positioned t-bacilli that are obviously in the process of splitting.
>
> 4) The t-bacilli are generated through degenerative growth of large rods in specific proteins and from streptococci. The smaller the t-bacilli, the more damaging they are to mice. The most harmful are those forms that barely have a visible size of 1 mm when viewed at 3600×

magnification. They can be immediately observed and verified through gram staining in fresh cancer preparations (magnification not less than 3000×). . . . The most important objectives that I use are:

Reichert 80× apochromatic lens, numerical aperture 1.10 for immersion in water,

Reichert 100× apochromatic lens, numerical aperture 1.10 for immersion in oil.

I examine t-bacilli in gram-staining with a 16x binocular, slanted lens barrel, which magnifies 50% more, and a 150× apochromatic lens object by Leitz [sic; see chapter 3 n130], which I special-ordered. I do not know the numerical aperture of the 150× object lens. These last configurations produce a superbly functional magnification of 3600×.

I have heard the objections of biologists and cancer researchers that examining anything at a magnification of more than 1–1500× makes no sense because no further structures can be resolved. This objection does not apply here because:

1. Examination of t-bacilli requires no better structure clarity, rather the ability to make structures visible that one could not otherwise see.
2. A magnification over 2000× allows one to see the finest movements in living preparations, which are not observable at a 1000× magnification.

33. Reich to du Teil, 15 Oct. 1937. Reich's guess about the bacterial classification is probably based upon the blue-green pigment characteristic of laboratory cultures of *P. aeruginosa* [*B. pyoceaneus*] and the color of his S-bacillus cultures. "This blue-green pigment is a combination of two metabolites of *P. aeruginosa*, pyocyanin (blue) and pyoverdine (green), which impart the blue-green characteristic color of cultures." This identification is unlikely (and Reich drops the idea) because *P. aeruginosa* cells are larger than the 0.2–0.6 μm rods Reich had obtained from sarcoma. It does stain gram negative, like the S-bacillus, and "Pyocyanin is a virulence factor of the bacteria and has been known to cause death in *C. elegans* by oxidative stress." This is interesting given Reich's attention to tissue suffocation as a hypothetical cause of cancer. For quotes, see http://en.wikipedia.org/wiki/Pseudomonas_aeruginosa. See also http://jem.rupress.org/content/4/5–6/627.abstract.

34. Reich, *BECP*, pp. 16–17. Reich later said that the t-bacilli could not be seen "in a phase [i.e., phase-contrast] microscope but only under darkfield at relatively low magnification, they appear as white, flittering points. With much higher magnification, you can see them with regular light [i.e., brightfield]." See Myron Sharaf, "Historical Notes," ca. 5 June 1950 (Reich archives).

35. Reich to E. Hval, 22 Dec. 1938 (RA, correspondence box 9, "1936–39" flds); Reich to Bon, 3 Dec. 1938 (RA, correspondence box 9, Bon flds), for example.
36. W. J. Kat to Reich, 23 Dec. 1938 and Reich reply, 30 Dec. RA, correspondence box 9, Bon flds); Homer Kesten, MD, Columbia U. College of Physicians and Surgeons, to Reich, 5 Jan. 1939 (RA, correspondence box 9, "1936–39" flds).
37. Wilhelm Reich, "Experimental Orgone Therapy of the Cancer Biopathy (1937–1943)," *IJSO* 2: 1–92 (1943): 25; included in *CB* (1973 FSG ed.), p. 244.
38. In a laboratory in Lancaster County, PA, microscopist Steven Dunlap is currently at work on just such a video setup, attempting to document the "natural organization of protozoa," as Reich called it, in grass and moss infusions. My thanks to Dunlap for showing me his equipment and for sharing some films he has made so far. With new halogen microscope lighting, Dunlap has found, the heat given off is sufficiently intense that only ciliates, no amoebae, will grow in infusions unless a heat-blocking filter is used.
39. Reich, *BECP*, p. 8.
40. Reich, *BP*, p. 113. Reich later observed that t-bacilli were produced simultaneously with PA bions in the incandescent carbon experiment, suggesting an alternate (or additional) explanation of why this preparation might induce tumor formation when injected into mice. Also see 9 Sept. 1937, Reich entry in bion lab ntbk 3 ("Basic Research III," OI box 7, ms. p. 37) Here Reich is beginning to guess at the role of excess carbon in blood bion experiments, by p. 40, the role of CO_2, on p. 41, cancer.
41. Reich, *BP*, p. 115.
42. Reich, "Erste Beobachtungen über den S-bacillus an Präparaten von Krebskranken" [Initial observations concerning S-bacilli in preparations from cancer-stricken patients] reports begun on 18 Oct. 1937, RA, mss. box 9, p. 4.
43. Ms. R-3152 re: "Bions and cancer" (RA, mss. box 9), p. 5a. Reich describes beginning to use sterile technique to examine fresh animal tissues for bions in Reich to du Teil, 26 Apr. 1937.
44. Ms. R-3152 re: "Bions and cancer" (RA, mss. box 9), p. 5a.
45. Ibid., pp. 5a–5b.
46. Ibid., pp. 8–9.
47. Ibid., pp. 10–11.
48. Ibid., p. 11.
49. See biographical entry for Kreyberg (1896–1984) in the *Norsk Biografisk Leksikon*, http://translate.google.com/translate?hl=en&sl=no&u=http://www.snl.no/.nbl_biografi/Leiv_Kreyberg/utdypning&ei=W2EBStfeDZLKMsXEqeYH&sa=X&oi=translate&resnum=1&ct=result&prev=/search%3Fq%3Dnorsk%2Bbiografisk%2Bleiv%2Bkreyberg%26hl%3Den%26client%3Dsafari%26rls%3Den. According to autobiographical materials, Kreyberg in the 1930s had "sought out contact with the group in the Labor Party called the Socialist

Physicians Association, as a sympathizer (but not a member). [Psychiatrist] Johann Scharffenberg did the same. Not long before the war, Scharffenberg joined as a full member. So I said to him: now there's only one 'sympathizer' i.e., fence-sitter, left" (namely, himself, Kreyberg). See ("Bakgrunn" section, "Krigserindringer [War Memories] (1940–1942)" fld, Kreyberg papers, PA #276, Staatsarkiv, Oslo). Here Kreyberg also notes that "Mussolini and Hitler came to power and in the summer of 1936 began the Spanish Civil War, which made a tremendous impression on me. I was then traveling to England and America to my first invited lecture at a foreign university [U. of Wisconsin, Madison]. I took along as travel reading Sinclair Lewis' book, *It Can't Happen Here*." See also Kreyberg to L. C. Dunn, 28 May 1936 (APS, Leslie C. Dunn papers B: D917, box 16). This and letter of 2 Aug. 1936 discuss plans for Kreyberg's visit to the Aug. 1936 (Genetics?) Cancer Congress at Madison, WI, on a Rockefeller Foundation fellowship. Kreyberg also visited Jackson Labs at Bar Harbor, ME—site of breeding experimental cancer mice—on 28 Aug. Kreyberg's papers given to the Staatsarkiv also include a pamphlet, "How and Why?" ["Hvorhen og Hvorfor?"] , written for his children before he left to go to war. It describes a bit of his personal, Christian philosophy, illustrated with lots of Viking motifs. The tone is rather sentimental, but his use of the behavior of an amoeba as a primitive model for behavior is quite interesting. This was very common among biologists at the time; however, his discussion here appears as though some parts of it could have been cribbed from Reich. Interestingly, Kreyberg displays himself in this pamphlet as the model of Reich's view that the mechanistic always must become mystical to "fill the gap" mechanistic science leaves, e.g., between living and nonliving matter.

50. For a good summary of what was known at the time about cancer in tar-painted mice, see Gustav Guldberg, *Experimental Researches on Precancerous Changes in the Skin and Skin Cancer* (Levin and Munksgaard: Copenhagen, 1931). The book opens, pp. 12–27, with a historical survey of work on tar-painted mice, and chap. 1 is an extensive description of the methods and findings to date (1931). Kreyberg charged Guldberg with plagiarizing his work in this book (see chapter 7, n8).

51. Reich to N. Hoel, Havrevold, 29 Mar. 1938 (RA, correspondence box 9, "1936–39" flds).

52. *Tidens Tegn* article "Gåten om livets opståen løst i Oslo?: Sløret trekkes fra dr. Wilhelm Reichs mystiske eksperimenter på Bergsløkken" [Has the genesis of life enigma been solved in Oslo? The veil is drawn from Dr. Wilhelm Reich's mysterious experiments at Bergsløkken], 21 Sept. 1937. A second subtitle was a quote from Reich: "—I do not create life, but reveal experimentally how the life process is originated." The reaction of Oslo University academics was prompted by the release of Reich's article "Dialectical Materialism in Bio-

logical Research," in ZPPS 4 (3): 137–148, published on 18 Sept. 1937, the first published report of any information about of his bion experiments.
53. Reich to du Teil, 23 Sept. 1937; most of this letter published in BP, pp. 111–113.
54. Reich to N. Hoel, Havrevold, 29 Mar. 1938 (RA, correspondence box 9, "1936–39" fld).
55. 25 Sept.–16 Oct. 1937, Blood drawn from 10 of Reich's colleagues, to produce bions: Tabelle der 10 Bione (OI box 7, "Basic Research III" ntbk, ms. p. 59), gives the age of each person plus details about the bions produced. See some notes on these bions, 5 Oct. 1937 in "Electrical Examination of Human Bions" (ms. p. 13) and on other pages in 30 Oct. 1937 Bion Culture Journal ms., entries followed by table (OI box 8, "Kultur Journal a" fld, ms. pp. 6–15, ms. pp. 22–24). Reich gave only a brief mention of the discovery of blood bions in BE, p. 119, since the discovery was made immediately before the book went off to press. By 1 Dec. 1937 he had prepared a formal protocol for preparation and culturing of human blood bions (OI box 8, Präparatherstellung und Protokoll" fld, "Vorlaufige Mitteilung uber die Herstellung von sterilen Menschenblutt-Bionen"). Most of Reich's observations and laboratory experiments on bions and cancer are recorded in three chronologically sequential lab notebooks in RA, Orgone Institute box 11, the first of which is a black notebook titled "Geschichte der Krebsforschung Nov. 1936–Nov. 1938.
56. Reich diary entry, 10 Oct. 1937 (Beyond, p. 115), emphasis in original.
57. Du Teil to Reich, 8 Oct. 1937.
58. Reich to du Teil, 14 Oct. 1937.
59. Kreyberg (from Radium Hospital) to Reich, 23 Sept. 1937 (RA, correspondence box 9, "1936–39" flds), "Jeg har gjort en almindelig bakterieenfarvning av den kultur på eggmedium hvorav jeg fikk et utstrykspreparat under mit besøk i forgårs. Billedet viser et utseende i full overensstemmelse med grampositive staphylo-coccer, og jeg vil ikke legge skjul på at jeg anser den opvekst som en banal medie-infeksjon."
60. Reich to Kreyberg, Tues., 5 Oct. 1937 (RA, correspondence box 9, "1936–39" flds).
61. Reich to du Teil, 13 Oct. 1937.
62. Reich later described Kreyberg's claim to have "controlled" his bion experiments as follows: "I never gave him any occasion to see whether I know anything about bacteriology or not, since I refused to be examined by him as he intended to do. It is a typical pest [i.e., Reich's concept of "emotional plague"] reaction since what actually happened is this: I showed Kreyberg in my laboratory cancer cells under 3000× magnification and also coal bions. He did not recognize any of these preparations. According to the pestilent function of projection, his ignorance appeared [to him] as my ignorance." Reich to Ola Raknes, 3 Mar. 1950 (Raknes papers, ms. 4° 2956: 8, National Library, Oslo).

63. Reich to du Teil, 8 Oct. 1937.
64. Reich to du Teil, 13 Oct. 1937, partial trans. in *BP*, pp. 119 (dated 15 Oct., but Reich told du Teil in his letter of 14 Oct. that it was written on 13 Oct.).
65. Mohr to L. C. Dunn, 5 Dec. 1937 (Dunn papers B:D917, APS, Mohr 1937–1939 file): "The last things from my period as a Dean. Of these I may mention the great joy that Kreyberg has now by the faculty been unanimously elected Harlitz's successor as Prof. of Pathological Anatomy and general pathology. . . . I don't doubt that this was the right result from every point of view, and am very glad to get this new blood into the faculty."
66. Reich to Kreyberg, 12 Oct. 1938 (RA, correspondence box 9, "1936–39" flds). The day after their meeting, 9 October, Reich recorded his account of what occurred in his diary (Personal box 2, fld "May 34-Feb. 37, ms p. 63) as well as in a formal "Protocol" document ("Protokoll 9.10.37," in black notebook "Geschichte der Krebsforschung Nov. 1936–Nov. 1938"OI box 11, RA). Kreyberg did not think himself stuck in an outdated, purely morphological approach to pathology. But for him the alternative to pure morphology was a genetic and (what would later be called) "molecular" approach. See Kreyberg to L. C. Dunn, 20 Feb. 1938, (Dunn papers B: D917, APS).
67. Reich, "Annahme zur Lösung der Granulationswucherungen zum Unterschied von Krebswucherungen" [Hypothesis addressing the question of how granulation tumors differ from cancer tumors], dated 6 Apr. 1938 (RA, ms. box 9). How widely or narrowly to cast the net in diagnosing a given tissue as "carcinoma" or "cancerous" is still a subject of dispute today among physicians and pathologists; see http://well.blogs.nytimes.com/2013/07/29/report-suggests-sweeping-changes-to-cancer-detection-and-treatment/?nl=todaysheadlines&emc=edit_th_20130730&_r=0.
68. Reich to du Teil, 14 Oct. 1937.
69. Kerrigan interview with Kreyberg (U.S. State Department investigation of Reich, Jerome Greenfield papers, New York U.).
70. Reich to Kreyberg, 12 Oct. 1938 (RA, correspondence box 9, "1936–39" flds).
71. Ibid.
72. S. A. Heyerdahl, MD (director of Radium Hospital) to Havrevold, 11 Oct. 1937 and Kreyberg to Reich, 15 Oct. 1937 (RA, correspondence box 9, "1936–39" flds).
73. Reich to du Teil, 15 Oct. 1937 (RA, correspondence box 10, "du Teil" flds).
74. Leland Rather, introduction to Virchow, *Cellular Pathology* (New York: Dover, 1971), pp. xxi–xxii. Rather, a practicing pathologist, remarks that the problem has only gotten worse since the 1930s in his field. Chester Raphael, MD, a student of Reich's, developed this point in more depth in a talk, "The Autonomic Nervous System from the Point of View of Reich's Functional Approach," July 1986 at Wilhelm Reich Museum. Transcript of talk available at the website accompanying this book. In the past thirty years, of course,

traditional pathologists have been contested with for ownership of the disease diagnosis, by molecular geneticists asserting that diseases are molecular pathology. See, e.g., Nicole C. Nelson, Peter Keating, and Alberto Cambrosio, "On Being 'Actionable': Clinical Sequencing and the Emerging Contours of a Regime of Genomic Medicine in Oncology," *New Genetics and Society* 32: 4, 405–428 (2013), DOI: 10.1080/14636778.2013.852010.

75. Reich to N. Hoel, Havrevold, 29 Mar. 1938 (RA, correspondence box 9, "1936–39" flds).
76. Kerrigan interview with Nic Waal (U.S. State Department investigation of Reich, Jerome Greenfield papers, New York U.), pp. 50–51, italics mine.
77. See, e.g., Reich to N. Hoel, 2 July 1936, (RA, correspondence box 9, "1936–39" flds; an edited version is in *BP*, pp. 67–69).
78. Nic Hoel to Ellen Siersted, June 1938, in Siersted, *Wilhelm Reich in Denmark*, *op. cit.*, p. 65.
79. Reich, *CB*, *op. cit.*, pp. 225.
80. Wilhelm Reich, "Experimental Orgone Therapy of the Cancer Biopathy (1937–1943)," *IJSO* 2: 1–92 (1943), pp. 16–17; *CB*, pp. 229–230.
81. Ibid.
82. Reich, *CB*, *op. cit.*, pp. 229–230, italics in original.
83. Chester Raphael, "Reich's Concept of the Cancer Biopathy," lecture at Wilhelm Reich Museum, 2 Aug. 1989, recording available from Reich Museum.
84. Ibid.
85. Christopher Lawrence and George Weisz, eds., *Greater Than the Parts: Holism in Biomedicine, 1920–1950* (New York: Oxford U. Press, 1998). See also Anne Harrington, *Reenchanted Science: Holism in German Culture from Wilhelm II to Hitler* (Princeton, NJ: Princeton U. Press, 1996).
86. Chester Raphael, "Reich's Concept of the Cancer Biopathy," lecture, *op. cit.* See also idem., "On the Air Germ Dogma," *Orgonomic Medicine* 1 (2): 159–161 (1956), in which Raphael critiques Rene Dubos's article "Second Thoughts on the Germ Theory," *Sci. Amer.* (May 1955). Dubos favors "disposition to disease," an ecological view of the "seed"/"soil" relationship rather than claiming that "germ = disease," but Raphael says he leaves the nature of the patient's "constitution" still vague, or perhaps confuses secondary chemical manifestations for primary energetic causes.
87. Raphael, "Reich's Concept of the Cancer Biopathy," lecture, *op. cit.*
88. Susan Sontag, *Illness as Metaphor* (New York: Farrar, Straus and Giroux, 1978).
89. See, e.g., Abigail Zuger, "A Fight for Life Consumes Both Mother and Son," *New York Times*, 29 Jan. 2008, http://www.nytimes.com/2008/01/29/health/29book.html?ex=1202274000&en=7f34dbd9c166d4f8&ei=5070&emc=eta1&_r=0: "Three decades of having cancer, being treated for cancer or waiting for

cancer to recur might bring out the inner philosopher in some. In Ms. Sontag, an inner adolescent seems to have emerged instead, with each battle and victory strengthening her determined appetite for life and her conviction that she was immortal."
90. Reich, *CB, op. cit.*, p. 319.
91. Reich to du Teil, 18 Dec. 1937, partial trans. published in *BP*, pp. 130–132, quotes from pp. 130–131.
92. See Lykknes et al., "Appreciated Abroad," *op. cit.*, p. 589n37.
93. Reich to du Teil, 18 Dec. 1937, *BP*, p. 131.
94. Croall, *op. cit.*, p. 263. Croall mentions this comment from Reich to justify his claim that "Reich was indeed extremely intolerant of anyone else's ability to judge his work, and quite convinced that his ideas would be unacceptable." However, one should note that this interpretation of the quote makes sense to Croall because he is certain that Reich's science was bogus—or at least wrong. Since Neill (Croall's subject) agrees with Reich, Croall feels compelled to conclude that Neill—whom he lauds as a brilliant iconoclast in education and in many other realms—must in this one instance have abandoned his critical judgment and become a "disciple" of Reich.
95. Reich to du Teil, 18 Dec. 1937, *BP*, p. 131. See also *Radium: Production, General Properties, Therapeutic Applications, Apparatus*, 346 pp., Union minière du Haut Katanga, Radium Department, http://books.google.com/books/about/Radium_production_general_properties_the.html?id=b9FsAAAAMAAJ.
96. Reich to du Teil, 18 Dec. 1937, *BP*, p. 131.
97. Ibid., p. 132.
98. See, e.g., Reich's repeated attempts to secure a patent on the orgone energy accumulator under a nonprofit medical research entity, with the express purpose of preventing it from being patented by someone else for profit. Reich, *American Odyssey, op. cit.*, pp. 60, 121, 133, 150, 154–155, 205–206, 217, 229, 249–250, 322–323, 356–357.
99. Reich, 18 Oct. 1938 report to French Academy of Sciences (RA, correspondence box 9, "1936–39" flds).
100. Reich, *BECP*, pp. 12–15, photos 24, 25.
101. Ibid.
102. Ms. of an Oct. 1938 early version of *Bion Experiments on the Cancer Problem* in German, which is in the next folder (in RA, correspondence box 9, "1936–39" flds), labeled "Vorlaufige und kurze Mitteilung über 'Bionexperimente zum Krebsproblem, vom Februar 1936 bis Oktober 1938," a four-page ms., exclusive of title page, dated 20 Oct. 1938. A French translation of the ms. is included, in a separate fld in Archives. In the next fld after this are included two addenda to that report, brief data tables dated 20 Dec. and 21 Dec. 1938.

103. Reich, "Cancer Model Experiment," OI box 8, "Präparatherstellung und Protokoll" fld, ms. pp. 57–59, quote on p. 57; Reich discusses this experiment in Reich to du Teil, 9 Apr. 1938, partial trans. in *BP*, p. 144.
104. "Cancel Model Experiment," *op. cit.*, p. 57.
105. Ibid.
106. Reich to du Teil, 19 Apr. 1938.
107. "Cancer Model Experiment," pp. 57–58.
108. Ibid., p. 58.
109. Ibid., p. 59.
110. Reich lab ntbk entry, Tues., 7 Feb. 1939, "*Die Zyankali-T-Probe*" (RA, OI box 8, "Research Preparations, 1937–39" fld, "Potassium Cyanide T-[bacilli] Trials"). In 1929, just as Reich's newly formed Socialist Association for Sex Counseling and Research opened the first of six clinics, known as Sexual Counseling Centers for Workers and Salarymen, the Communist play to revolutionize abortion politics, *Cyankali* [Potassium cyanide] by Friedrich Wolf, was a big hit, so much so that it was made into a feature film in 1930. Thus, Reich's choice of the term "Cyankali T-trials" for his experiment ten years later contained a reference to an episode he knew was very important for his sex-political work in the past, not merely to KCN. See http://movies.amctv.com/movie/137689/Cyankali/overview; see also http://www.letmewatchthis.ch/watch-46909-Cyankali and http://de.wikipedia.org/wiki/Cyankali_(Wolf).
111. This may be the reason Reich changed his initial thinking (discussed in Chapter 1, Table 1.1, note b), through Sept. 1937, that H+ ion belonged with the parasympathetic and OH- ion with the sympathetic.
112. Reich lab ntbk entry, Tues., 7 Feb. 1939, "*Die Zyankali-T-Probe*" (RA, OI box 8, "Research Preparations, 1937–39" fld), "Potassium Cyanide T-Trials."
113. Ibid.
114. Reich, "Experimental Orgone Therapy of the Cancer Biopathy (1937–1943)," *IJSO* 2: 1–92 (1943): 33.
115. Reich lab ntbk entry, Tues., 7 Feb. 1939, "*Die Zyankali-T-Probe*" (RA, OI box 8, "Research Preparations, 1937–39" fld), "Potassium Cyanide T-Trials." For packet-amoeba PA bions paralyzing S-bacilli, see Reich to du Teil, 13 Sept. 1937; also a ts. titled "Eigenschaften des S-bacillus" [Properties of the S-bacillus], n.d. (but ca. late 1937 or early 1938 since he is still calling them "S-bacilli"), ts. pp. 10–11 (RA, mss. box 9).
116. "Eigenschaften des S-bacillus" [Properties of the S-bacillus], n.d. (but ca. late 1937 or early 1938 since he is still calling them "S-bacilli"), ts. p. 11 (RA, mss. box 9).
117. *BP*, p. 110, diary entry of 15 Sept. 1937.
118. Reich, *BECP*, *op. cit.*, pp. 11–12, esp. photos 16 and 17.

119. Reich, "Experimental Orgone Therapy," *op. cit.*, pp. 39–44; CB, *op. cit.*, pp. 295–305.
120. Reich, CB, *op. cit.*, pp. 252–253, quotes on p. 253.
121. Ibid., pp. 298–305.
122. Reich, CB (1973 FSG ed.), pp. 305–306, 309. Reich reported the same cause of death for most of the human cancer patients who later volunteered for experimental treatment with the orgone accumulator—a striking fact given the widespread false rumor that Reich claimed to "cure cancer."
123. Chester Raphael and Helen McDonald, *Orgonomic Diagnosis of the Cancer Biopathy* (New York: Orgone Institute Press, 1952); originally published as *Orgone Energy Bulletin* 4 (2): 65–128 (Apr. 1952). This booklet summarized the entire content of a course Reich taught on bions and cancer in July–Aug. 1950.
124. See Anon., "Pneumocystis pneumonia—Los Angeles," *MMWR Morb. Mortal. Wkly. Rep.* 30: 250–252 (1981); see also Aimee Wilkin and Judith Feinberg, "*Pneumocystis carinii* Pneumonia: A Clinical Review," *Am. Fam. Physician* 60 (6): 1699–1708 (15 Oct. 1999), http://www.aafp.org/afp/1999/1015/p1699.html.
125. Chester Raphael, "*Pneumocystis carinii* Pneumonia as a Possible Endogenous Infection," talk given at Wilhelm Reich Museum conference, 12–14 July 1982, recording courtesy of Wilhelm Reich Museum.
126. Alan Cantwell, *The Cancer Microbe* (Los Angeles: Aries Rising Press, 1990); Harris Coulter, *AIDS and Syphilis: The Hidden Link* (Berkeley, CA: North Atlantic, 1987); Lynn Margulis et al., "Spirochete Round Bodies: Syphilis, Lyme Disease and AIDS; Resurgence of the 'Great Imitator'?," *Symbiosis* 47: 51–58 (2009).

6. Opposition to the Bion Experiments

1. Odd Hølaas, "Reichs påstand om å ha skapt liv er nonsens" [Reich's claim of discovering life is ridiculous], *Tidens Tegn*, 22 Sept. 1937.
2. "Redaktion von *Tidens Tegn*" (RA, correspondence box 9, "1936–39" flds). According to Norwegian historian Håvard Nilsen (personal communication), "*Tidens Tegn* (TT) initially was a conservative organ with broad views and excellent writers. In the 1930s, however, the editor Rolf Thommesen had connections with the Fatherland Society, which developed into a group of fascist sympathizers. Thommesen recruited Vidkun Quisling as a columnist and printed the first statement from newly formed NS-Party in TT before the Party had an organ of its own. So TT was viewed as pro-fascist in the 1930s, although it did not have an outspoken pro-fascist policy or platform. Many people who do not know about the Quisling-connection believe it to have been a liberal-conservative magazine, and it was revived as such in the 1980s. So to be precise: TT was NOT a fascist journal, as it did not have such a

platform. But that the editor Thommesen sympathized with the fascists because he wanted a united conservative front against the labour movement is clear."

3. Reich to du Teil, 23 Sept. 1937. From what Reich says, the patient could possibly be Rolf Stenersen or the chemist Odd Wenesland. Reich described an episode like this ten years later in his book *Listen, Little Man!* (New York: Orgone Institute Press, 1948), p. 87. If it was Wenesland, his crisis of confidence may have been triggered when he was unable to replicate results of some of du Teil's control experiments; see OI box 7, "Journal, Odd Wenesland" fld, entries of 29 Sept.–18 Oct. 1937. Reich thought the journalist for *Tidens Tegn*, Odd Hølaas, wrote a "hatchet job" article on his work because he was a Communist and Stalinist who hated Reich's sex-economy. See "Die soziale Macht des Gerüchts in der Bürokratie," *op. cit.*, p. 5, new sec., pp. 12–13 (after 9 pp. of this ms., there begins a new section, written later, after 1947; thus, I have begun anew paginating from the tenth page of the entire ts.).

4. Reich, BE, p. 6. See Espen Søbye, *Rolf Stenersen. En biografi* (Oslo, 1995) for more on Stenersen (1889–1978); see also http://www.snl.no/.nbl_biografi/Rolf _E_Stenersen/utdypning, retrieved 24 July 2010. In Dec. 1936, Stenersen personally paid for an expensive piece of equipment Reich needed (RA, bion lab ntbk 1, ms. p. 128). On Christensen, see "Lars Christensen," *Store norske leksikon*. Kunnskapsforlaget. http://www.snl.no/Lars_Christensen.

5. Reich, BP, pp. 99–100. Original letter RA, correspondence box 9, "1936–39" flds, Reich to Frank Blair Hansen and H. M. Miller, Rockefeller Foundation Office, Paris. Thirty thousand kroner was equivalent to about $6,800 in 1936–1937; it is equivalent to over $100,000 today (2013). Monthly expenses of $300 to $400 in 1937 would be equivalent to $4,500 to $6,200 in 2015 dollars.

6. Ibid.

7. Reich to Rockefeller Foundation, 19 Oct. 1937 (RA, correspondence box 9, "1936–39" fld); the letter was not sent at that time.

8. See Staatsarchiv, Oslo; Kreyberg papers PA no. 276, box 14, bound vol. "Guldberg disputas." In Feb.–Mar. 1932, Kreyberg initiated proceedings at the university claiming plagiarism of his work by Gustav Guldberg (who was at the Pathological-Anatomical Institute of the University Hospital, Oslo, working under Prof. Francis Harbitz) and his *Experimental Researches on Precancerous Changes in the Skin and Skin Cancer* (Copenhagen: Levin and Munksgaard, Copenhagen, 1930). Mohr, Harbitz [at that time Oslo U. prof. of pathological anatomy; Kreyberg replaced him upon his retirement at end of 1937], and Schreiner appear to support Kreyberg's claim [e.g., Mohr on ms. p. 60]. This controversy tells us something about Kreyberg's combative style of conflict, his reliance on highly technical abilities to back up his position, and his allies at the university. Kreyberg has marked pp. 44–68 of Guldberg's book as the pages

involving plagiarism. The book is published as *Acta Pathologica et Microbiologica Scandinavica, Supplementum VI*. Pt. 2 is supp. VIII (1931).

9. W. E. Tisdale Officer's Diary, 19 May 1938, RG 12.1, Rockefeller Foundation Archives, Rockefeller Archive Center, North Tarrytown, NY (hereafter, RAC). On Otto Mohr, see biographic entry in *Norsk Biografisk Leksikon*, http://translate.google.com/translate?hl=en&sl=no&u=http://www.snl.no/.nbl_biografi/Otto_Lous_Mohr/utdypning&ei=CisESurSO8OrtgeyjPj8Bg&sa=X&oi=translate&resnum=1&ct=result&prev=/search%3Fq%3Dnorsk%2Bbiografisk%2Btove%2Bmohr%26hl%3Den%26client%3Dsafari%26rls%3Den . The details of Mohr's career are surveyed in an obituary by A. M. Dalcq, "Notice Sur La Vie et l'ouvre de M. O. L. Mohr, Correspondent Étranger," *Bull. de l'Acad. Royale de Médicine de Belgique* 7: 691–698 (1967). On Tove Mohr, see http://translate.google.com/translate?hl=en&sl=no&u=http://no.wikipedia.org/wiki/Tove_Mohr&ei=CisESurSO8OrtgeyjPj8Bg&sa=X&oi=translate&resnum=3&ct=result&prev=/search%3Fq%3Dnorsk%2Bbiografisk%2Btove%2Bmohr%26hl%3Den%26client%3Dsafari%26rls%3Den.

10. W. E. Tisdale Officer's Diary, 19 May 1938, RG 12.1, Rockefeller Foundation Archives, RAC. Ragnar Nicolaysen (1902–1986) was professor of nutrition (whose biochemical position the RF had lobbied to create) at the university. Both men in the next few years received significant RF support. The full quote is "Dr. Wilhelm Reich—émigré psychoanalyst who has had anthropological training, now author of a book entitled 'binos' [sic; should be bions], the formula for things living, recommended to the RF by Malinowski, is in the opinion of Mohr, Langfeldt, and Nicolaysen a charlatan. In connection with his stay in Norway, the medical faculty was recently asked to pass on R's work. They reported that from a medical basis, no reason to extend his stay. (TBK) [i.e., Tracy B. Kittredge]." Einar Langfeldt (1884–1966) the physiologist should not be confused with Gabriel Langfeldt (1895–1983) the psychiatrist, who later came to the United States and began spreading the rumor that Reich had been institutionalized as insane; see *Reich Speaks of Freud* (New York: Farrar, Straus and Giroux, 1967), pp. 234–235; see also http://translate.google.com/translate?hl=en&sl=no&u=http://www.snl.no/.nbl_biografi/Gabriel_Langfeldt/utdypning&ei=-BEASounJ9WEtwfBtPWMBw&sa=X&oi=translate&resnum=1&ct=result&prev=/search%3Fq%3Dnorsk%2Bbiografisk%2Beinar%2Blangfeldt%26hl%3Den%26sa%3DG . On R. Nicolaysen, see http://translate.google.com/translate?hl=en&sl=no&u=http://www.snl.no/.nbl_biografi/Ragnar_Nicolaysen/utdypning&ei=TaABSs3nDY_EMsLCreAH&sa=X&oi=translate&resnum=6&ct=result&prev=/search%3Fq%3Dragnar%2Bnicolaysen%26hl%3Den%26client%3Dsafari%26rls%3Den%26sa%3DG.

Albert Fischer of the Rockefeller Institute in Copenhagen would also likely have given Reich a thumbs down if he were consulted, based on the disagreeable exchange he had with Reich after Reich visited his lab in December 1936 to demonstrate bion preparations. Reich, *People in Trouble, op. cit.*, p. 269.

11. Reich, *BP*, pp. 99–101. Original document RA, correspondence box 9, "1936–39" flds. Reich applied to the Paris RF office on 2 Mar. 1937; he was turned down in a letter from RF Officer Daniel P. O'Brien on 24 Mar. (RA, same flds).

12. Robert Kohler, *Lords of the Fly: Drosophila Genetics and the Experimental Life* (Chicago: U. of Chicago Press, 1994), pp. 98, 139, 145–146; also Garland Allen, *Thomas Hunt Morgan: The Man and His Science* (Princeton, NJ: Princeton U. Press, 1978), pp. 279–281, 376–378ff.

13. On Tove Mohr (1891–1981), Norwegian campaigner for women's rights and birth control education, see http://translate.google.com/translate?hl=en&sl=no&u=http://no.wikipedia.org/wiki/Tove_Mohr&ei=CisESurSO8OrtgeyjPj8Bg&sa=X&oi=translate&resnum=3&ct=result&prev=/search%3Fq%3Dnorsk%2Bbiografisk%2Btove%2Bmohr%26hl%3Den%26client%3Dsafari%26rls%3Den; on Mohr's views on eugenics, see Nils Roll-Hansen, "Eugenics in Norway," *Hist. Phil. Life Sci.* 2: 269–298 (1980). H. J. Muller still believed in eugenic ideas (and Mohr still opposed them) strongly enough in 1960 that Mohr took issue with him on the subject in a private letter; see Mohr to Muller, 2 Jan. 1960 (Muller papers, Lilly Library, Indiana U., ser. 1, box 25) in which Mohr concludes, "It may be daring for me to object to an authority of your dimensions. Still I feel like doing it in this case. Perhaps it is my rigid training in medicine which makes me feel so strongly here. I have not forgotten the difficulties in attaining exact diagnoses." Mohr was discussing the vagueness of such terms as "genius," in response to Muller's suggestion that sperm from "geniuses" be collected and used to inseminate women.

14. O. L. Mohr to Max Hodann, 12 Jan. 1937 (Mohr papers, Brevs 413, National Library, Oslo). On Hodann, see Wilfried Wolff, *Max Hodann (1894–1946): Sozialist und Sexualreformer* (Hamburg: von Bockel, 1993). In his book *History of Modern Morals* (Eng. trans. Stella Browne [London: Heinemann, 1937]), Hodann discusses Reich's work on pp. 133, 151–156; Reich's bioelectric experiments are discussed on p. 316. Mohr discusses at length Tove's struggles with reactionary people in Oslo in her attempt to liberalize abortion laws, as well as his own battles with the school board to try to institute sex education in public schools; see Mohr to L. C. Dunn, 28 May 1936 (Dunn papers, APS, box 22): "By Jove, the priests and the male school teachers finally seemed to consider me the very devil. And matters did not improve when I succeeded in preventing our University from sending a personal representative to the Heidelberg festivals [i.e., as a boycott of the Nazis]." In this same letter, Mohr describes his wife Tove's struggles to make abortion legal and the social forces that oppose her, especially "an old sadist bachelor by the name of Natvig, chief

physician at the maternity ward of the city," who tried to drive away all Tove's patients to intimidate her from working on changing abortion law and who was not above publishing a book "full of the most perfidious lies. But what can you do? A legal suit is out of the question in such situations, of course."

15. See, e.g., Hodann to Reich, 29 Dec. 1936 (RA, correspondence box 9, "1936–39" flds): "I still feel very strongly the impression of your film demonstrations. I am convinced that these [bion] experiments will assume an enormous importance, for the development of biological theory and thus for the biological basis of the 'orgasm formula,' as well as for pathology and therapy." Swiss sex reform advocate Fritz Brupbacher was also a longtime supporter of Reich.

16. H. B. from Socialdepartment, Medizinalabteilung, Oslo to Reich, Lindenberg, and Motesiczky, 25 [Apr.] 1938 [sic; original Norwegian version in next fld is dated Oct. 1938 and placed in that order in the collection] (RA, correspondence box 9, "1936–39" flds, Reich's German trans.).

17. See biographical entry for Kreyberg (1896–1984) in the *Norsk Biografisk Leksikon*, http://translate.google.com/translate?hl=en&sl=no&u=http://www.snl.no/.nbl_biografi/Leiv_Kreyberg/utdypning&ei=W2EBStfeDZLKMsXEqeYH&sa=X&oi=translate&resnum=1&ct=result&prev=/search%3Fq%3Dnorsk%2Bbiografisk%2Bleiv%2Bkreyberg%26hl%3Den%26client%3Dsafari%26rls%3Den.

18. Mohr to Dunn, 5 Dec. 1937 (L. C. Dunn papers B:D917, American Philosophical Society, Philadelphia, Mohr 1937–1939 file). The internal Oslo U. deliberations leading to Kreyberg's appointment as professor can be found at U. of Oslo archives, Medical Faculty, Professors, Pathological Anatomy and Forensic Medicine 1937–1939, Staatsarchiv, Oslo, 3A 198 67. Deliberations over the choice for the job are under Church Ministry records (KUD), 1 Skolekontor [First School Office] D, Staatsarchiv, box 448. In June 1929, radiochemist Ellen Gleditsch was appointed to full professor at the U. of Oslo, despite misogynist opposition from many faculty, including her predecessor in the post. In this story, Goldschmidt's attempt to get the job for his protégé Hassel could be seen as a model for Mohr's attempts on behalf of Kreyberg, in this sense: the openings were *very* limited in a "small town" and thus were hotly contested; see Annette Lykknes et al., "Appreciated Abroad," *op. cit.*

19. Reich to A. Schjødt, 16 June 1938 (RA, correspondence box 9, Eng. trans. in Reich, *BP*, pp. 156–158).

20. W. E. Tisdale RF Officer's Diary, 16 Nov. 1936, *op. cit.*

21. H. M. Miller RF Officer's Diary, 12–13 Jan. 1937, RF Archives, RAC. Thjøtta's institute was called the "Kaptein W. Wilhelmsen Og Frues Bakteriologiske Institutt," apparently named after his shipping magnate patron. At the July–Aug. 1947 International Congress of Microbiology in Copenhagen, a session on electron microscopy of bacteria was chaired by Thjøtta and

attended by Stuart Mudd, American bacteriologist and evangelist for electron microscopy in bacteriology (Thjøtta to Wyckoff, 4 Apr. 1949, Mudd papers, box MMB 18, file J/1948–9; on Mudd's evangelism, see N. Rasmussen, *Picture Control, op. cit.*).

22. See Robert Kohler, *Partners in Science: Foundations and Natural Scientists, 1900–1945* (Chicago: U. of Chicago Press, 1991), pp. 265–406; also Pnina Abir-am, "The Discourse of Physical Power and Biological Knowledge in the 1930s: A Reappraisal of the Rockefeller Foundation's 'Policy in Molecular Biology,'" *Social Studies of Science* 12: 341–382 (1982). See also Glenn Bugos, "Managing Cooperative Research and Borderland Science in the National Research Council, 1922–1942," *Historical Studies in the Physical and Biological Sciences* 20: 1–32 (1989), in which mitogenetic radiation as "borderland science" is discussed.

23. RF, H. M. Miller Officer's Diary, 19–21 Jan. 1937, *op. cit.*

24. RF, W. E. Tisdale Officer's diary, RG 12.1, 5 pp. ts. See also Tisdale to Mohr, 6 July and 31 Aug. 1938 (Mohr papers, Institute of Genetics, Oslo U.), in which Tisdale is very solicitous, attempting to get Mohr to send his formal request with the implication that it is likely to be immediately granted. Mohr later became rector of Oslo U., 1945–1951.

25. See Gunnar Broberg and Nils Roll-Hansen, *Eugenics and the Welfare State* (East Lansing: Michigan State U. Press, 2005), pp. 158, 172–175. For more on Scharffenberg (1869–1965), see N. J. Lavik, *Makt og Galskap* (Power and Madness) (Oslo: Pax Forlag, 1990); also see the recent biography by Espen Søbye, *En Mann Fra Forgangne Århundrer: Overlege Johan Scharffenbergs Liv Og Virke, 1869–1965: Enarkivstudie* (Oslo: Oktober Forlag, 2010). A charming caricature of Scharffenberg is in "De steller med mange ting, herr Scharffenberg," *Dagbladet* (nr. 304), 31 Dec. 1977, p. 15. See also Norsk Biografisk Leksikon entry: http://translate.google.com/translate?hl=en&sl=no&u=http:// www.snl.no/.nbl_biografi/Johan_Scharffenberg/utdypning&ei =z5cBSsvSO5qeMs_r7ekH&sa=X&oi=translate&resnum=3&ct=result&prev =/search%3Fq%3Dnorsk%2Bbiografisk%2Bjohan%2Bscharffenberg%26hl%3 Den%26client%3Dsafari%26rls%3Den%26sa%3DG.

26. Nils Roll-Hansen, "Norwegian Eugenics: Sterilization as Social Reform," in *Eugenics and the Welfare State, op. cit.*, p. 158.

27. Ibid., p. 172. Scharffenberg's original argument was "Forslag til sterilisasjonslov for Norge," *Arbeiderbladet* (30 Jan. 1932), pp. 2–3.

28. Søbye, *En Mann, op. cit.*

29. See, e.g. Loren Graham, "Science and Values: The Eugenics Movement in Germany and Russia in the 1920s," *Amer. Hist. Rev.* 82: 1133–1164 (1977). On the Norwegian case, see also Nils Roll-Hansen, "Eugenics in Norway," *Hist. Phil. Life Sci.* 2: 269–298 (1980).

30. Roll-Hansen, "Norwegian Eugenics," *op. cit.*, pp. 153–154.
31. Diane Paul, *Controlling Human Heredity*, *op. cit.*
32. Roll-Hansen, "Norwegian Eugenics," *op. cit.*, pp. 173.
33. Ibid., p. 174.
34. Scharffenberg to Nic Hoel, 9 Mar. 1938 and 25 Mar. 1938 (RA, correspondence box 9, "1936–39" flds).
35. See, e.g., Scharffenberg to Nic Hoel, 9 Mar. 1938 and 25 Mar. 1938 (RA, correspondence box 9, "1936–39" flds); Reich to Havrevold (and N. Waal), 29 Mar. 1938 (*Beyond*, pp. 138–139, orig. in RA, correspondence box 9). Oslo U. Dean K. Schreiner had written to Havrevold after Scharffenberg's provocation, also in a threatening tone. See Reich to du Teil, 19 Apr. 1938.
36. A summary of all the articles published is in Sverre A. Eriksen, *Bioner eller Roquefort? En Bibliografi til tredveårenes Reich-debatt i Oslo-avisene* (Oslo: Sandvika, 1973). See, e.g., 13 Apr. 1938, Arbeiderbladet article by Scharffenberg, "Are Dr. Reich's Experiments Scientifically Valid?"; Psychologist Ola Raknes replies 20 Apr. in same paper. Ola Raknes was the only Norwegian student of Reich's who kept up with (and publicly declared himself persuaded of the validity of) Reich's later physical orgone energy work, including use of the orgone energy accumulator with his patients, who included Sean Connery in the 1960s. See Jan Olav Gatland, *Ord og Orgasme: Ein Biografi om Ola Raknes* (Oslo: Norske Samlaget, 2010).
37. For example, 10 Mar. 1938, P. O. von Törne (prof. at the Åbo Akademi, Åbo Finland).
38. Håvard Friis Nilsen, "Sexual Politics in Norway: Wilhelm Reich and Leon Trotsky, 1933–1936," in Marta Kuzma and Pablo Lafuente, eds., *Whatever Happened to Sex in Scandinavia? Art and the Politics of Emanicipation* (New York: Walther Koenig/OCA, 2011), pp. 136—145. Also idem, "The History of the Norwegian Psychoanalytic Society," in Peter Loewenberg and Nellie Thompson, *100 Years of the IPA: The Centenary History of the International Psychoanalytical Association 1910–2010; Evolution and Change* (London: Karnac Books, 2012), pp. 140–148.
39. See, e.g., *Fritt Folk* attack and J. Quisling's (brother of V. Quisling) attack on Reich in *Morgenbladet*, 23 Apr. 1938, BP, pp. 147–148. Vidkun Quisling called Reich a Jewish pornographer; see H. Nilsen, "Widerstand in der Therapie und im Krieg 1933–1945: Die Psychoanalyse vor und während der Besatzung Norwegens durch die Nationalsozialisten," in Mitchell G. Ash, ed., *Psychoanalyse in totalitären und autoritären Regimen* (Vienna: Brandes and Apsel verlag, 2010), pp. 177–208, published in Eng. as "Resistance in Therapy and War: Psychoanalysis Before and During the Nazi Occupation of Norway 1933–1945," *Int. J. Psychoanal.* (2013), pp. 725–746.

40. Reich to N. Hoel, Havrevold, 29 Mar. 1938 (RA, correspondence box 9); edited version in *BP*, pp. 138–139.
41. See Reich to Gjessing, 21 Nov., 11 Dec., and 23 Dec. 1936; also Gjessing to Reich, 22 Jan. 1937 RA, correspondence box 9, "1936–39" fld).
42. Reich to N. Hoel, Havrevold, 29 Mar. 1938, *op. cit.*: *BP*, pp. 138–139.
43. W. Hoffmann, "Dr. Reich og vi andre" [Dr. Reich and we others: Reply to Annaeus Schjødt], *Dagbladet*, 21 June 1938.
44. "Correction," *Dagbladet*, 22 June 1938.
45. Reich to du Teil, 19 Apr. 1938 (RA, Reich-du Teil correspondence). The passage from Scharffenberg's article reads, "In France control experiments by 'Professor' Roger du Teil from the University of Nice are supposed to begin now. This was reported in Dr. Reich's magazine 1937 (pages 206–207). Mr. Roger du Teil came to Oslo himself in summer to work 10 days in the laboratory, and the bion-publication contains two papers from him in which he is expressly referred to as 'professor from the University of Nice.'

However, there is no university in Nice. In the well-known calendar, 'Minerva' [i.e., Abteilungen Universitäten und Fachhochschulen], (Section Universities and Colleges 1937, page 1200) only a 'University Center of the Mediterranean,' constructed in 1933 and part of the University in Aix-Marseille, is cited. The purpose of this institution is merely to provide courses for foreigners visiting Nice.

Mr. Roger du Teil can hardly be a university or college professor because his name is definitely not in the 'Minerva' register."
46. Du Teil to Reich, 26 Apr. 1938.
47. Schjødt to Scharffenberg, 12 May 1938 (RA, correspondence box 9,"1936–39" flds).
48. Gunnar Leistikow, "The Fascist Newspaper Campaign in Norway," *IJSO* 1: 266–273 (1942), p. 273.
49. Ibid.
50. *Dagbladet* and *Arbeiderbladet* published the lists of signatories on 27 Apr. 1938 (on page 1 in *Arbeiderbladet* under the title "Får dr. Reich opholdstillatelse ut fra asylretts-prinsippet?" [Will Dr. Reich gain a residence permit based on the right to asylum principle?]); see *BP*, p. 148. Originals of these signed letters, dated 8–9 Mar. 1938, plus copies of several drafts of the letter, can be found in the Sigurd Hoel papers, ms. 2351, "Ad W. Reich: diverse" fld, Norway National Library. See also Arne Stai, "Oppgjøret omkring psykoanalysen og Wilhelm Reich!," in *Norsk kultur- og moraldebatt i 1930-årene* [Norwegian culture and the morality debates of the 1930s], 2nd ed. Oslo, Gyldendal (1954), pp. 99, 151 [1978 ed. in "Fakkel" ser.]; *BP*, pp. 126, 135; Stai married Elsa Lindenberg after she and Reich separated and Reich departed Norway for America. Christian

August Lange (1907–1970) was a friend of Reich's (*BP*, p. 126); a drawing by his wife Lizzie [Elise Marie Getz (1908–1983)], "Adolescents in Trouble," is displayed in the Wilhelm Reich Museum. Lange published in Reich's journal, e.g., "Zum heutigen Geschlechtsleben der Jugend," ZPPS 4: 162–176 (1937). Lange's father, Christian Lous Lange (1869–1938), also a signer, was a Nobel Peace Prize winner. Halvard Manthey Lange (1902–1970) later became Norway's minister of foreign affairs for many years (c.f. "The Lange Room" in NATO's headquarters).

51. Quote from Leistikow, *op. cit.*, p. 272.
52. Malinowski statement, 12 Mar. 1938, *Reich Speaks of Freud* (New York: Farrar, Straus and Giroux, 1967), p. 219 (original in RA, correspondence box 9, "1936–39" flds).
53. Quoted in Sharaf, *Fury on Earth*, *op. cit.*, pp. 231–232.
54. Leistikow, *op. cit.*, pp. 272–273.
55. Similarly, did Reich "assemble a group of disciples," as many, including Sharaf describe it? Or did numerous talented people seek Reich out because they found his ideas important? The latter seems more borne out by the historical record about Sigurd Hoel, Harald Schjelderup, Nic Hoel, Arnulf Øverland, Theodore Wolfe, A. S. Neill, and others.
56. K. E. Schreiner (U. Oslo prof. of anatomy, dean of medical faculty) to Havrevold, 11 Apr. 1938 (RA, correspondence box 9, "1936–39" flds, Reich's Ger. trans. plus the original letter in Norwegian).
57. Ibid.
58. Ibid.
59. Reich diary entry, 22 Dec. 1937 (Personal box 2, fld "May 34–Feb. 37, ms p. 79; edited version in *BP*, p. 132).
60. Ibid., 15 Feb. 1938 (Personal box 2, fld "May 34–Feb. 37, ms p. 81; edited version in *BP*, p. 136).
61. Reich to du Teil, 19 Apr. 1938, partial trans. in *BP*, pp. 144–146 (original in RA, correspondence box 10, du Teil flds): "From Dean Schreiner's letter to Havrevold, you can see what kind of attitude we are up against."
62. Ibid., pp. 144–145.
63. Ibid., p. 145.
64. Ibid.
65. Ibid., p. 146.
66. Ibid.; this section excerpted from the published version in *BP*.
67. See, e.g., Reich, *PT*, *op. cit.*, p. 271: "Havrevold had made this urgent suggestion after a conversation in which Schreiner scared him into thinking that was necessary." See also Apr. 1938, K. E. Schreiner to Havrevold (RA, correspondence box 9, "1936–39" flds); and 11 Apr. 1938, K. Schreiner to Havrevold, *op. cit.*

68. Reich to Annaeus Schjødt (Reich's lawyer), 16 June 1938, *BP*, pp. 156–158, quote on 156.
69. Ibid., pp. 156–158, quote on 156. Most of this account of Kreyberg and Dysthe's machinations is taken from this official statement to Reich's lawyer; see also Lars Christensen to Havrevold, 4 Apr. 1938 (RA, correspondence box 9, "1936–39" flds). Regarding Reich's reaction to the 12 Mar. *Anschluss*, Sigurd Hoel showed up at Reich's apartment that evening to see how he was holding up, because of the horror he knew the Nazi annexation of Austria must have been causing his dear friend (M. Sharaf, *Fury on Earth, op. cit.*, p. 253). Hoel: "I never saw Reich cry. But he was close to tears that night."
70. Reich to du Teil, 30 Mar. 1938 (*BP*, pp. 139–140). Reich's level of concern is suggested by the fact that on 30 Mar. he also wrote to Walter Briehl and O. Spurgeon English, two former American students of his, now practicing psychiatrists in the United States, to ask whether they would be able to help him in the event he decided to relocate to America before the coming war broke out.
71. P. Neergaard to Reich, 21 Apr. 1938 (RA, correspondence box 9, Neergaard flds).
72. Reich to Neergaard, 23 Apr. 1938 (RA, correspondence box 9, Neergaard flds).
73. Birger Bergersen (1891–1977) was sympathetic to Reich's work; he was a Norwegian anatomist and politician for the Labor Party. He served as a professor (1932–1947) and rector (1938–1945) of Norges tannlegehøgskole; see http://en.wikipedia.org/wiki/Birger_Bergersen.
74. Reich to Palmgren, 23 Apr. 1938 (RA, correspondence box 9, "1936–39" flds); Reich to du Teil, 23 Apr. 1938. Bergersen had made the suggestion in conversation with Sigurd Hoel, who told Reich (Reich to du Teil, 25 Apr. 1938).
75. Reich to du Teil, 21 Apr. 1938. Reich's sense that the general public's interest was connected to deep, mystical yearnings was reinforced by a letter he received: "When the press campaign about bions began in Norway around Easter, 1938, I received an unusual letter. A lonely woman lamented that her dog, her only companion, had died. Could I bring him back to life? The belief in miracles is not limited to lonely people. For this reason, I must to begin this introduction with an attempt to clarify my work." Introduction to "From Sexual Hygiene to Cancer Research" [Von der Sexualhygiene zum Krebsproblem], unpublished ms. (RA, mss. box 9); this ms. was completed by 28 Apr. 1938 according to Reich in his letter to du Teil of that date.
76. Reich to du Teil, 25 Apr. 1938.
77. Du Teil to Reich, 26 Apr. 1938.
78. Reich to du Teil, 28 Apr. 1938.
79. Schreiner in his 11 Apr. 1938 letter to Havrevold says, "It is a tremendous surprise that the control experiments by the French researchers should be

treated as strictly confidential because two of these gentlemen, Dr. Martiny and Dr. Bonnet, have already discussed them with my source very openly and without the least reservation." Dysthe seems likely to be Schreiner's unnamed source at this time.
80. Reich to du Teil, 28 Apr. 1938.
81. Reich to Schjødt, 16 June 1938, *BP*, quote on p. 157.
82. Ibid., quote on p. 158.
83. Ibid., p. 158, 17 June 1938.
84. Reich to du Teil, 15 June 1939: "Our Dr. Kreyberg here in Oslo is therefore nothing more or less than a common liar and scoundrel [*Lügner und Schuft*]. I do not envy him the position in which he will find himself a few years from now." *BP*, p. 223. In his diary, Reich previously remarked, on 19 June 1938, "Kreyberg, that pathological anatomist!" (Personal box 2, fld "Feb. 37–Oct. 38, ms p. 86). Kreyberg had convinced bacteriologist Vivi Berle to leave Reich's lab, for instance. (See Reich diary, 28 Feb. 1938, RA, Personal box 2, fld "Feb. 37–Oct. 38, ms p. 81). In Nov. 1941, reflecting back on the episode, Reich wrote, "Where objective argument fails, they resort to rumor. They are curious but they take pains to avoid straightforward and simple contact with my laboratory and go asking questions of others instead. It is a fact that Norwegian 'authorities' sent emissaries to Malinowski in London, to Bonnet in Paris and to du Teil in Nice, in order to find out what I was doing. They could have reached my laboratory in Oslo by a fifteen minutes' journey on the streetcar." Reich, "Biophysical Functionalism and Mechanistic Natural Science," *IJSO* 1: 97–107 (1942), quote on p. 103.
85. Reich to du Teil, 28 Apr. 1938 (partial trans. in *BP*, pp. 149–151).
86. See G. Holmgren, "Reidar Gording" (obituary), *Acta Otolaryngol.* 42 (4–5): 439 (Aug.–Oct. 1952). Gording was Oslo's most preeminent ear, nose, and throat doctor. As an indicator of the "small town" Oslo was, Gording's daughter Elisabeth, an actress in the National Theater, was married to Reich's colleague Odd Havrevold. See http://translate.google.com/translate?hl=en&prev=/search%3Fq%3Dnorsk%2Bbiografisk%2Bjohan%2Bvogt%26hl%3Den%26sa%3DG&sl=no&u=http://www.snl.no/.nbl_biografi/Elisabeth_Gording/utdypning.
87. Reich to Schjødt, 16 June 1938, *BP*, p. 156.
88. Reich to du Teil, 8 June 1938. Partial trans. in *BP*, pp. 154–155, quote on p. 154.
89. 2 July 1938, Henri Bonnet to Reich (RA, correspondence box 9, "Bonnet" flds).
90. Reich, *BP*, p. 154n.
91. Du Teil to Reich, 5 June 1938.
92. Reich to du Teil, 8 June 1938, partial translation in *BP*, pp. 154–155, quote on p. 154.

93. Re Annaeus Schjødt (1888–1972), see Reich, "The Social Power of Gossip in Bureaucracy," RA, Personal box 4, ms. p. 19. See also his entry in the Norsk Biografisk Leksikon.
94. Reich to Schjødt, 16 June 1938, *BP*, p. 157.
95. Ibid.
96. Raknes to René Allendy, 14 June 1938 (RA, box 10, du Teil correspondence fld).
97. Ibid.
98. Du Teil to Reich, 21 June 1938.
99. Ibid.
100. Ibid.
101. Ibid.
102. Ibid.
103. Ibid.
104. Raknes to du Teil, 23–24 June 1938 (RA, du Teil flds).
105. Ibid.
106. Du Teil to Reich, 25 June 1938.
107. Ibid.
108. Ibid.
109. Ibid.
110. Reich to du Teil, 30 June 1938 (*BP*, pp. 161–164, not sent, correspondence box 10, du Teil flds), quote on p. 161.
111. Ibid., pp. 161–162.
112. Ibid., p. 162.
113. Ibid.
114. Ibid.
115. Ibid.
116. Ibid.
117. Ibid.
118. Du Teil to Reich, 28 Sept. 1938. Reich learned at second hand just after the war that du Teil had been imprisoned for "frauds," that is, probably failure to pay his debts. Thus, the relative roles of his financial problems and du Teil's caution about involvement with Reich are probably impossible to disentangle. See L. Delpech to Reich, 13 Aug. 1945 and Reich's reply, 24 Aug. 1945 (RA, correspondence box 10, Delpech fld).
119. See Marie Bardy, "Reich and du Teil," *op. cit.*, and du Teil, "The Bions: An Afterword," *op. cit.*
120. This interpretation is supported by the final paragraph of the 27 Apr. 1938 article in *Arbeiderbladet*, "Får dr. Reich opholdstillatelse ut fra asylretts-prinsippet?" [Will Dr. Reich gain a residence permit based on the right to asylum principle?], p. 1.

121. Reich to Neill, 4 June 1938, *Record of a Friendship, op. cit.*, p. 14.
122. Neill to Reich, 8 June 1938 (ibid., p. 15).
123. Ibid.
124. Reich to Neill, 10 June 1938 (ibid., pp. 15–16).
125. Reich to du Teil, Emil Walter, and Paul Neergaard, 20 June 1938 (du Teil flds, partial trans. in *BP*, p. 159).
126. Reich to Emil Walter, 25 June 1938, *BP*, pp. 159–160. Walter published his book *Unser naturwissenschaftliches Weltbild : sein Werden vom Altertum bis zur Gegenwart* (Zurich: Niehans, 1938) at this time.
127. Reich to Emil Walter, 25 June 1938, *op. cit.*
128. Sharaf, *Fury on Earth, op. cit.*, p. 214.
129. Wilhelm Reich, *Selected Writings* (New York: Farrar, Straus and Giroux, 1961), pp. 535–539.
130. Ibid.
131. "Die am 1.10.38 vorhandenen Bion-Kulturen" (OI box 8, "Präparatherstellung und Protokoll" fld).
132. See, e.g., 21 Sept. 1938, Reich to A. Palmgren (RA, correspondence box 9, "1936–39" flds), Reich to French Academy of Sciences 16 Sept. 1938, and their reply of 27 Sept. (RA, correspondence box 9, unlabeled fld containing French Academy correspondence); Reich to du Teil, 16 Sept. 1938; Reich to Neill, 16 Sept. 1938 and Neill reply 19 Sept. 1938 plus following exchanges (*Record of a Friendship, op. cit.*, pp. 16–19). Bernal and Needham steered Neill to a Prof. Spooner at Cambridge University. See Neill to Reich, 19 Oct. 1938 (RA, correspondence box 9, "1936–39" flds), including enclosed letter from E. J. C. Spooner (Cambridge U. Dept. of Pathology) to Neill, 17 Oct. 1938 (same fld), in which Spooner tells Neill, "I am afraid that I must have given [Joseph] Needham a wrong impression, for I cannot remember saying that I could do anything for Reich. As a matter of fact, the impression that I had from Needham was that Reich's cultures were somewhere in the London area and, as the [Munich] crisis was then in full flower, that he feared for their safety. I am sorry that I cannot either accept cultures of this kind or investigate them."
133. See Lars Amund Waage, *Den Framande Dyen* [The strange/foreign town], a 1999 novel about Reich's first four months in Norway. My thanks to Dr. Karl Fossum for this reference.
134. Deliberations over Thjøtta's choice as professor of bacteriology and serology, 27 Sept. 1935, are in University of Oslo archives, Medical Faculty, Professors, Staatsarchiv, Oslo, 3A 198 67, Church Ministry records (KUD), 1 Skolekontor [First School Office] D, Staatsarchiv, box 447.
135. Reich diary, 9 March 1937 (Personal box 2, fld "May 34–Feb. 37, ms p. 48; partial translation in *BP*, p. 102).

136. Reich diary, 9 March 1937 (Personal box 2, fld "May 34–Feb. 37, ms p. 48). Noted in Bion lab ntbk 2, ms. p. 88, 8 Mar. 1937, i.e., the same day Thjøtta examined the cultures and told Havrevold they were nonsense.
137. Bion lab ntbk 2, ms. p. 96, "Partial English translation" ts. p. 54, 15 Mar. 1937.
138. Reich diary, 8 March 1937 (Personal box 2, fld "May 34–Feb. 37, ms p. 49; partial translation in *BP*, p. 102).
139. Hoffmann, "Dr. Reich og vi andre," *Dagbladet*, 21 June 1938. See n86 above for further personal ties Havrevold had with officials at Oslo University.
140. Sigurd Hoel to Ellen Siersted of 4 June 1937 (Sigurd Hoel papers, National Library of Norway, Brevs 355). On Scharffenberg's pressuring; see, e.g., Scharffenberg to Nic Hoel (RA, correspondence box 9, "1936–39" flds). After Reich's departure from Norway, Havrevold became a neurology and endocrinology specialist. See, e.g., Odd Havrevold to Sigurd Hoel, 23 May 1944 (Havrevold papers, Brevs 350, National Library of Norway).
141. In a 1939 "getting things off the chest" letter to Scharffenberg (shortly before he left for New York), which Reich decided not to send, Reich remarked upon Scharffenberg's social eugenic views: "Even Freud battled against the all-too-easy trend to 'explain away' the sexual troubles of youth and the nightmares of frustrated women by unexplored genes. From the 'theory of degenerative genetic substances' to Hitler's 'racial theory' it is only one step." Higgins and Raphael, eds., *Reich Speaks of Freud, op. cit.*, pp. 223–224.
142. Reich to du Teil, Thurs., 23 Sept. 1937 (*BP*, p. 111–113). By Mohr himself, see *Heredity and Disease* (New York: W. W. Norton, 1934).
143. Mohr to Bentley Glass, 10 Apr. 1933 (Glass papers, ms. collection 105, APS).
144. On Konstad, see http://www.nuav.net/ns.html; on Hansen (1895–1971), see Mohr to L. C. Dunn, 20 June 1945 (Dunn papers, B:D917, APS Library): "Professor Klaus Hansen—you remember how much time I spent in order to castrate this criminal—of course proved to be a very dangerous traitor from the very first day—well, I am sure he started much earlier. He was the only university professor who joined the [Nazi] party. Recently he tried to escape, but was caught in Copenhagen. He kept close personal contact with the leading Gestapos during the whole war." U.S. Embassy official William M. Kerrigan was unable to interview Hansen in 1952 because he was in prison for his Nazi activities. N.b.: discussion of Hansen's role in bion controversy, p. 6 of 30 June 1952 Foreign Service Dispatch from Greenfield archives. See also http://translate.google.com/translate?hl=en&sl=no&u=http://www.tidsskriftet .no/%3Fseks_id%3D1312556&ei=9aIBSvyHGJaeMuTryNsH&sa=X&oi =translate&resnum=3&ct=result&prev=/search%3Fq%3Dnorsk%2Bbiografisk %2Bklaus%2Bhansen%26hl%3Den%26client%3Dsafari%26rls%3Den.
145. Reich, *BP*, pp. 164, 209–210. Konstad was probably the official making it difficult for H. J. Muller to get a visa for his new wife Dorothea in Oct.–Nov.

1939; see Mohr to Muller, 20 Nov. 1939 (Muller papers, Lilly Library, Indiana U.). See also Reich to Schjødt, July 1938 and to Walter Briehl, 28 Nov. 1938, *BP* pp. 164–165, 179.

146. S. Hoel to H. Schjelderup, 15 Nov. 1940 (Sigurd Hoel papers, National Library of Norway, Brevs 355), Hoel says that attacking the public straw man of Reich as if it were true could lead to serious consequences that Schjelderup may not have considered. "In the worst case, the book may indirectly put the aforementioned analysts in an unfavorable position, or even directly in danger. . . . And *if* the book were to be exploited, beyond your wishes and control, how would you appear then? You cannot avoid that to many people's minds, you will appear as the informer. Such a thing will be long remembered." People react strongly, said Hoel, "correctly in my opinion, against anything that could be understood as treachery. . . .

These were my arguments. I think they are reasonable arguments. But [Nazi member] Klaus Hansen gave the most essential argument. As you will remember, you mentioned a conversation you had had with him, coincidentally. He warned you against publishing a book on psychoanalysis now. But when he heard the tendency in the book [against Reich], he changed his mind. Such a book, you could safely publish! That is the kind of safety I do not think you should strive for." Despite Hoel's argument, Schjelderup published the book, *Nevrosene og den nevrotiske karakter* (Oslo: Gyldendal, 1940). This letter is also cited in H. Nilsen, "Resistance in Therapy and War: Psychoanalysis Before and During the Nazi Occupation of Norway 1933–1945," *Int. J. Psychoanal.* 94: 725–746 (2013).

147. Evang (1902–1981), later director of the Health Directorate of the Norwegian government, became internationally known for his work in the World Health Organization. See his entry in *Norsk Biografisk Leksikon:* http://translate.google.com/translate?hl=en&sl=no&u=http://www.snl.no/.nbl_biografi/Karl_Evang/utdypning&ei=iKYBSvjcDZSoMbqKzdwH&sa=X&oi=translate&resnum=2&ct=result&prev=/search%3Fq%3Dnorsk%2Bbiografisk%2Bkarl%2Bevang%26hl%3Den%26client%3Dsafari%26rls%3Den.

Nic and Evang were in a romantic relationship in the 1920s before she married Hoel, and Sigurd Hoel's 1927 novel *Sinners in Summertime* was based upon Nic's summer flings with Trygve Braatøy (1904–1953), Evang, and possibly also Wilhelm Hoffmann, according to Helge Waal (Nic's son); Strick oral history interview with Helge Waal (son of Nic Waal), 17 Apr. 2009, Oslo. In the 1930s Evang was "a young medical doctor and prominent Socialist intellectual" who found the 1934 Norwegian sterilization law acceptable (Broberg and Roll-Hansen, *op. cit.*, p. 174). In 1945 Evang spread the rumor that Reich was insane; see Reich, *American Odyssey* (New York: Farrar, Straus and Giroux, 1999), p. 281.

148. Reich to Raknes, 24 Dec. 1949 (Raknes papers, ms. 4° 2956: 8, National Library, Oslo). Reich discussed his 1928 organization in Vienna in "The Socialistic Society for Sexual Advice and Sexual Research," in Margaret Sanger and Hannah Stone, eds., *The Practice of Contraception* (Baltimore: Williams and Wilkins, 1931), p. 271.
149. Some of the FDA's files on the Reich case can be found at the National Library of Medicine, beginning in 1947, e.g., 23 Oct. 1947, FDA memo to chief of Eastern District, from Cyril C. Sullivan, chief of Boston Station (NLM FDA Reich papers, box 2, fld 1 of 2). I found no material relevant to the bion experiments in this collection, however.
150. The entire file is in Jerome Greenfield papers, Taminent Library, and Robert F. Wagner Labor Archives, New York U. My thanks to Dr. Philip Bennett for bringing this file to my attention. Kerrigan later went on to serve as head of the Economic Section of the U.S. Embassy in Israel; see http://archive.org/stream /aahoo10.1965.002.umich.edu/aahoo10.1965.002.umich.edu_djvu.txt.
151. Kerrigan, "Introduction" to report on his interviews, Foreign Service Despatch 1252, 811.557/6–3052, Greenfield papers, p. 1.
152. Ibid.
153. Ibid., pp. 9–10.
154. Ibid.
155. Ibid., interview with Thjøtta 1952, p. 8.
156. Ibid., p. 21.
157. Ibid., p. 47.
158. Ibid., interview with Mohr, 20 May 1952, pp. 31–33.
159. Ibid., interview with Langfeldt, 20 May 1952, p. 29.
160. Ibid.
161. Kerrigan's "The lady doth protest too much, methinks" instinct about anti-Semitism behind Mohr's comments has not been substantiated in any way about Mohr by my research. On Reich's suspicion of Thjøtta's anti-semitism, see his 20 May 1937 note, *BP*, p. 106; original in bion lab ntbk 2, ms. p. 155. Dr. Frøydis Langmark, a trainee of Kreyberg's, says (pers. comm., 25 Apr. 2009) that Kreyberg delighted in telling the story for decades of how he defeated Reich. She relates that in the private correspondence he let her read, Kreyberg said the contamination in Reich's experiments probably came from "skitt under neglene" ("filth under his fingernails," reminiscent, as she pointed out to me, of the common [at that time] expression "skittne jøde"- "filthy Jew"). Note Kreyberg's genealogical interest, in proving that his family name goes back as far as 1468 in Germany, so much interest that he left this information in the material he donated to the State Archives in 1973. Langmark adds, "As most professors at that time (1930's, 40's and 50's and maybe more recently) also LK probably was arrogant and self-assertive, thinking he knew the truths in

most fields outside his field and specialty, he was often referred to as frightening and with a formidable temperament. He just spoke freely to me, obviously thinking he served a good cause concerning WR and Nic Waal [i.e., by opposing them]. I don't think he was more anti-Semitic than any of his colleagues, rather the opposite."

162. Ibid. The same source also denied that Reich was a member of the Norwegian Communist Party (ibid., p. 4). For a much more detailed discussion of this, see Philip Bennett, "Wilhelm Reich, the FBI, and the Norwegian Communist Party: The Consequences of an Unsubstantiated Rumor," *Psychoanalysis and History* 16: 95–114 (Jan. 2014).

163. Personal communication from Langmark, 5 May 2009. See also Helge Waal, *Nic Waal: Det urolige Hjerte, op. cit.*, pp. 226–227.

164. The anti-Nazi activities of Mohr, Scharffenberg, Kreyberg, and others are detailed in Maynard Cohen, *A Stand against Tyranny: Norway's Physicians and the Nazis* (Detroit: Wayne State U. Press, 1997).

165. On Nic Waal's equally heroic resistance, see ibid., pp. 143–144; see also oral history interview with Helge Waal, 17 Apr. 2009, Oslo. Mohr describes at great length the situation under Nazi occupation in Norway in a letter to L. C. Dunn, 20 June 1945 (Dunn papers, B:D917) and in a letter to Curt Stern (n.d., 1945, Stern papers, ms. collection 5, "Genetics Institute, Oslo, Norway" fld, both letters at Amer. Phil. Soc. Library, Philadelphia). A similar account of conditions in occupied Holland is given in A. J. Kluyver to Charles N. Frey, 16 Nov. 1945 (ASM Archives, Barnett Cohen papers, Hollinger box 1, U. of Maryland, Baltimore County).

166. Christian August Lange was arrested by the Germans in 1941, along with his brother Halvard. Both spent the rest of the war years in Sachsenhausen. See S. Hoel to C. A. Lange, 8 May 1945, just after Lange was freed (Hoel papers, Brevs 355, National Library, Oslo). Lange struggled with health problems for the rest of his life as a result of the camp years, especially nerve problems, and his family believes he committed suicide by drowning himself. On Øverland, see http://translate.google.com/translate?hl=en&sl=no&u=http://www.snl.no /.nbl_biografi/Arnulf_%25C3%2598verland/utdypning&ei=iA4AStHxFsrJtgfpyv 2IBw&sa=X&oi=translate&resnum=1&ct=result&prev=/search%3Fq%3Dnorsk %2Bbiografisk%2Barnulf%2B%25C3%2598verland%26hl%3Den%26sa%3DG.

167. Motesiczky (1904–1943) was a wealthy Austrian count who trained as a therapist under Reich. For more on him, see Christiane Rothländer, *Karl Motesiczky: Eine biographische Rekonstruktion* (Vienna: Turia and Kant, 2009). On his participation in the bioelectric experiments, see Reich to W. Hoffmann and H. Löwenbach, 6 Sept. 1935 (*BP*, pp. 48–50). An example of his publishing in Reich's journal is "Religöse Ekstase als Ersatz der sexuellen Auslösung," *ZPPS* 4: 23–34 (1937).

168. See Michael Fend, "Erna Gál," http://www.lexm.uni-hamburg.de/object/lexm_lexmperson_00004622?wcmsID=0003. See also Croall, ed., *op. cit.*, *Neill of Summerhill*, p. 146.
169. In none of Mohr's correspondence with many other scientists I examined, not even with frequent correspondents like L. C. Dunn or Bentley Glass, with whom he discussed every small detail of life in Oslo, does Mohr ever mention the Reich bion controversy. This is striking given the prominence of the controversy in Oslo, but it is consonant with the idea that Mohr viewed the entire episode as unimportant pseudoscience from the outset.
170. Nils Roll-Hansen, *The Lysenko Effect: The Politics of Science* (Amherst, NY: Prometheus, 2005), p. 242.
171. Kohler, *Lords of the Fly, op. cit.*, p. 140.
172. Fleming, "Emigré Physicists," *op. cit.* A similar push, driving ecology in the direction of ecosystems analysis using radioactive tracers to quantify nutrient flows, was provided by Atomic Energy Commission funding in the years after World War II. See Joel Hagan, *An Entangled Bank: The Origins of Ecosystem Ecology* (New Brunswick, NJ: Rutgers U. Press, 1992); see also Angela Creager, *Life Atomic: A History of Radioisotopes in Science and Medicine* (Chicago: U. of Chicago Press, 2012), chap. 10.

7. SAPA Bions and Reich's Departure for the United States

1. Bon to Reich, 2 Nov. 1938 (RA, correspondence box 9, "Bon" flds).
2. Reich to Bon, 11 Mar. 1939 (RA, correspondence box 9, Bon flds); Nevlunghavn is in Larvik, Vestfold, Norway; see http://www.visitnorway.com/en/Product/?pid=30585.
3. Reich report, "Beobachtungen über Strahlungsphänomene bei SAPA-Bionen," 10 Mar. 1939 (RA, mss. box 9, a 70-p. report), p. 2. See also Reich, *CB, op. cit.*, p. 82 (originally published in Reich, "The Discovery of the Orgone," *IJSO* 1: 108–130 [1942], p. 117). By Feb. 1941, Reich wrote Einstein that he had now obtained the SAPA bions eight out of sixteen times attempted; see Reich to Einstein, 20 Feb. 1941, in Mary Boyd Higgins, ed., *American Odyssey: Letters and Journals, 1940–1947* (New York: Farrar, Straus and Giroux, 1999), p. 74. It is interesting to compare the role of serendipity in Reich's discovery of SAPA radiation with that involved in the 22 Oct. 1934 discovery by Enrico Fermi about slowing of neutrons [see Richard Rhodes, *The Making of the Atomic Bomb* (New York: Simon and Schuster, 1987), pp. 218–220].
4. Reich report, "Beobachtungen über Strahlungsphänomene," *op. cit.*, p. 1. This seems to be a SexPol bulletin that was sent out to Havrevold, Theodore Wolfe in New York, Emil Walter in Zurich, Willem Bon in Amersfoort, and the director of medicine, Oslo; it was notably *not* sent to du Teil, from whom Reich had heard nothing in months.

5. Ibid., p. 1.
6. Reich entry in Bion lab ntbk 3 ("Basic Research III, ms. pp. 162–163, 3 Feb. 1939, OI box 7).
7. Reich report, "Beobachtungen über Strahlungsphänomene," *op. cit.*, p. 1. The original notes on these tiny light rays are in Reich entry in Bion lab ntbk 3, 4 Feb. 1939 ("Basic Research III," ms. pp. 162–163, OI box 7). On 19 March, Reich tried out some specialized Reichert equipment, which helped visualize the rays more clearly. He proposed to Reichert of Vienna that he would supply SAPA bion cultures if the firm would develop "special optical microscope lenses" for the study of this new type of radiation (letter in OI Box 11, fld "Bion und Krebsforschung 29.X.38–15.XII.39").
8. Ibid. Reich describes the conjunctivitis in a letter to Tage Philipson on 18 Feb. 1939 (*BP*, p. 189); also in a letter to A. S. Neill of 19 Feb. 1939 (mistakenly, Reich typed 19.I.39 instead of 19.II.39), published in *Record of a Friendship, op. cit.*, pp. 25–26.
9. Reich report "Beobachtungen über Strahlungsphänomene," *op. cit.*, p. 2.
10. Reich diary entry, 16 Feb. 1939 (Personal box 3, fld "Pers. Tagebuch 29.X.38–9.V.39," ms. p. 25; *BP*, p. 188).
11. Reich to Philipson, 18 Feb. 1939 (RA, correspondence box 9, "1936–1939" fld; *BP*, p. 189).
12. Reich report, "Beobachtungen über Strahlungsphänomene," *op. cit.*, p. 2.
13. In spectrophotometry, for example, a sample can be placed in a quartz cuvette if one wants to measure its absorbance of wavelengths in the ultraviolet range. This is important with DNA samples, with their strong absorption at 260 nanometers, among others.
14. First mentioned in Reich to Neill, 19 Feb. 1939 (again, note Reich's typo error of "19.I.39") (RA, correspondence box 9, Neill flds), published in *Record of a Friendship, op. cit.*, pp. 25–26. The first lab notes on SAPA bions paralyzing t-bacilli are in Bion lab ntbk 3, 3–4 Feb. 1939, ms. pp. 162–163. Reich describes this and says the bions "killed amoeboid crawling cancer cells at a distance. These events were filmed. Unadulterated, deep-reaching ulcers similar to x-ray ulcers appeared on two mice, which had been re-inoculated with T-bacillus cultures. SAPA I prevented the otherwise fatal effect of the T-bacillus." See Reich report, "Beobachtungen über Strahlungsphänomene," *op. cit.*, p. 5.
15. Reich to Philipson, 18 Feb. 1939 (*BP*, p. 189).
16. Reich to Bon, 18 Feb. 1939 (RA, correspondence box 9, "Bon" flds).
17. Ibid., 27 Feb. 1939 (RA, correspondence box 9, "Bon" flds), partial trans. in *BP*, pp. 191–192.
18. Ibid.
19. For the "field" around RBCs, see Raphael and MacDonald, *Orgonomic Diagnosis, op. cit.*, p. 70.

20. Paulo and Alexandra Correa discuss the "array envelope" or capsule in their article "PA and SAPA Bion Experiments," *op. cit.*, discussed much further in the Epilogue to this book.
21. See Murphy, *Fundamentals of Light Microscopy*, *op. cit.*, pp. 64–65.
22. Raphael and MacDonald, *Orgonomic Diagnosis*, *op. cit.*
23. See, e.g., Marcel Bessis, *Living Blood Cells and Their Ultrastructure* (New York: Springer, 1973). Other issues with erythrocytes possibly relevant to interpretation of the Reich Blood Test are discussed in Makoto Nakao, Toshiko Nakao, and Saburo Yamazoe, "Adenosine Triphosphate and Maintenance of Shape of the Human Red Cells," *Nature* 187: 945–946 (10 Sept. 1960), doi:10.1038/187945a0; Eric Ponder, *Hemolysis and Related Phenomena* (New York: Grune and Stratton, 1948); Robert F. Furchgott and Eric J. Ponder, "Disk-Sphere Transformation in Mammalian Red Cells: II; The Nature of the Antisphering Factor," *J. Exp. Biol.* 27: 117–127 (1940). Furchgott was a biochemist whose later work included explaining the mechanism of action of Viagra. See http://en.wikipedia.org/wiki/Robert_F._Furchgott.
24. See, e.g., Richard Rhodes, *The Making of the Atomic Bomb* (New York: Simon and Schuster, 1987), pp. 47–48.
25. Reich, *CB*, *op. cit.*, p. 86 (originally published in Reich, "The Discovery of the Orgone," *IJSO* 1: 108–130 (1942), p. 119. During the war, Reich used aliases—in this case "Dr. F" for Havrevold—to protect his European colleagues.
26. Ibid. Bohr was in America on a trip that later became famous because when he disembarked in New York he let slip to a Princeton physics colleague the importance of the new discovery of uranium fission. On Sat., 28 Jan. 1939, that news broke to wire services all over the world. Since Reich had not realized Bohr was in America, it is likely that—being absorbed in his SAPA and Cyankali experiments—he had not yet heard this news from three weeks earlier.
27. Reich, *CB*, *op. cit.*, p. 87 (originally published in Reich, "The Discovery of the Orgone," *IJSO* 1: 108–130 [1942], p. 120).
28. Reich to Philipson, 23 Feb. 1939 (*BP*, pp. 190–191).
29. Reich report "Beobachtungen über Strahlungsphänomene," *op. cit.*, pp. 2–3.
30. Ibid., p. 3.
31. Ibid.
32. Ibid., p. 4.
33. On the recognition of carcinogenic effects, see A. R. Bleich, *The Story of X-rays* (New York: Dover, 1960), chap. 4. Muller's discovery was announced in *Science* 66: 84–87 (22 July 1927); Muller's full data were given in a talk late summer at the 5th International Genetics Congress in Berlin.
34. Reich diary entry, 6 Mar. 1939 (Personal box 3, fld "Pers. Tagebuch 29.X.38–9.V.39," ms p. 27v; edited version in *BP*, p. 193).

35. Reich, "The Discovery of the Orgone," *IJSO* 1: 108–130 (1942), p. 119.
36. Reich report, "Beobachtungen über Strahlungsphänomene," *op. cit.*, pp. 5–6.
37. Ibid., p. 6.
38. Ibid., p. 118; emphasis in original.
39. Reich, "The Discovery of the Orgone," *IJSO* 1: 108–130 (1942), p. 120. See also Reich to Bon, 14 Mar. and 17 Mar. 1939, *BP*, pp. 193–195, where Reich describes how his understanding of the electroscope's behavior is emerging.
40. There is some evidence to suggest that Moxnes, like Kreyberg (the two men knew one another from the Radium Hospital), had an irrational degree of negativity toward Reich. More than ten years later, when Reich's student Ola Raknes gave an invited talk about Reich's recent (1950) work on orgone energy to the Medical Students' Association in Oslo, Moxnes and Kreyberg both attended. Moxnes insisted his electroscope measurement proved SAPA radiation was a fantasy. Raknes told Reich that Moxnes embellished his account (perhaps his memory had been embellished as well since 1939) to say that he had had a long discussion with Reich about different forms of radiation and how to measure them. Reich replied to Raknes, "Dr. Moxnes simply lied when he told the audience that the different kinds of radiation and ways of measuring them were discussed at our brief meeting. The only thing that actually happened was that I brought with me a culture of SAPA bions, that he approached this culture to his radiumscope, and that no reaction occurred. Whereupon he, in an utterly unscientific manner declared that 'there is no radiation.' Thereupon I corrected him, saying that the only thing he can say is that there is no *radium* activity, but that the negative result does not exclude radiation of some new kind. I was proved correct by later events. There was no discussion of any techniques or theories, of radium examination, etc. This is a simple lie, made up for the occasion of your lecture." Reich to Raknes, 3 Mar. 1950 (Raknes papers, ms. 4° 2956: 8, National Library, Oslo). In such a "he said/she said" case, the historian cannot judge conclusively which account is closer to the truth. But the discussion of Kreyberg's similar behavior by the U.S. Embassy official Kerrigan in 1952 (in Chapter 6) seems suggestive.
41. Bon to Reich, 22 Feb. 1939 (RA, correspondence box 9, Bon flds).
42. Reich to Bon, 11 Mar. 1939 (RA, correspondence box 9, Bon flds).
43. Ibid., 27 Feb. 1939 (RA, correspondence box 9, Bon flds), partial trans. in *BP*, pp. 191–192, quote on p. 191.
44. Ibid. For magnetization of metal objects, see Reich to Bon, 8 Mar. 1939; and Reich, "The Discovery of the Orgone," *op. cit.*, p. 118.
45. Reich to Bon, 27 Feb. 1939, *op. cit.*
46. Bon to Reich, 3 Mar. 1939.
47. Ibid.
48. Reich to Bon, 7 Mar. 1939.

49. Ibid., 22 Mar. 1939, partial trans. in *BP*, pp. 197–198.
50. Ibid.
51. Reich report, "Beobachtungen über Strahlungsphänomene," *op. cit.*, p. 5.
52. Reich to Bon, 8 Mar. 1939.
53. Ibid.
54. Ibid.
55. Ibid.
56. Reich report, "Beobachtungen über Strahlungsphänomene," *op. cit.*, p. 3.
57. Ibid., p. 4.
58. Emphasizing this point and extending this research, Reich wrote Bon on 24 May 1939 to report, "In diagnostic terms, the prickling and burning feeling on the skin is very important. Three women introduced test tubes [containing SAPA cultures] into their vaginas and left them there for only one to three minutes. One woman exhibited an extremely strong reaction and experienced a prickling and burning sensation deep inside her. These were women in whom I had found amoeboid cells in samples of vaginal secretion taken following intermenstrual bleeding. These cells were promptly destroyed when they were exposed to the radiation. The radiation is strong when the SAPA are cultured in broth + KCl. The also become significantly larger in that medium." Reich later reported that the radiation killed *Trichomonas vaginalis* as well.
59. Reich to Bon, 11 Mar. 1939. Reich was in consultation with a technical specialist at Reichert of Vienna regarding designing the special equipment. See Reich to Reichert, 13 March 1939 (RA, OI Box 11, fld "Bion und Krebsforschung 29.X.38–15.XII.39").
60. Ibid., p. 4. The Faraday cage later played an important role in the development of the orgone energy accumulator.
61. Ibid., p. 5.
62. Reich diary entry, 21 Aug. 1938, *BP*, p. 165.
63. Reich to Bon, 22 Mar. 1939, *BP*, pp. 197–198. The differing ways metallic versus insulating materials react to the radiation later became the basis for an "orgone energy accumulator" made of alternating layers with a metallic layer innermost. See Reich, *CB*, (1973 FSG ed.) pp. 90–93, 104–126.
64. Reich to Bon, 22 Mar. 1939, comment at end of letter, left out of *BP* partial trans.: "I can let you have new cultures any time, but do not keep the cultures anywhere near metals, because of the reflected radiation problem."
65. Reich to Bon, 17 Mar. 1939, where he uses the term "orgonicity." In Reich to Bon, 18 Mar. 1939, Reich says, "Bearing in mind the origin of its discovery, I have called this radiation 'orgone'" (*BP*, p. 196). Reich explains the etymology of the term more explicitly in "The Discovery of the Orgone," *op. cit.*, p. 121.
66. Reich diary entry, 28 Mar. 1939, *BP*, p. 199.
67. Reich to Bon, 18 Mar. 1939, *BP*, p. 195.

68. Ibid., Fri., 17 Mar. 39 (RA, correspondence box 9, Bon flds, partial trans. in *BP*, p. 194).
69. Reich to Bon, 18 Mar. 1939 (*BP*, pp. 195–196).
70. This experiment and the two subsequent ones are described in Reich, "Drei Versuche mit Gummi am statischen Elektroskop," pp. 26–27 in Reich, *Klinische und Experimentelle Berichte* no. 7 (Oslo: SexPol Verlag, 1939); Eng. trans. M. Sharaf published in *Orgone Energy Bulletin* 3: 144–145 (1951).
71. Reich to Bon, 4 Apr. 1939 (*BP*, pp. 201–202).
72. Ibid.
73. Reich, "Drei Versuche," Eng. trans., *op. cit.*, p. 145. Reich reports more electroscope experiments in his 17 Mar. 1939 letter to Bon, as well as recording some of his developing ideas on the same date in his diary; for both, see *BP*, pp. 194–195.
74. Reich to Bon, 17 Mar. 1939 (*BP*, p. 194).
75. Ibid., 18 Mar. 1939 (*BP*, p. 195).
76. Ibid., 11 Mar. 1939.
77. Ibid.
78. Reich to Bon, 22 Mar. 1939 (*BP*, p. 198).
79. Bon to Reich, 23 Mar. 1939, emphasis in original.
80. Ibid.
81. Reich to du Teil, 10 Apr. 1937.
82. Table on display in Wilhelm Reich Museum exhibit on bion experiments: "Eigenschaften der Bionkulturen." Nine bion types are described here: PA, BOPA, HAPA, SEKO I, SEKO II, BLUKO I, BLUKO III, Preparation 11, and SAPA.
83. Bon to Reich, 23 Mar. 1939.
84. Reich to Bon, 27 Mar. 1939.
85. Ibid.
86. Ibid.
87. Ibid.
88. Reich to Bon, 30 Mar. 1939, partial trans. in *BP*, pp. 199–200.
89. Bon to Reich, 2 Apr. 1939.
90. Ibid.
91. Bon to Reich, 12 May 1939.
92. For example, in Reich to Bon, 11 Mar. 1939, from which this quote is taken.
93. Reich to Bon, 17 May 1939; partial trans. in *BP*, pp. 211–215, quote on p. 215.
94. In a comparable case, Reich notes (*BE*, p. 68) that his "culture medium d" occasionally showed a reddish growth. Standard bacteriology would no doubt pronounce that this must be "just" *Serratia marcescens*, a reddish bacterium not infrequently found growing on polenta (which has led to some claims of Christian miracles, the red smear being interpreted as the blood of Christ; see

http://astro.temple.edu/~tomfeke/serratia_marcescens.htm). But this, again, begs the question of the origin of the bacteria. In the era of prions and cancer viruses, endogenously originating microbes sound at least somewhat less farfetched than they did in the 1930s.

95. Reich discussed the "Sarcinae" matter further in the unpublished ms. "Misslingen von Versuchen in Amersfort, Holland," (RA, Personal box 4, n.d., ca. 1941 or 1942); quote from p. 1.

96. Ibid., p. 3.

97. One recent analysis, however, reasserts the "sarcina" interpretation, claiming that the marine cyanobacteria *Myxosarcina* "and a variety of closely related oceanic sarcina also found in the SAPA experiment cultures present an unsurmised resistance to high heat and . . . exist as inclusions in carbonaceous and vitreous (quartz and quartzites) ocean sand." See Paulo Correa and Alexandra Correa, "The PA and SAPA Bion Experiments and Proto-Prokaryotic Biopoiesis," *J. Biophys. Hematol. Oncol.* 1 (2): 1–49 (2010), http://www.aetherometry.com/publications/direct/JBioHemOnc/JBHO01-02-01.pdf. This work is discussed further in the Epilogue.

98. Ibid., p. 2. Reich's analysis here reminds one of Ludwik Fleck's (1934, *op. cit.*) description of how "words become battle cries," representing whole "party platforms" of associated paradigms, conceptual structures, or belief systems.

99. Ibid., pp. 2–3.

100. Herrera to Reich, 12 July 1939. In Herrera to Reich, 14 Aug. 1939, Herrera suggested Reich's bions were identical with what Herrera called *Micrococcus brownianus*. In his 12 July letter, Herrera wrote, "La culture de vos biontes devra avoir une grande importance pour l'humanite. Je les trouve partout et ils devront avoir une grande influence sur la vie, la santé, et la mort." (The culture of your bions is bound to be of the greatest importance for humanity. I find them everywhere, and they will undoubtedly have a great influence on matters of life, health, and death.)

101. Herrera to Reich, 12 July 1939.

102. Reich, "Die soziale Macht des Gerüchts in der Bürokratie" [The social power of gossip in bureaucracy], RA, Personal box 4, new sec., p. 11. Reich discusses these matters with Herrera in Reich to Herrera, 25 Sept., 9, 13, and 24 Oct. 1939; and Herrera to Reich, 3 Oct. (all in RA, general correspondence fld, box 11). Also Herrera to Reich, n.d., sometime between 10 and 24 Oct. 1939 (RA, correspondence box 9, "1936–39" flds, misfiled in Mar. 1939 letters); Reich to A. Herrera, 25 Oct. and 14 Nov. 1939; and Herrera to Reich, 2 and 8 Nov. 1939 (RA, correspondence box 9, "1936–39" flds). In his 25 Oct. letter, Reich said he now felt certain that his T-bacilli were identical with Herrera's *Micrococcus brownianus*.

103. Bon to Reich, 2 Apr. 1939.

104. For background on early theories of two separate "vitreous" and "resinous" electrical fluids, preceding Benjamin Franklin's persuasive case for a single electrical fluid, see, e.g., Thomas Hankins, *Science and the Enlightenment* (New York: Cambridge U. Press, 1985), pp. 61–62. See also Benjamin Franklin, *Experiments and Observations on Electricity*, ed. I. B. Cohen (Cambridge, MA: Harvard U. Press, 1941), pp. 42–43.
105. Reich to Bon, 12 Apr. 1939, partial trans. in *BP*, pp. 202–203.
106. Ibid., p. 203.
107. Reich to Bon, 24 May 1939, *BP*, pp. 216–219, quote from p. 218.
108. Ibid., pp. 216–217.
109. Ibid., p. 219.
110. Bon to Reich, 30 May 1939, emphasis in original.
111. Ibid.
112. Reich, "Misslingen von Versuchen in Amersfort, Holland," p. 4 (RA, Personal box 4). See also a Related document, "The Effects of Humidity on Orgone Trials" (late 1939?) in mss. box 9.
113. Ibid., emphasis in original. Not incidentally, this inability to measure orgone phenomena during the humid summer in New York is the reason Reich moved his laboratory to the low humidity of the mountains of western Maine during the summers by the early 1940s. The permanent laboratory he eventually built there became, after his death, the Wilhelm Reich Museum.
114. On the days immediately after the Nazi occupation of Norway, see Kreyberg to Dunn, 16 July 1940, sent from Toronto (L. C. Dunn papers, APS).
115. Albert J. Kluyver to Chas. N. Frey, 16 Nov. 1945 (ASM Archives, Barnett Cohen papers, Hollinger box 1).
116. See, e.g., W. F. Bon, *Physisch-Chemisch Onderzoek van Het Ooglenseiwit α-crystalline: Academisch Proefschrift* (Amsterdam: Drukkerij Holland, 1955); his popular science works include *Wat Weet Ik van "Aardstrahlen"?* (Amsterdam: Nederlandsche Keurboekerij, 1949) and *Wat Weet Ik van Magnetisme?* (Amsterdam: Nederlandsche Keurboekerij, 1950). I am grateful to Olga Amsterdamska for help in tracking down information about Bon.
117. The originals for Einstein's side of the correspondence are in the Jewish National and University Library, Jerusalem; for example, Einstein to Reich, 7 Feb. 1941 (55 838–1 and 2, ms. original 839–1 and 2, Einstein file). The full correspondence was published by Reich, titled *The Einstein Affair* (Rangeley, ME: Orgone Institute Press, 1953), available as a reprint from the Wilhelm Reich Museum.
118. Warren Weaver to Julius Weinberger, 26 May 1941 (RA, correspondence box 12, "Weinberger" fld).
119. Reich to Weinberger, 29 May 1941, *American Odyssey, op. cit.*, pp. 100–101.
120. Reich to Weinberger, 6 June 1941, *American Odyssey, op. cit.*, pp. 104–106.

121. Otto Rahn, "The Disagreement in Mitogenetic Experiments, a Problem in Bacterial Physiology," presented 29 Dec. 1933 at 35th annual S.A.B. mtg., Philadelphia, published a few months later in *J. Bact.* 28: 153–158 (1934).
122. Alexander Hollaender, "The Problem of Mitogenetic Rays," in B. M. Duggar, ed., *Biological Effects of Radiation*, vol. 2 (New York: McGraw-Hill, 1936), pp. 919–959. Hollaender (1898–1986) became a major figure in radiation biology, particularly after World War II when he worked at Oak Ridge National Laboratory.
123. Ibid., p. 920. This was quite similar to Beutner's 1938 description of why George Crile's "autosynthetic cell" experiments did not get a fair hearing because of sensationalist publicity. See Beutner, 1938, *op. cit.*, pp. 174–186, esp. 182–186, "Artificial Cells and Politics." On Gurwitsch, see pp. 209–213. Reinhard Beutner was a physiological chemist, "whose life-long work on bioelectric potentials" involved much work on the action potential of nerves, which he believed by 1952 he could explain "in terms of the electromotive effect of acetylcholine." Beutner was originally brought to work on this problem to the Rockefeller Institute by Jacques Loeb; he spent much of his career at the University of Pennsylvania College of Pharmacy. See Stuart Mudd to Carl F. Schmidt, 6 June 1952, box 21, file "S 1951–52," Mudd papers, U. of Pennsylvania Archives.
124. Hollaender, *op. cit.*, pp. 919–920.
125. Ibid., p. 920.
126. R. A. Harman, "Current Research with SAPA Bions," *J. Orgonomy* 21: 42–52 (1987), claims to have produced SAPA bions; more work on these is reported by K. Carey and S. Dunlap, "Culturing SAPA Bions," *J. Orgonomy* 22: 68–75 (1988). However, there are sufficient problems with these experiments that this author remains skeptical of their claims to have produced SAPA bions. The cultures did not grow on the inoculation streak, for example, but rather near the edge of an agar plate where one would most likely expect to find air contaminant organisms. So these might well be *Sarcinae* that happen to have similar morphology to SAPA bions. Also, no controls were reported in either article.
127. Hollaender, *op. cit.*, p. 920.
128. Ibid., p. 924.
129. A. Hollaender and W. D. Claus, *An Experimental Study of the Problem of Mitogenetic Radiation*, NRC Bulletin no. 100 (Washington, DC: National Research Council, 1937). The study was dated 30 June 1937. For press reaction see, e.g., *New York Times*, 9 Jan. 1938, "Tests Cast Doubt on Theory of Rays; Gurwitsch Claim That Living Matter Propels Ultra-Violet Radiation Is Disputed, Controversy Is Revived; Wisconsin University Research Fails to Find Support for 'Physical Phenomenon.'" The article opens, "The rays

which, according to the Russian physiologist, Alexander Gurwitsch, are given off by plants and animals, must be regarded with a skeptical eye, conclude Drs. Alexander Hollaender and Walter D. Claus of the University of Wisconsin in a report submitted to the NRC." http://www.nytimes.com/top/classifieds/realestate/columns/living_in/index.html?query=HOLLAENDER%20,%20ALEXANDER&field=per&match=exact.

130. Kurt H. Stern, "Mitogenetic Radiation: A Study of Authority in Science," *J. Washington Acad. Sci.* 65 (3): 83–90 (1975). See also Bryan Wynne, "Between Orthodoxy and Oblivion: The Normalisation of Deviance in Science," in Roy Wallis, ed., *Sociol. Rev. Monog.* 27, *On the Margins of Science: the Social Construction of Rejected Knowledge* (1979), pp. 67–84, about "J-rays."

131. Rahn's unpublished 1952 autobiography (ts., American Society for Microbiology Archives), pp. 53–54.

132. See C. Frederick Rosenblum [Courtney Baker], "An Analysis of the U.S. Food and Drug Administration's Scientific Evidence against Wilhelm Reich: Part 2: The Physical Evidence," in Jerome Greenfield, *Wilhelm Reich vs the USA* (New York: Norton, 1977), pp. 358–367.

133. John R. Lion, MD, to James DeMeo, 18 Jan. 1982: "I recall quite well that my father [MIT physicist Kurt Lion] was called upon to prove that the box was just a box and that Dr. Reich was a fraud." This pretty clearly implies bias, going into the experiments. DeMeo's reply of 22 Jan. 1982 points out that Lion's test of the temperature differential between an orgone accumulator and a control box "did not control for very important parameters of the local meteorology [relative humidity most of all], which influence the accumulator's functioning. Hence, without such controls. . . , his study loses meaning." My thanks to Dr. DeMeo for sharing his correspondence with Lion.

134. See Richard Blasband, "An Analysis of the U.S. Food and Drug Administration's Scientific Evidence against Wilhelm Reich: Part 1: The Biomedical Evidence," pp. 343–357 in Greenfield, *Wilhelm Reich vs the USA, op. cit.*

135. See, e.g., Courtney Baker and Robert Dew, "Bion Migration," *Annals of the Inst. for Orgonomic Science (AIOS)* 1: 24–32 (1984). See also C. Baker, Byron Braid, Robert Dew, and Louisa Lance, "The Reich Blood Test: 105 Cases," *AIOS* 1: 1–11 (1984); idem, "The Reich Blood Test: Clinical Correlation," *AIOS* 2: 1–6 (1985); and C. Baker and R. Dew, "Studies of the Reich Blood Test in Cancer Mice," *AIOS* 3: 1–11 (1986); C. F. Baker and P. S. Burlingame, "The Effects of Calcium on Preparation 6," *AIOS* 3: 12–17 (1986).

136. A remark is in order, lest the reader think I am being partisan on behalf of Reich and that I apply the methodologies of judgment asymmetrically to Reich and to his opponents. I try to make clear throughout that it is, in fact, rather the archival record that is asymmetrical. Reich left a voluminous archive. But in all of the preserved correspondence I could find of his opponents in

numerous different archives in Norway and in the United States, there were only about a hundred letters total. And in none of them was Reich's work mentioned. Thus, they left very little archival record of their views, except in the popular press, and I was forced to use more inference to reconstruct those views. Where I was able to locate a useful third-party source, the U.S. Embassy official's interviews of many of Reich's opponents, the opinions that the official expresses are his own. This material casts interesting light on the players, but I am not imposing any opinions on the official or on the reader.

137. In the recent excellent survey of our knowledge of cancer, Siddhartha Mukherjee's *The Emperor of All Maladies: A Biography of Cancer* (New York: Scribner, 2010), for example, Reich's story has disappeared altogether from the story of cancer research. So established is the pseudoscience narrative, Reich's theory apparently is no longer considered important enough to include even to criticize it.

Epilogue

1. Franklin Harold, *The Way of the Cell: Molecules, Organisms and the Order of Life* (New York: Oxford U. Press, 2001); Harold Hillman, *The Case for New Paradigms in Cell Biology and in Neurobiology* (Lewiston, NY: Edwin Mellen Press, 1991) and idem, *Evidence-Based Cell Biology* (Maastricht: Shaker Pub., 2008); Gerald Pollack, *Cells, Gels and the Engines of Life: A New, Unifying Approach to Cell Function* (Seattle: Ebner and Sons, 2001); Stephen Rothman, *Lessons from the Living Cell* (New York: McGraw-Hill, 2002). Also relevant are Gilbert Ling, *Life at the Cell and Below-Cell Level: The Hidden History of a Fundamental Revolution in Biology* (New York: Pacific Press, 2001); and Stanley Shostak, *Death of Life: The Legacy of Molecular Biology* (London: Macmillan, 1998). A more extensive bibliography can be found in Marie Kaiser, "The Limits of Reductionism in the Life Sciences," *Hist. Phil. Life Sci.* 33: 453–476 (2011). More recently, Pollack applies his work to the origin-of-life problem in G. Pollack, X. Figueroa, and Q. Zhao, "Molecules, Water, and Radiant Energy: New Clues for the Origin of Life," *Int. J. Mol. Sci.* 10: 1419–1429 (2009).

2. Lynn Margulis and Dorion Sagan, *Slanted Truths: Essays on Gaia, Symbiosis and Evolution* (New York: Springer Verlag, 1997), chap. 20.

3. James A. Shapiro, *Evolution: A View from the 21st Century* (Chicago: FT Press Science, 2011).

4. Robert Chambers and Edward L. Chambers, *Explorations into the Nature of the Living Cell* (Cambridge, MA: Harvard U. Press, 1961); Paul Weiss, *The Science of Life: The Living System—A System for Living* (Mt. Kisco, NY: Futura, 1973); Hans Selye, *In Vivo: The Case for Supra-Molecular Biology* (New York: Liveright, 1967); Gilbert Ling, *A Physical Theory of the Living State: The*

Association-Induction Hypothesis (Waltham, MA: Blaisdell, 1962); Harold Hillman, *Certainty and Uncertainty in Biochemical Techniques* (Henley on Thames, UK: Surrey U. Press, 1972). By Hillman, see also "The Unit Membrane, the Endoplasmic Reticulum, and the Nuclear Pores are Artifacts," *Perception* 6: 667–673 (1977); "Biochemical Cytology: Has It Advanced in the Last 35 Years?" *The Biologist* 65: 1–16 (1983); "Some Microscopic Considerations about Cell Structure: Light versus Electron Microscopy," *Microscopy* 36: 557–576 (1991); also Harold Hillman and Peter Sartory, *The Living Cell: A Re-examination of Its Fine Structure* (Chichester, UK: Packard, 1980).

5. Marston Bates, *The Forest and the Sea* (New York: Random House, 1960).
6. Disciplinary tensions over turf between physiology and cell biology are clearly also at stake in this case, but they do not negate the importance of the dispute about experimental evidence.
7. Franklin Harold, "Postscript to Schrödinger: So What Is Life?," *ASM (Am. Soc. for Microbiol.) News* 67: 611–616 (2001), p. 613.
8. Franklin Harold, "Review of *Life at the Cell and Below-Cell Level* by Gilbert Ling," *Cell Biol. Int.* 26: 1007–1009 (2002), pp. 1007, 1009.
9. F. Harold to Strick, pers. comm., 3 Jan. 2002; Jan Sapp, "Cytoplasmic Heretics," *Perspect. Biol. Med.* 41: 225–290 (1998).
10. Rothman, *Lessons from the Living Cell*, pp. 20–21.
11. Strick, "Pijper," *op. cit.*, pp. 300, 302–304.
12. See, e.g., Joan Acocella, "The Empty Couch," *New Yorker*, 8 May 2000, pp. 112–118; Erica Goode, "Sharp Rise Found in Psychiatric Drugs for the Very Young," *New York Times*, 2 Feb. 2000; Marcia Angell "The Epidemic of Mental Illness: Why?" in http://www.nybooks.com/articles/archives/2011/jun/23/epidemic-mental-illness-why/?pagination=false and "The Illusions of Psychiatry," http://www.nybooks.com/articles/archives/2011/jul/14/illusions-of-psychiatry/?pagination=false.
13. Strick oral history interview with Smith, 30 Jan. 1997. See Chapter 2, note 63 for publications by Smith on the origin of life.
14. Roger Straus to Mary Higgins, 23 July 1973 (courtesy of Higgins).
15. Bernard Grad, "Wilhelm Reich's Experiment XX," *CORE (Cosmic Orgone Engineering)* 7: 130–143 (1955).
16. Bernard Grad, "Calcium Carbonate and Pleomorphism," in G. Jensen, ed., *Pleomorphic Microbes in Health and Disease, Proceedings of the First Annual Symposium Held June 18–19, 1999 in Montreal, Quebec, Canada* (Holger N.I.S.: Port Dover, Ontario, Canada, 1999), pp. 55–59; also B. Grad, "Studies on the Origin of Life: The Preparation of Primordial Cell-like Forms," *Pulse of the Planet* 5: 79–87 (2002).
17. Grad, "Studies on the Origin," *op. cit.*, p. 82.

18. J. Činátl, "Inorganic-Organic Multimolecular Complexes of Salt Solutions, Culture Media and Biological Fluids and Their Possible Significance for the Origin of Life," *J. Theor. Biol.* 23 (1): 1–8 (Apr. 1969), quotes on p. 1.
19. Ibid., p. 8.
20. Ibid., p. 9.
21. A. Smith and D. Kenyon, "Is Life Originating *De Novo?*," *Persp. Biol. Med.* 15: 529–542 (Aug. 1972). At this time, Kenyon had not yet switched his views from being a supporter of evolution to a supporter of creation science/intelligent design, which had happened by 1980 or so.
22. G. H. Pollack, Xavier Figueroa, and Qing Zhao, "Molecules, Water and Radiant Energy: New Clues for the Origin of Life," *Int. J. Molec. Sci.* 10: 1419–1429 (2009), pp. 1423–1424.
23. Ibid., p. 1425.
24. C-Y. Wu, L. Young, D. Young, J. Martel, and J. D. Young, "Bions: A Family of Biomimetic Mineralo-Organic Complexes Derived from Biological Fluids," *PLoS ONE* 8 (9): e75501. doi:10.1371/journal.pone.0075501 (25 Sept. 2013), http://www.plosone.org/article/info:doi/10.1371/journal.pone.0075501, p. 1.
25. Jan Martel, David Young, Hsin-Hsin Peng, Cheng-Yeu Wu, and John D. Young, "Biomimetic Properties of Minerals and the Search for Life in the Martian Meteorite ALH84001," *Ann. Rev. Earth and Planet. Sci.* 40: 167–193 (2012). For more recent news on this problem, see also http://online.liebertpub.com/doi/full/10.1089/ast.2013.1069.
26. However, Reich did appear by 1950 to be reconsidering the importance of genes. See Chapter 4, note 55.
27. See Dick and Strick, *Living Universe, op. cit.*, chap. 3, on the ongoing "gene first" versus "metabolism first" debate. More recently, see Neeraja Sankaran, "How the Discovery of Ribozymes Cast RNA in the Roles of Both Chicken and Egg in Origin-of-Life Theories," *Stud. Hist. Phil. Biolog. Biomed. Sci.* 43: 741–750 (2012).
28. Wu et al., "Bions," *op. cit.*, p. 13.
29. Ibid., p. 12.
30. Ibid., p. 15.
31. Correas, "The PA and SAPA Bion," *op. cit.*, p. 6.
32. Ibid., p. 13.
33. Ibid., p. 12.
34. Ibid., pp. 14–16, quote on p. 7. On the discoveries of extremophiles, see Dick and Strick, *Living Universe, op. cit.*, pp. 65–67, 106–109; see also Lynn Rothschild and Rocco Mancinelli, "Life in Extreme Environments," *Nature* 409: 1092–1101 (22 Feb. 2001).
35. Correas, "The PA and SAPA Bion," *op. cit.*, p. 27.

36. Ibid., pp. 5–6.
37. Ibid., p. 6.
38. Reich, *CB* (1973 FSG ed.), pp. xvi–xvii.
39. See Chapter 4, note 55 on this issue.
40. Correas, "The PA and SAPA Bion," *op. cit.*, p. 31.
41. Ibid., p. 35.
42. Ibid., p. 41.

ACKNOWLEDGMENTS

The primary source documents consulted in the research for this book have been in German, English, French, and Norwegian. Thus, I must express my gratitude for translation assistance from a number of able colleagues: with the German, to Sander Gliboff, Tina Lindemann, Manfred Fuckert, Conny Huthsteiner, Otto Alber, and especially Diane Schnellhammer; with the French, to Lisa Gasbarrone and especially to Evelyn Sellers; with the Norwegian, to Bodil Erikson, Håvard Nilsen, and Karl Fossum. I am grateful for feedback from many colleagues after presentations of parts of this material at the History of Science Society Annual Meetings (2005, 2009, and 2013), the Countway Library of Medicine (2010), and the Ischia Summer School for History and Philosophy of the Life Sciences (2013).

Håvard Nilsen showed me warm hospitality and assisted me in many ways during the research in Oslo. Nils Roll-Hansen generously helped with many contacts and background on the Oslo story, particularly on Otto Lous Mohr. Dr. Carl van der Hagen gave me access to the correspondence of Mohr at Oslo University. Dr. Frøydis Langmark shared recollections of her mentor, Leiv Kreyberg, and Dr. Helge Waal did likewise regarding his mother, Nic Waal (Nic Hoel), one of Reich's students in Norway. Dr. Stephen Nagy helped me understand countless important points about light microscopy technique and generously supplied some microphotos of bion preparations, including Figure 2.4. He also helped me learn the techniques of the bion experiments at a course at the Wilhem Reich Museum in 2002. Dr. Dan Friedman provided me with much helpful background on the story of mitogenetic radiation. Kevin Hinchey of the Reich Trust and Jack Eckert at the Countway Library of Medicine were extremely helpful with access to the Reich Archives. Heidi Stoa was very helpful at the National Library of Norway archives. The staff of the Rockefeller Archive Center made working there easy and pleasant. Susan Goldsmith was an extremely capable and critical research assistant, particularly in intelligently transcribing interviews on a wide range of subject matter from microscopy to psychology.

Dr. David Brahinsky introduced me to Reich's work and has helped me with too many conversations to count and offered more support than I can repay. Dr. C. Grier Sellers demonstrated making Reich's egg medium IV under sterile conditions and helped clarify many technical points in numerous exchanges. Dr. Philip Bennett shared helpful ideas and sources, including a comprehensive catalog of Reich's personal library. Mary Henderson

ACKNOWLEDGMENTS

was extremely helpful with all matters at the Wilhelm Reich Museum. The late Dr. Chester Raphael provided much useful information about Reich and about his microscopy work, particularly regarding the Reich Blood Tests and Reich's theory of cancer. The Wilhelm Reich Infant Trust generously granted permission to reproduce photos and to quote from Reich's works and letters, both published and unpublished, as well as to post photos, film, and unpublished manuscripts on the website accompanying this book (http://wilhelmreichbiologist.org). I had much technical help from Phil Eskew in creating the website. The trust also gave me access to the late Dr. Charles Oller's microphotos and text from 1970–1971 work replicating Reich's "natural organization of protozoa" experiments. Dr. Morton Herskowitz shared personal papers and his recollections of Reich.

My wife, Alison Glick, has been a keen-eyed editor as well as supplying steady love and encouragement in a task drawn out over years. My children, Rachel and Alexander, and my stepdaughter Tamar inspired me daily in this work and in life in general. They give me astute advice and hope for the future.

Most of all, I wish to thank Mary Higgins, longtime trustee of Reich's estate, who has for many years helped me find archival documents, allowed me to examine Reich's microscopes and filming equipment, time-lapse films and library, and done everything possible to facilitate the work of a scholar in this long process. Ms. Higgins deserves credit for preserving Reich's scientific equipment in the Wilhelm Reich Museum, for preventing further illegal book burning by the U.S. Food and Drug Administration, and, with the help of publisher Roger Straus, for bringing Reich's work back into print, beginning in 1960, after its destruction by the U.S. government. Without those long labors of hers, I might never have come across this story. Many conversations with her have contributed to my sense of Reich as a scientist and as a human being. Responsibility for any errors that remain here is, of course, mine alone.

Funding for translation help came from the Franklin and Marshall College Office of College Grants Resource Fund. Funding for travel to archives in Oslo during a Franklin and Marshall sabbatical year, 2008–2009, was provided by the Carnegie Mellon Foundation. I am most grateful for both.

In 2003, as my research at the Wilhelm Reich Museum progressed for this book, Ms. Higgins asked if I would be willing to advise her in setting up a committee to review scholarly applications for access to Reich's archives, once those opened in November 2007. There was a need for an outside scholar to explain to those in Reich's Trust the typical procedures scholarly archives use to regulate access. As these discussions progressed, I was invited to be a member of a three-person committee that would evaluate applications for their scholarly merit, once the archives were opened. Thinking I could contribute a voice from a scholar educated about Reich but from outside their organization, I agreed. I recused myself from ruling on my own application for access, of course, but it has been a great opportunity to learn of other scholars' projects on wide-ranging aspects of Reich's work. The criteria and procedures for application to conduct research in the Reich archive are posted at the Wilhelm Reich Infant Trust website.

INDEX

Boldface page numbers reference figures

achromatic lenses, 112, 128, 411n12
Addison, William, 188
AIDS, 216–217
ALH84001, 77, 319
Allen, F. J., 27–30
Allendy, René, 116–117, 247–249, 388n60
amoeba, 11–12, 74–83; natural organization of, 81–83; parallel with cancer cells, 82–83, 190
anti-Semitism, 6, 108, 259, 263–264, 437n161
apochromatic lenses, 128, 275, 304, 377n71, 411n12, 413–414n32
apolitical science, 156–157
Arbeiderbladet, 230, 232, 253
Archaea, 323, 352n14
Aristotle, 24, 34
armor, 72–74. *See also* character armoring; muscular armoring
Arrhenius, Svante, 151, 283, 398n14
Astbury, W. T., 150, 161
Austria, 14, 16, 239, 259; Anschluss by Nazis 1938, 239
autonomic nervous system, 59–64, 70, 73–74; contractility of, 60–61, 70, 74; Reich's functional clarification of sympathetic vs. parasympathetic innervations, 60–63. *See also* biopathies
autumnal moss or grass, 77, 82–83, 189
Avery, Oswald, 92, 225, 320, 325–326

Bacillus proteus (now *Proteus vulgaris*), 107, 384n26
Bacillus subtilis, 118–122, 144
bacterial cyclogeny, 91–92

"Basic Antithesis of Vegetative Life, The," 13, 51, 58–63, 76, 161, 174, 320, 329; role of choline vs adrenaline, 13, 59–63; mistranslated as "vegetable life," 59, 371n150; ; role of Ca^{+2} vs K^+ ion, 59–63, 76, 213, 329; role of lecithin vs cholesterin, 59–63, 76–77, 86, 101–102, 132, 150–151; role of H^+ vs OH^- ion, 62, 372n164, 421n111
Bastian, Henry Charlton, 90–91, 114, 132, 380–381n114, 402–403n56; conflict with French Academy of Sciences, 90, 122; conflict with John Tyndall, 118, 402–403n56
Bates, Marston, 312
Beale, Lionel, 188, 390n94
Béchamp, Antoine, 114
Bennett, John Hughes, 114, 188, 411n12, 412n23
Bergersen, Birger, 240–241
Berggrav, Kari, 82–83, **131**, 180, 377n72, 381–382n14
Bergson, Henri, 21, 24, 35
Berle, Vivi, 129, **131**, 180, 258; difficulty seeing motility of bions, 129; bacteriology training, 180
Bernal, John Desmond, 4, 45–48, 51, 150, 155, 159–164, 168, 253–254, 266–268, 401–402n51, 402n53, 403n61, 404n66, 434n132; *The Social Function of Science*, 159; opinion of *Die Bione*, 159–163, 401–402n51; defense of Lysenko, 162–163, 166
Bernfeld, Siegfried, 60, 368–369n129
Bertalanffy, Ludwig von, 41–42, 44, 364–365n94, 365n98
Beutner, Reinhard, 256, 375n48, 385n42, 447n123
Bibring, Edward, 17

455

INDEX

Bibring, Grete Lehner, 17
Bildungstrieb, 23, 358–359n33
bioelectrical experiments, 64–72; relevance for personal relationships, 68, 407n92; Reich conflict with Hoffmann, Löwenbach, 71–72, 231–232
biological detectors, 45–46, 305–306
"Bion cookbook," 130–133, 255, 346, 391nn105–106
Bion Experiments on the Cancer Problem, 132, 210, 420n102
bions, 79–329; "explained away" semantically, 34, 111–112, 293, 385n43, 444–445n94; as synthetic ("gemischer") approach to origin of life, 76–77, 160, 376–377n63, 402n53; lifelike properties of, 80, 84–88, 94–95; migration in electrical field, 80, 89–90, 94, 111–112, 308; ability to grow in pure culture, through successive generations, 94–95, 99–101, 105–110, 118–127; same as histological "molecules," Brown's "Active molecules," "microzymas," 114, 188, 411n12, 412–413n23; distinguished from decay bacteria, 143; opposition of PA vs. T-bacilli, 144, 146, 397n2; energy field surrounding, 272, 275, 326; subject of artwork, 354n34. *See also* cataphoresis; *Sarcinae lútea*; *Sarcina ventriculi*
bion types, 89–90, 333–334; preparation 6, 86–88, 131, 333; earth (Prep. 1), 80–81, **85**, 333; packet amoebae, **93**–98; 3e (soot), 133–**136**, 196–197, 333; incandescent carbon (soot, blood charcoal, etc.), 133–138, 210–212; iron, 137, 289, 320, 334, 395n125; BOPA, 144–145, 197, 333, 444n82; BLUKO, 144–145, 210–212, 333–334, 444n82; HAPA, 197, 210, 444n82; blood, 197, 333, 417n55; SEKO, 210–212, 333, 396–397n148; SAPA, **214**, **273**, 270–299, 334, 444n82; grass, moss (Prep. 5), 333; SEPA, 334
biopathies, 63, 70, 204–207, 217, 372n165; parasympatheticotonia, 63, 70; sympatheticotonia, 63, 70; examples: cancer, asthma, paranoid schizophrenia, 63, 70, 204–207; disturbance of basic pulsatory functioning, 63, 201, 206; as diseases of the total organism, 70, 204–207, 217; as "self-infection" diseases, 117, 119–122, 190, 217

biosignatures, 77, 319–320
biotic energy, 31–33
blue field entoptic phenomenon, 111–112
Blumenthal, Ferdinand, 170, 214, 406–407n88
Bohr, Niels, 274, 276, 441n26
Bon, Willem F., 270–302; belief that life is a radiation phenomenon, 270, 291–292; interpretation of SAPA radiation as negative electricity, 295–296
Bonner, John Tyler, 43
Bonnet, Henri, 142–143, 239, 242–252
Bourdet, Claude, 117, 388n60
Brady, Mildred E., 3
Brandt, Willi, 65, 373n5
Brefeld, Oskar, 91–92
Breuer, Josef, 25–26
Brown, Robert, 110–114
Brownian movement, 5, 98, 109–118, 122–127, 134–138, 151, 182, 319, 330; contrasted with motility and pulsation of bions, 110, 115–116, 127, 138, 399n39; interpreted as a spectrum of kinds of movements, 110–111, 385n39, 385n42, 388n53; Brown's discovery of, 110–114; as a fetish term, in Reich's opinion, 111; "black-boxing" of, 112, 388n53; Buffon and Needham anticipate Brown's discovery, 113
Brücke, Ernst, 23
Brupbacher, Fritz, 426n15
Büchner, Ludwig, 50, 368n119
Buffon, Comte de, 113
Bülow-Hansen, Aadel, 373n5
Burke, J. Butler, 31, 187, 279
Bütschli, Otto, 61, 76, 372n161

calcium metabolism, 317, 319–321
cancer: Reich falsely accused of claiming ability to cure, 2–3, 208, 215, 219, 321, 422n122; examination of cells in living state, 5; cancer viruses, 6, 190, 214–215, 309, 444–445n94; Reich's concept of biopathy, 70, 204–208, 372n165; role of t-bacilli in, 94, 191–194, 196–197, 205–206, 212–215, 331; bion injection therapy experiments, 108, 213–214; role of mitosis in, 186, 204–206; parallel with grass and moss infusions, 187, 190; role of tissue suffocation in, 188–197, 212–213, 414n33; spindle, caudate shaped cells, 190,

INDEX

412–413n23; tumor only a late manifestation of a systemic disease, 206–208; death by clogging of kidneys, liver, spleen, 208, 215, 321, 422n122; Reich 1950 course on, 376n63, 402n55

cancer "disposition," 197, 207

Candolle, Augustin de, 23

Cannon, Walter B., 41–42

Cantwell, Alan, 217

carbol fuchsin stain, 392n109

Carrell, Alexis, 170, 189, 406n88

cataphoresis, 80, 89–90, 94, 111–112, 308; reversal of current as control, 80, 111–112, 308

cell model experiments, 76–77, 316–317

character analysis, 1, 18–19, 55, 73, 223

character armoring, 18–19, 57, 73

Christensen, Lars, 222, 239–240, 335, 423n4

Činátl, Jaroslav, 316–318

Claus, Walter, 306, 447–448n129

coacervates, 160, 318, 320

Cohn, Ferdinand, 118

colloidal turbidity, 86–87, 89–90, 134–135, 150, 393–394n119, 398n11

colloid chemistry, 33–35, 46, 89–90, 124, 149–152, 154–155, 211, 267–268, 302, 360n51; Ben Moore links with specific life energy idea, 33; lyophilic colloids, relation to "germs," 109, 150; growth from 1900–1940 as a field of research, 149–152; Whiggish "Dark Age of Biocolloidology" discourse, 150, 154–155, 267–268, 398n9

consciousness and sensation, related to origin of life, 49–56, 171–172, 282, 286–287, 367n111

conservation of energy, 12–13, 17–19, 69, 148, 353n21. *See also* sex-economy

Correa, Paulo and Alexandra, 310, 321–326, 381n125, 441n20, 445n97

Coulter, Harris, 217, 379n106

Crile, George Jr., 76, 447n123

Crowther, James G., 45, 410–411n8

culturability of bions, 86, 88–90, 93–97, 99, 102–105, 109–110, 114–115, 118–119, 122–127; colloidal turbidity of solution as a precondition of, 86–87, 89–90, 134, 150, 393–394n119; electrical charge as a precondition of, 87, 94, 134; morphological changes due to aging of cultures, 91–92, 281; blue-gray, pure cultures of soot bions, 118–120; possible contamination of a culture by *B. subtilis*, 118–121; difficulties replicating, use of Sy-Clos system, 120, 139–141; Lapicque downplays significance of, 122–127; small, key technical details, 131–133, 281, 297–298, 304–306, 415n38

culture media: nutrient agar, 91, 94, 191, 281; plain agar, 91, 128, 272; egg white, 91, 213; bouillon, 94, 120, 134; blood agar, 118, 130, 272, 281; malt agar, 128; egg medium IV, 130, 132, 144, 212, 132, 210–212, 272, 333–334; medium e, 132; importance of freshness, 133, 281; polyvalent, 136, 394n121; autoinoculation of, 444–445n94

"Cyankali," 421n110

cyanogen, 29–30, 151, 213–214, 398

Czernowitz, 16, 407n90

Dagbladet, 230, 232

Dallinger, William, 130, 390n94

darkfield condenser, 272

darkfield observations, 103, 213, 272, 275, 281–282, 326, 413–414n32, 414n34; visibility of t-bacilli, 413–414n32, 414n34

dark room experiments, 274–278, 287, 297–299

Darwin, Charles, 8, 30, 33, 39, 355n6, 361n60

death instinct. *See* Todestrieb

Debré, Robert, 142, 239, 243–250, 335, 389n66

Decourt, Philippe, 141–142

Deel, Dr., 105, 114–115

dehydration, 53, 57, 60, 62, 76, 86, 376–377n63; calcium as agent of, 57, 60, 62, 76

dialectical materialism, 46–54, 146–166; as via media between mechanism and neo-vitalism, 21, 46–54, 162; vs. "crude mechanical materialism," 22, 50, 368n119; largely discredited in the West by association with Lysenko, 40, 46, 162–163; basic principles, 46–47; three dialectical laws, 47–48; Reich's version compared with Oparin, 153–155, 160–161; Reich's version compared with Bernal, 153–155, 160–166

Dialectical Materialism and Psychoanalysis, 14, 20–21, 49–53

Dialectics of Nature, 44, 47, 50, 366n109, 402–403n56

INDEX

"Der dialektische Materialismus in der Lebensforschung," 155
Dictyostelium discoideum, 43
Dikemark mental hospital, 65, 72, 230–233, 286
Diplococcus pneumonia ("pneumococcus"), 92
"disposition to disease," 197, 207, 419n86
DNA, 92, 268, 316, 324–326
Dobzhansky, Theodozius, 40
Driesch, Hans, 24, 34, 44, 159, 358–359n33
Drosophila, 40, 268; "fly network," 223, 258, 266
Drysdale, William, 130, 390n94
DuBois-Reymond, Emil, 23, 25–26, 64, 113
Dubos, Rene, 117, 419n86
Dunlap, Steven, 415n38
du Noüy, Pierre LeComte, 99, 157, 399n39
du Teil, Roger, 9, 91, 93–97, 99–100, 103–109, 114–127, **131**, 133–144, 147–148, 173–185, 191, 197–199, 208–211, 200, 237–255, 290–291, 301–302, 335; disagreement with Reich over terminology, 96–97, 118–122, 182–183; July-August 1937 visit to Reich's lab in Oslo, 115, **131**, 137, 394–395n124; experimental demonstrations in Paris, 9–10 Aug. 1937, 116–118; invention of h-tube, SyClos system, 139–141; disagreement with Reich over strategy, 175–185, 243–252; believed Reich's personality triggers much opposition, 181–185; dismissed from his job, June 1938, 243–251; question of status as "professor," 251–252, 429n45; Reich loses contact with, 252, 301–302
Dyson, Freeman, 324
Dysthe, Roald, 242–249

earth bions, 80–81, **85**, 333
egg medium: formula for producing, 132; dissolution of, by BLUKO, 210–212
Ehrenberg, Christian, 113
Einstein, Albert, 302, 366n104
élan vital, 24, 358–359n33
electron microscopy, 5, 102–103, 267–268
electrophoresis, 226, 267–268
electroscope, 279, 283–287, 295–296, 300
Enderlein, Gunther, 92
endogenous infection, 91, 117, 119–122, 188–190, 293, 295, 309, 330, 444–445n94; autoinoculation of culture media as model for, 210–212, 444–445n94; new model of, 321

endosymbiosis, 92, 322–323
energetic functionalism (also called biophysical functionalism, later orgonomic functionalism), 16–17, 50–52, 58, 157, 397n6, 399n23
energia, 30, 360n51
energy, conservation of. *See* conservation of energy
energy field around bions, 275, 326
energy principle, 10–12, 17–19, 148, 162, 269, 302
Engels, Frederick, 41, 47, 153, 161, 402–403n56; *Dialectics of Nature*, 41, 49, 161, 366n109; critique of Feuerbach, 50, 366n109
entelechy, 24, 34–35, 44, 358–359n33
Entwicklungsmechanik, 24, 41
Eosin-haematoxylin stain, 145
erective potency/impotence, 13, 17, 67–68, 371–371n156
Ernst, Norbert, 395n127
erythrocytes. *See* red blood cells
eugenics, 20, 37–41, 156, 159, 223, 228–230; appeal all across the political spectrum, 229
Evang, Karl, 233, 259–260, 436n147
evasion of a problem, using words, 34, 111–112, 293, 385n43, 444–445n94
"The Expressive Language of the Living," 171, 407n91, 407–408n93

false narratives about Reich, repeated uncritically, 3–4, 308–310, 350n4
Faraday, Michael, 23, 274
Faraday cage, 69–70, 274–283, 443n60
Feitelberg, Sergei, 60
Fenichel, Otto, 17, 20, 54, 166–167; fabricated rumor that Reich was in mental hospital, 167
"field theory," 41–43
Fischer, Albert, 99–103, 128, 138, 224–225, 335, 381–382n4, 382n5, 424–425n110
Fleck, Ludwik, 7, 92, 445n98
fluorite lenses, 103, 396n130; ability to transmit ultraviolet radiation, 396n130
Food and Drug Administration, U.S. (FDA), 1–4, 307–310, 437n150
Four-beat formula. *See* orgasm formula; tension-charge formula

458

INDEX

Fränkel, Ernst, 214
French Academy of Sciences, 8–9, 90, 99, 122–127, 210, 220, 340–341
Freud, Anna, 167; involvement in engineering Reich's expulsion from the IPA, 167
Freud, Sigmund, 1, 17–18, 53–54, 56, 355nn6–7, 356n9, 368–369n129; underwent Steinach operation, 362–363n74
Fromm, Erich, 20
Function of the Orgasm, The (1942), 372n163, 421n111
Funktion des Orgasmus, Die (1927), 19, 53

Gaasland, Gertrud, 65–66, **131**, 300; wife of Willy Brandt, 65; moves Reich's lab to New York City, 300
Gál, Erna, 265, 439n168
Galton, Francis, 40
Geison, Gerald, 4, 9, 308
gelatin, 83–84, 86–87, 131, 152; red gelatin in Prep. 6, 131, 384–385n33
"gene first" approach to origin of life, 320, 324–326, 451n27; vs. "metabolism first" approach, 320, 451n27
genes: Reich skepticism of eugenic excesses, 20, 37–41, 161, 258, 320, 324–326, 402n55, 435n141; Reich reconsiders skepticism about, 402n55
Geschichte des Materialismus, 49, 367n111
Giemsa stain, 100–102, 340
Gjessing, Rolf, 65, 226, 230–233
Gleditsch, Ellen, 409–410n6, 426n18
Goethe, Johann von, 23
Goldschmidt, Richard, 44
Gording, Reidar, 243, 243, 250
Grad, Bernard, 9, 316–319, 376n63; calcium carbonate "primordial form" experiments, 316–319
Graham, Loren, 21, 46–48
Graham, Thomas, 30, 33
gram stain, 87, 96–97, 143–145, 191, 413–414n32; related to electrical charge of bions, 143; recipe for, 392n109
granulation tissue, 199–201, 236, 418n67
Greenfield, Jerome, 260, 350n3
Griffith, Frederick, 92
Guernica, 156–157

Gurwitsch, Alexander, 42–43, 45–46, 152, 303–307, 364n87, 410n8

Hadley, Philip, 92
Haeckel, Ernst, 31, 49, 49, 76, 375–376n52
Hahn, Arthur, 80, 377n64
Haldane, John B. S., 4, 159–161, 163, 253–254, 366n109, 401n49; opinion of *Die Bione*, 159, 253–254; ambivalence about Lysenko, 163, 403–404n65
Hansen, Frank Blair
Hansen, Klaus, 196, 218, 259, 436n144, 436n146; arrest for Nazi collaboration, 436n144
Harold, Franklin, 312–313
Hartmann, Max, 13–14, 61, 75, 143, 152, 187–188, 398–399n19. *See also* "relative sexuality" theory
Harvey, William, 22
Hasvold, Nina, 265
Havrevold, Odd, 105–108, **106**, 119, **131**, 180–184, 195, 198, 201–202, 219–220, 222, 230–231, 235–237, 239–41, 250, 257–258, 264, 273, 275, 299, 335, 432n86, 435n140
Hegel, G. W. F., 148
Helmholtz, Hermann, 23, 357n23, 358n28
Henderson, L. J., 41–42, 364n87
Hering, Ewald, 25, 30–31, 358n28, 367n111
Herrera, Alfonso, 76, 160–161, 293–295, 342, 375–376n52, 376n53, 402n53, 445n100
Hertwig, Richard, 204–205
Hessen, Boris, 45–46, 365n100, 366n104
Higgins, Mary Boyd, 10, 315–316, 347, 454
Hillman, Harold, 312–313, 449n1, 449–450n4
Hodann, Max, 223, 258, 425–426n14, 426n15
Hoel, Nic, 55, 65, **203**, 202–204, 230, 250, 264–265, 369n137, 430n55, 428n34. *See also* Waal, Nic
Hoel, Sigurd, **55**, 65, 208–209, 220–221, 233–234, 241, 262, 253, 335, 369n137, 430n55, 431n69, 436n147; circulates open letter in support of Reich's visa, 233, 429–430n50; novel *The Troll Circle*, 369n137; challenges Schjelderup, 436n146
Hoffmann, Wilhelm, 65, 69, 71–72, 169, 232, 258, 436n147
Hogben, Lancelot, 45–46, 164–166
Hølaas, Odd, 195–196, 218, 423n3

459

INDEX

holism, biomedical, 5, 41–46, 63, 117, 207
holistic materialism, 41–46, 356n15
Hollaender, Alexander, 304–306, 410–411n8, 447n122
Holmes, Frederic ("Larry"), 4
hormone research, 13, 59, 353–354n25
h-tube, 140–**141**
Humboldt, Alexander von, 408n94
humidity: role in absorbing orgone radiation, 300, 446n113; low in Oslo, 300
Huxley, Julian, 41, 45, 400n45
Huxley, Thomas H., 30, 76, 110, 114, 400n45, 402–403n56
hydrogen ions, acidity, 62, 421n111

immanent criticism, Reich's demand for, 173–180, 183, 251, 253
incandescence, heating to the point of, 86, 133–137, 156, 211–212, 256, 269–270, 281, 289, 292–293, 309–310, 333
independent scientist, 164–168, 405n79
intuition. *See* Reich, Wilhelm: intuition in research

Janus Green stain, 111
Jennings, Herbert S., 76, 364n82, 364n87, 372n161
Joffe, A. F., 45–46
Jones, Ernest: attempts to rewrite Freud's views on neo-Lamarckism, 41; involvement in engineering Reich's expulsion from the IPA, 167, 369n131; attempts to write Reich out of the history of psychoanalysis, 406n81
J-rays, 306

Kallikak family study
Kammerer, Paul, 21, 33–40, 44, 48, 187, 284, 361n61, 362–363n74; as Reich's professor, 21, 33–36, 187; support for neo-Lamarckism, 33, 37–41; concept of a specific biological energy, 33–35, 284; belief in abiogenesis, 37; suicide of, 37, 39; experiments with Steinach, 37–40, 362–363n74
Keller, Philip, 64
Kendall, Arthur I., 92
Kerrigan, William M., 5–6, 260–264; suspicions of anti-Semitism, 437–438n161

Kittredge, Tracy B., 222–223, 424–425n10
kleptomania, 178
Koch, Robert, 111; dogmatic belief in monomorphism, 92, 111
Koch's bacillus (BK, *Mycobacterium tuberculosis*), 94, 116, 190
Koestler, Arthur, 40, 163
Konstad, Leiv, 259, 435n144
Kramer, Sol, 9, 376n63, 402n55
Kraus, Friedrich, 13, 57–59, 174, 354n27; "fluid theory" of life, 57–58; support for Reich, 174, 408n100
Kreyberg, Leiv, 5–6, 178, **196**, 225, 250–251, 261–262, 264, 336, 415–416n49, 416n50, 423–424n8; protégé of Mohr, 195–196, 200, 218, 222–227; appointment as professor, 195–196, 200, 218, 222–227, 423–424n8, 426n18; Kerrigan's opinion of his objectivity, 261; sabotages Nic Waal's appointment as professor, 264; and Socialist Physicians Association, 415–416n49
Kuhn, Thomas, 7–8, 184–185, 352n14

laboratory notebooks, 3, 9–10, 157–**158**, 308; agreement between notebooks and published account of bion experiments, 10, 86–87, 134–137, 189, 308–309; listing of Reich's notebooks examined for this book, 345–346
Lamarck, Jean Baptiste de, 33, 37–41
LaMettrie, Julien Offray de, 50
Lange, Christian August, 65, 233, 265, 429–430n50, 438n166
Lange, Christian Lous, 429–430n50
Lange, Friedrich, 49, 367n111
Lange, Halvard Manthey, 233, 429–430n50
Langfeldt, Einar, 196, 218, 226–227, 336, 424–425n10
Langfeldt, Gabriel, 231, 264, 336, 424–425n10
Langmark, Frøydis, 437–438n161
Lapicque, Louis, 122–127
lava, volcanic, 137, 395n127
Lebenskraft, 23, 358–359n33
lecithin, 61, 64, 79–80, 89, 100–101, 128, 146, 148, 318
LeDuc, Stefan, 76, 105
Leistikow, Gunnar, 233–235

INDEX

Lenin, Vladimir I., 41, 49, 366n104
Lepeschinskaia, Olga, 163, 317, 404n67
Leunbach, Jonathan H., 99–103, 336, 381–382n4, 382n5
Levy, Hyman, 45, 164–165
Lewis, Warren and Margaret, 157, 377n71, 400n40
Libido: Freud's initial concept, 10–13, 17–19, 26–27, 36, 52, 353–354n25; chemical idea of, 13, 26–27, 52; dammed-up river metaphor, 13, 26–27, 353–354n25; quantitative measurement of, 29; electrical idea of, 54; sending out "toward the world," 60–61
Liebeck, Lotte, 56
Liebig, Justus von, 23
life energy, specific, 31–35, 284
"life rays," 186–188, 302–307; Reich's earliest hunches, 187–188
Ling, Gilbert, 312–313
Lion, Kurt, 307
living state, observations on, 5, 154–155, 157–158, 200–202, 309, 312–313, 400n41, 409n120, 418–419n74
Loeb, Jacques, 24, 41–42
Löhnis, Felix, 92, 379–380n107
Lovelock, James E., 6, 166–167, 184–185
Löwenbach, Hans, 65, 71–72, 169
Ludwig, Carl, 23, 66
Lysenko, Trofim D., 40, 46, 162–163, 223, 266, 403–404n65

Mach, Ernst, 366n104
macromolecule concept, 149–150, 267–268, 302, 400n45
magnification: vs. resolution, 102–103, 127–129; above 2000x, 102–103, 115–117, 127–129, 138, 382n10, 390n94; upper limit, 128–129, 390–391n99; Reich's lenses used, 383n13, 413–414n32, 414n34
Malinowski, Bronislaw, 174, 222, 234–235, 405n76
Margulis, Lynn, 217, 312, 385n43
Martel, Jan, 318–321
Martiny, Marcel, 116–117, 242
Marx, Karl, 49–56
masochism: Reich's theory of, 53; Freud's theory of death instinct, 53; Freud's opposition to Reich's theory, 53, 368–369n129
Masse und Staat, 163, 231
Mass Psychology of Fascism, The, 1, 20, 403n63; burned by Nazis, proscribed by Communist Party, 20
Masters and Johnson, 70–71
materialism, crude mechanistic, 22, 50–51, 368n119
Mayr, Ernst, 40
McClintock, Barbara, 7, 171
mechanism, 32, 409n120. *See also* materialism, crude mechanistic
mechanism-vitalism controversy, 22–36, 75, 360n53; Reich's sense of what he had contributed to its resolution, 75, 147–148
mechano-electrical leap, problem of, 151–152
memory as a property of matter, theories of, 367n111
"memory traces," 27, 49, 52, 353n25
Mendel, Gregor, 38–42
Menninger Clinic, 3
"metabolism first" approach to origin of life, 320, 324–326, 451n27
methylene blue stain, 87
microcinematography, time lapse technique, 81–82, 377n71
Micrococcus brownianus, 445n100, 445n102
Miller, H.M., 225
Misch, Walter and Käthe, 13
Mitchell, Peter, 6, 166, 185
mitochondria, 111, 322–323
mitogenetic radiation, 43, 186–188, 205, 279, 303–307, 364n87, 410–411n8
mobility of scientific discovery (limited by high quality microscopes), 116, 138, 395n129
modern synthesis (neo-Darwinian), 40–41
Mohr, Otto Lous, 195–196, 200, 218, 219, 222–227, 258, 262–266, 401n49, 424–425n10, 425n13, 425–426n14; September 1937 press release, 195–196, 218; and Max Hodann, 223; member of "fly network," 223, 258, 266; opposition to eugenics, 223, 263–265; and 1937 Moscow Genetics Congress, 223, 266; Kerrigan's opinion of his objectivity, 262–264
Mohr, Tove, 223, 263–264, 425n13, 425–426n14
Monod, Raoul-Charles, 116, 388n60

monomorphism, 92, 111
Moore, Ben, 31–33, 149
moralism in biology, 37–40, 152, 363n77. *See also* genes: Reich skepticism of eugenic excesses
Morgan, Thomas H., 40, 402n55
Mot Dag, 55, 259
Motecitzky, Karl (also pen name Karl Teschitz), 265, 438n167
Möwus, Franz, 370–371n146
Moxnes, N. H., 279, 285, 442n40
Muller, Hermann J., 159, 259, 278, 320, 403–404n65, 405n75; eugenic ideas of, 159, 259, 425n13; research on genetic effects of ionizing radiation, 278; proponent of "gene first" origin of life approach, 320
Müller, Johannes, 22
Müller, L. R., 372n163
Munich crisis (Sept. 1938), 103, 252, 395n129, 434n132
muscular armoring, 1, 57, 70, 73–74, 190, 197
Mycobacterium tuberculosis, 111, 183. *See also* Koch's bacillus
Myxosarcina, 310, 322–326

Nägeli, Carl, 31
Nagy, Stephen, 85, 396n130
Nanney, David, 370–371n146
nanobacteria, nanoparticles, 6, 318–321
natural organization of protozoa, 78–83, 189, 323; parallel with development of cancer cells, 82–83, 189; spring grass vs. autumnal grass, 82–83, 189
Naturphilosophie, 36, 357n23
Needham, John Turberville, 113
Needham, Joseph, 44–46, 164, 365–366n101, 434n132
Neergaard, Jørgen, 127
Neergaard, Paul, 127–130, 336
Neill, Alexander S., 159–160, 234–235, 238, 253–254, 349n2; and Summerhill School, 159; attempts to interest scientists in bion experiments, 159–160, 253–254
neo-Lamarckism: Kammerer's experiments on, 37–41; Kammerer's suicide causes widespread perception of invalidation, 39; Freud's interest in, 39–41

neo-vitalism, 21, 24–27, 34–36, 45, 160, 268–269, 358–359n33, 360n63
Netherlands, under Nazi occupation, 300–301
New School for Social Research, 342
Nice (France); Natural Philosophy Society of, 103; University (Centre Univers.-Mediterranean), 99, 232, 253; Bacteriological Laboratory, 105
Nicolaysen, Ragnar, 226–227, 336, 424–425n10
Nissen, Ingjald, 221
Nordberg, Dr., 113
Norway, under Nazi occupation, 265, 438nn165–167
Norwegian Labor Party, 223, 229, 230–231, 259, 343
N-rays, 306

objective lens, 150x, 103, 128–129, 383n13, 396n130, 413–414n132
Oller, Charles I., 345, 376n63
"On the Energy of Drives," 12, 19, 49
Oparin, Alexander I., 4, 31, 46–48, 160–162, 320; support for Lysenko, Lepeschinskaia, 163
Oparin-Haldane hypothesis, 160–161, 400–401n45
opposition of PA vs T-bions, 144, 146, 397n2; Reich's recognition of the antithesis as experimental fact, 144; Reich's hypothesis based on dialectical-materialism, 146
"Orgasm as an Electrophysiological Discharge, The," 58–59
orgasm formula, 59, 149
orgastic potency, 13, 17–19, 59, 174; as the basis of all Reich's subsequent research, 18
"orgone box," as derisive label, 3, 309, 350n5. *See also* orgone energy accumulator
orgone energy, 2, 4, 6, 35, 270–300, 309, 330; specific biological energy, 35, 284; discovery of, in SAPA bion cultures, 270–299; effects on photographic plates, 274, 279–280, 296–297; visual impressions, especially in darkroom, 276–278; derivation of name "orgone," 284, 287, 350n6; distinct from electromagnetism, 284–287; absorbed by humidity, 300, 446n113

INDEX

orgone energy accumulator, 2–3, 181, 291, 307–309, 349n2, 420n98, 422n122, 428n36, 443n60, 443n63; false rumors that Reich claimed it could cure cancer, give sexual potency, 2–3, 208, 215, 321, 422n122; derisively called "orgone box," 2–3, 307, 309, 350n5

Orgonomic Diagnosis of the Cancer Biopathy, 215–216, 402n55, 422n123, 440n19

org-tierchen (Org-protozoa), 79, 376n57, 376n61

origin of life, one time only vs continuous, 153, 161

Oslo, as "small town," 8, 223, 227, 257–260, 434n133

Oslo University: Psychological Institute, 12; medical faculty, 106, 222, 227, 235, 254; O. L. Mohr, as rector of (1945–1951), 336

Ostwald, Wilhelm, 149

Ostwald, Wolfgang, 149

"outsiderness," 164–168

Øverland, Arnulf, 55–**56**, 65, 233, 265, 337, 369n137, 430n55, 438n166; in concentration camp, 265; "You Must Not Sleep!," 369n137

PA bions, 93–94, 144, 146, 322, 397n2; contrasted with staphylococci, 96–97, 108–109; injection experiments, 108, 201, 213–214, 288, 302; paralyze cancer cells, 195

packet amoebae, **93**–94, 96–**97**; characteristics shared with *Sarcina* spp., 96–98

Palmgren, Axel, 240, **78**

Pantostat, 77

paradigm-shifting science, 7–8, 174–175, 180, 184–185, 352n14, 391n103

parasympathetic, 59–63, 70

Pasteur, Louis, 8–9, 27, 88–90, 132, 402–403n56

pathology: examination of tissues in dead, fixed, stained condition, 5, 200–202, 309, 418n66, 418–419n74

Pauli, Wolfgang, 149–151, 397n8

Pflüger, Eduard, 29–30, 151, 213, 359n45, 398n14

Philipson, Tage, 271, 274

phospholipids. *See* lecithin

physicians, vs laboratory scientists, 174, 198

physicists, intellectual migration to biology, 157–158, 267, 311

Pijper, Adrianus, 157, 409n120

plastids, 322

plastidules, 31, 49

pleomorphism, 91–92, 120, 129, 319, 380n109

pneumococcus. See *Diplococcus pneumonia*

Pneumocystis carinii, 216–217

Pollack, Gerald, 312–313, 318; speculates on possibility of current-day origin of life, 318

polyvalent culture medium. *See* culture media: polyvalent

Pouchet, Felix, 90–91, 122; hay infusions, dust as ingredient for spontaneous generation, 90–91; infusions opened on glaciers, 90–91

prions, 6, 309, 321, 444–445n94

protocells, 320, 324–326

protoplasm theory of life, 30

protozoa, natural organization of, 83, 182, 310, 321, 415n38

pseudo-amoebae, 83–84

pseudopod: metaphor of extended pseudopod as "reaching out," 60–61, 165, 352–353n18, 371–372n156

"pseudoscience," 4–6, 8, 307–310, 350–351n6; as "epistemological wastebasket" category, 4, 8, 307–310; Reich's bion experiments mistakenly called, 4–6, 8, 256, 307–310

psychophysical parallelism, 21, 52–54, 60

quartz lenses, condenser, 297

quartz slide, 43, 273

Quisling, Jørgen, 231

Quisling, Vidkun, 231, 244

radiobes, 31, 187, 279, 303

Raditsa, Leo, 172, 350n2, 408n95

radium, 187, 208–209, 287–288, 442n40; radium industry, 187, 208–209, 287–288

Radium Hospital, Oslo, 194, 201, 279

Rahn, Otto, 303, 307, 364n91, 410n8

Raknes, Ola, 242, 247–250

Raphael, Chester M., 206–208, 216–217, 275, 350–351n6, 418–419n74, 419n86, 441n23, 448n135; on the "subjective factor" in irrational responses to Reich's work, 172–173; on cancer, 206–208; on Reich Blood Test, 216, 275, 441n23, 448n135; on *Pneumocystis* pneumonia, 216–217; critical response to Rene Dubos, 419n86

INDEX

Rather, Leland J., 202, 400n41, 411n12, 412–413n23, 418–419n74
red blood cells, 193–194, 197, 214–216
reductionism, physical-chemical, 44, 357n19
Reich, Annie, 167
Reich, Wilhelm, 1–15, **11**, 16–39, 48–98, **100**, 99–145, **131**, 132–310, 337; critical of Stalin, earlier than most leftist sympathizers, 1, 15, 163, 165–168, 231, 404n69, 405n75; intuition in research, 7, 168–174; interested in movement, more than structure, 11, 170–171, 399n39, 407n91, 407–408n93; critical of Marxist parties, 15, 20, 167–168; familiarity with animal life from farm upbringing, 16, 171; training under Freud, 17, 356n9; training under Wagner von Jauregg, 17, 356n9; leader of psychoanalytic technical seminar, 18; as gifted clinician, 18, 57, 174, 355n7; concern to understand energy behind drives, 19; skepticism about genes, 20, 37–41, 161, 258, 320, 324–326, 402n55, 435n141; expelled from Communist Party, 20, 167–168, 235; not a vitalist, 35–36, 358n33, 362n68; break with Freud, 53–54, 355n6, 368–369n129; expulsion from International Psychoanalytic Association, 54, 167, 369n131; father fixation on Freud, 56; ability to see what others overlooked, ignored, 57, 171–172, 255–256, 259; criticized as aggressive, provocative, 71, 174–185, 203–204, 241, 250, 260, 264–265; attempts to measure, quantify perceptive ability, 73–74, 171–172, 282, 286–287, 407–408n93, 408n94; reasons for resisting bacteriological terminology, 93–98, 111–112, 118–122, 288–295; balance between skepticism of authorities and openness to contact, 165–168, 237–238, 250–251, 253–257, 289–291; financial support, 166, 222, 423nn4–5; desire for immanent criticism, 173–180, 183, 253; and "formal" scientific writing style, 183, 255–257; concern over profiteering, 208–209, 287–288, 420n98; views on adolescent sexuality, 223–224; regarded as a Trotskyite, 230–231, 404n69; talent for training new analysts, 355n7
Reich Blood Test, 197, 216, 275, 441n23, 448n135
Reichert model Z microscope, 103–104, 116–117, 127–129, 138, x, 383nn12–13, 390n94

"Reichian," "disciple" begs the question, 235, 308, 420n94, 430n55
Reilly, James, 142, 396n138
"relative sexuality" theory, 13–14, 61, 143, 152, 187–188, 398–399n19 (*see also* Hartmann, Max); "baby thrown out with the bathwater," 370–371n146
resolution vs. magnification (microscopy), 102–103, 127–129, 390–391n99, 413–414n32
respiration: impeded by muscular armoring, 73–74, 190, 197; aerobic respiration impeded in cancer cell metabolism, 188–194, 205, 213
Rhumbler, Ludwig, 74–76
Richter, C. P., 64
Rife microscope, 129, 390–391n99
Ristelhueber, Rene, 133
Rivers, Thomas, 92
Roberts, William, 132, 392n110
Rockefeller Foundation, 105, 222–227, 265–269, 291, 424–425n10; "old boy" network, 222–227, 265–269, 424–425n10; as cause of divergence between life science research trajectories, 225–226, 267–269
Rockefeller Institute: Copenhagen, 99–103; New York, 225
Roll-Hansen, Nils, 227–230, 265–266
Ronchese, Angel Denis, 105, 114–115, 138
Rothenberg, Michael, 349n2
Rothman, Stephen, 312–314
Rous sarcoma virus, 190, 214–215, 309
Roux, Wilhelm, 24, 41

SAPA bions, 131, **214**, 270–300, **273**, 330; difficulties replicating, 131–132, 305, 308, 447n126; biological effects of, 213–214, 277–278, 282, 286–287; ability to paralyze cancer cells, *T. vaginalis*, 213–214, 294, 440n14, 443n58; experimental injection into cancer mice, 214, 288, 302; first creation of, 271–273; radiation, properties of, 272–288, 295–300; visual energy field, 275, 326; radiation described as orgone energy, 283–288, 295–300; Correas' experiments with, 310, 321–326
Sarcina lutea, 93–98, 288–295
Sarcina ventriculi, 93, 380–381n114
s-bacilli (later called t-bacilli, q.v.), 94, 191–195

INDEX

Scharffenberg, Johann, 221, 227–233, **228**, 251, 258, 262, 337; supporter of social eugenics, sterilization, 227–230; declared Hitler a psychopath, 228–229, 258; and Socialist Physicians Association, 230, 415–416n49; started false rumors about sex between mental patients, 230–233; questions credentials of du Teil, 232; questions credentials of Reich, 232–233; accused of ulterior motives, 233, 235; Hoel's speculations about his motives, 235, 262; fears of heritable mental problems in his family, 262; Kerrigan's opinion of his objectivity, 262

Schaxel, Julius, 41, 44, 162, 364–365n94, 365n96

Scheerer's phenomenon, 111–112

Schjelderup, Harald, 11–12, 72, 108, 241, 259, 265, 352n16; support for Reich, 11–12, 108; later distances himself from Reich, 259; challenged by Hoel, 436n146

Schjødt, Annaeus, 241, 243–244; responds to Scharffenberg's challenge of Reich's credentials, 234; as prosecutor of V. Quisling, 244

Schreiner, Anaton, 108

Schreiner, Kristian, 106, 222, 235–237, 242–243, 430n61

Schrödinger, Erwin, 49, 158, 268, 450n7

Schwann, Theodor, 163, 411n12, 412–413n23

"seed vs soil" debate over infectious disease, 117, 207, 419n86

self-active matter, 113, 386–387n50, 387n51, 412–413n23; psychological defensive attitude toward, 172, 289–290

Semon, Richard, 21, 49, 367n111

Serratia marcescens, 444–445n94

sex-economy, 6, 19, 51–52, 54, 177–178, 183, 233

Sexualität im Kulturkampf, Die, 1, 163, 231

Sharaf, Myron, x, 71, x, x, 405n77, 444n70

Sharpey-Shafer, Edward, 32, 36

Siersted, Ellen, 189, 204, 258

simulacra, 76–77, 375n48. *See also* cell model experiments

simultaneous identity and antithesis, concept of, 10, 53–57, 60, 330–331; psychosomatic identity and antithesis, 21

Slavet, Eliza, 41

Smith, Adolph, 9, 315, 318, 376n63

Smith, Theobald, 92

Smits, J. M. M., 280, 289

Socialist Association for Sex Counseling and Research (Vienna), 259–260, 421n110, 437n148

Socialist Physicians Association (Oslo), 230, 259, 415–416n49

Soddy, Frederick, 187

Sontag, Susan, 208

soot: bions from, 133–138, **136**, 192, 196–197, 210–212, 333; as cause of scrotal cancer in chimney sweeps, 192; relation to respiratory disturbance (Warburg), 192–194

spectrum, of forms between living and nonliving, 110–111, 317–319

Spemann, Hans, 43

Spinoza, Baruch, 352n16

Spooner, E. J. C., 434n132

spores: heat-resistant, 108, 118–122; of protozoa, supposed, 77–78; Reich's explanation of, as bions, 137, 147–148

Stai, Arne, 429–430n50

staphylococci, 108–109, 179, x, x, 325; differences from PA bions, 96–97, 108–109; Kreyberg and Thjøtta use this label for PA bions, 96–97, 108–109, 198; Reich possibly reconsiders differences, 325

State Department, U.S., 260

Staudinger, Hermann, 150, 267

Steig, William, 349n2

Steinach, Eugen, 37; experiments with Kammerer, 37, 362–363n74; rejuvenation surgery, 37, 362–363n74; Reich's impression of, 37

Stenersen, Rolf, 222, 337, 423nn3–4

Straus, Roger, 1, 315–316, 349n2

streptococci, 97, 110, 189, 309

Svedberg, The, 150, 162, 400–401n45

Sverdrup, Harald, 6, 261–262

"swan-necked" flasks, 90

swelling, 59–63, 78–81, 160; potassium as agent of, 62, 76, 81

Sy-Clos system, 120, **139**–140, **141**

sympathetic (branch of autonomic nervous system), 59–63, 70

Tarchanoff, Ivan, 65
tar-painted mice, tumors of, 195, 416n50, 423–424n8; relation to respiratory disturbance (Warburg), 192–194, 210–213
t-bacilli, 94, 146–147, 191–194, 196–197, 205–206, 212–215, 331; isolated from sarcoma tissue, 94, 191–194; grown from putrefied protein material, 191, 213; mouse injection experiments, 191, 213–214; characteristics shared with *Pseudomonas aeruginosa*, 191, 414n33; role in cachexia, death from cancer, 205–206, 331; magnification required, 413–414n32, 414n34
teleology in biology, 37–41
tension-charge formula, 58–59, 61, 79, 86; as distinction between life and non-life, 76, 142, 155–156
Thjøtta, Theodor, 106–109, 107, 225–226, 257, 261–262, 264, 337, 437–438n161; Appointment as Oslo U. prof. of bacteriology, 106, 257, 261, 434n134; financial support, 225, 426–427n21; Kerrigan's opinion of his objectivity, 261
Tidens Tegn, 195–196, 202, 218–221, 341, 343, 422–423n2
time-lapse filming, 81–82, 377n71
Tisdale, W. E., 224–226
Todestrieb (death instinct), 53
Tracey, Constance, 222
Traube, Moritz, 76
Treponema pallidum, 217
Trichomonas vaginalis, 443n58
Trotsky, Leon (Lev), 230–231, 375n52, 405n75
tuberculosis, 117, 175, 190, 209; belief in cause as endogenous infection, 117, 190; sanatoria for, 209
Tyndall, John, 88–91, 118, 132, 402–403n56; infusions opened in glacier air, 90–91, 379n103; fractional sterilization (Tyndallization), 118

Ullevål Hospital, 194
ultracentrifuge, 5, 150, 162, 226, 267–268
Union Minière du Haut-Katanga, 208, 287–288
Üxküll, Jakob von, 35, 361n64

vaginal secretions, 58, 65, 157, 309, 443n58; examination of cells in the living state, 157, 309
Van den Areud, Guy, 125–126, 242–245, 248–249
vegetative (i.e., autonomic) nervous system, 59–64, 70, 73–74; term has nothing to do with plant life or vegetables, 7, 52, 59, 371n150
vegetotherapy, 55, 73–74, 101–102, 264, 373n5, 407–408n93
Veraguth, Otto, 65
Verworn, Max, 49
Vienna: hypocritical sexual mores, 356n10; Red, 20, 187
Vincent, Dr., 116
Virchow, Rudolph, 157, 400n41, 412–413n23, 418–419n74
viruses, 6, 92, 190, 214–15, 309, 380n108, 380n110, 444–445n94
vitalism, 9, 21, 24–27, 34–36, 45–46, 62, 77–78, 99, 160, 268–269, 311, 314, 358–359n33, 360n53, 362n68, 364n84, 365–366n101; use as a label of dismissal, 8, 49, 314, 364n84; Reich's view distinguished from, 358–359n33, 360n53, 362n68
Vogel, Steven, 115–116
Vogt, Ragnar, 221
Vorticella, 79. *See also* org-tierchen
Vygotsky, Lev, 46

Waal, Helge (son of Nic), 436n147, 438n163, 438n165
Waal, Nic, 55, 202–204, 230, 264–265, 336, 374n34, 436n147, 438n163, 438n165; relationship with Reich, 202–204; Oslo U. professorship sabotaged by Kreyberg, Langfeldt, 264, 436n147, 438n163, 438n165; rescues children from Jewish orphanage, 265. *See also* Hoel, Nic
Waddington, C. H., 45–46
Wagner von Jauregg, Julius, 17, 356n9
Wallin, Ivan, 322
Walter, Emil, 105, 255–257, 337, 434n126
Warburg, Otto, 188–194
Watase, Shôsaburô, 92
Watson and Crick DNA structure, 268–269, 311–312
Weart, Spencer, 186–187, 306–307

Weaver, Warren, 225–226, 267–269, 302–303, 307, 311; agenda of importing physical-chemical methods into biology, 5, 225–226, 267–268, 427n22; director of Natural Sciences grants at Rockefeller Foundation, 225–226, 267; view on "life rays," 302–303
Weinberger, Julius, 302–303
Weiss, Paul, 41–44, 312, 364n90, 400n40
Wells, H. G., 158, 164, 400–401n45
Wenesland, Odd, **131**, 180, 220–221, 394n121, 423n3
Werskey, Gary, 46, 164, 365n100, 366n104
Whig history, 150, 154–155, 267–268, 398n9
Whitehead, Alfred North, 42
Wilhelm Reich Museum: contains Reich's library, microscopes, 82, 129–130, 345
Woese, Carl, 352n14

Wöhler, Friedrich, 22–23
Wolfe, Theodore P., 18, 21, 58, 439n4
Woodger, J. H., 45, 365n98

X-ray diffraction, use in crystallography, 150, 161–162, 267, 401n48

Young, John D., 6, 310, 318–321

Zavadovsky, B., 45–46
Zeitschrift für Politische Psychologie und Sexual-ökonomie (ZPPS), 6, 177–178
zinc sulfide discs, 275–276
Zinsser, Hans, 92
Zondek, Samuel G., 57
"Zur Triebenergetik," 12, 19, 49. *See also* "On the Energy of Drives"